TIME AND COMPLEXITY IN HISTORICAL ECOLOGY

HISTORICAL ECOLOGY SERIES

THE HISTORICAL ECOLOGY SERIES

WILLIAM BALÉE AND CAROLE L. CRUMLEY, EDITORS

This series explores the complex links between people and landscapes. Individuals and societies impact and change their environments, and they are in turn changed by their surroundings. Drawing on scientific and humanistic scholarship, books in the series focus on environmental understanding and on temporal and spatial change. The series explores issues, develops concepts that help to preserve ecological experiences, and hopes to derive lessons for today from other places and times.

William Balée, editor, *Advances in Historical Ecology*

David L. Lentz, editor, *Imperfect Balance: Landscape Transformations in the Precolumbian Americas*

Roderick J. McIntosh, Joseph A. Tainter, and Susan Keech McIntosh, editors, *The Way the Wind Blows: Climate, History, and Human Action*

Laura M. Rival, *Trekking Through History: The Huaorani of Amazonian Ecuador*

Loretta A. Cormier, *Kinship with Monkeys: The Guajá Foragers of Eastern Amazonia*

Kenneth M. Bauer, *High Frontiers: Dolpo and the Changing World of Himalayan Pastoralists*

Edited by William Balée and Clark L. Erickson

TIME AND COMPLEXITY IN HISTORICAL ECOLOGY

STUDIES IN THE NEOTROPICAL LOWLANDS

COLUMBIA UNIVERSITY PRESS   NEW YORK

Columbia University Press

Publishers Since 1893

New York   Chichester, West Sussex

Copyright © 2006 Columbia University Press

All rights reserved

Library of Congress Cataloging-in-Publication Data

Symposium on Neotropical Historical Ecology (2002 : Tulane University)

Time and complexity in historical ecology : studies in the neotropical lowlands / edited
   by William Balée and Clark L. Erickson.

   p.   cm. — (Historical ecology series)

Papers originally presented at the Symposium on Neotropical Historical Ecology at the
   Neotropical Ecology Institute of Tulane University in October 2002.

Includes bibliographical references and index.

ISBN 978-0-231-13562-7(cloth : alk. paper)—ISBN 978-0-231-50961-9 (electronic)

1. Human ecology—Latin America—Congresses. 2. Human ecology—Tropics—
   Congresses. 3. Rain forest ecology—Latin America—Congresses. 4. Ethnobiology—
   Latin America—Congresses. 5. Agriculture—Tropics—Congresses. 6. Land use—
   Latin America—Congresses. 7. Landscape changes—Latin America—Congresses. I.
   Balée, William L., 1954– II. Erickson, Clark L. III. Title. IV. Series.

GF514.S96   2005

304.2'098—dc22   2005049370

⊗

Columbia University Press books are printed on permanent and durable acid-free paper.

Printed in the United States of America

c 10 9 8 7 6 5 4

Title page art by Daniel Brinkmeier.

To our friend and colleague Jim Petersen (1954–2005)

CONTENTS

Preface   ix

Contributors   xi

Time, Complexity, and Historical Ecology   1
WILLIAM BALÉE AND CLARK L. ERICKSON

PART 1

1. The Feral Forests of the Eastern Petén   21
    DAVID G. CAMPBELL, ANABEL FORD, KAREN S. LOWELL, JAY WALKER,
    JEFFREY K. LAKE, CONSTANZA OCAMPO-RAEDER, ANDREW TOWNESMITH,
    AND MICHAEL BALICK

2. A Neotropical Framework for *Terra Preta*   57
    ELIZABETH GRAHAM

3. Domesticated Food and Society in Early Coastal Peru   87
    CHRISTINE A. HASTORF

4. Microvertebrate Synecology and Anthropogenic Footprints
    in the Forested Neotropics   127
    PETER W. STAHL

PART 2

5. Pre-European Forest Cultivation in Amazonia   153
    WILLIAM M. DENEVAN

6. Fruit Trees and the Transition to Food Production in Amazonia   165
CHARLES R. CLEMENT

7. The Historical Ecology of a Complex Landscape in Bolivia   187
CLARK L. ERICKSON AND WILLIAM BALÉE

8. The Domesticated Landscapes of the Bolivian Amazon   235
CLARK L. ERICKSON

9. Political Economy and Pre-Columbian
Landscape Transformations in Central Amazonia   279
EDUARDO G. NEVES AND JAMES B. PETERSEN

10. History, Ecology, and Alterity: Visualizing Polity in
Ancient Amazonia   311
MICHAEL HECKENBERGER

11. Between the Ship and the Bulldozer: Historical Ecology of
Guajá Subsistence, Sociality, and Symbolism After 1500   341
LORETTA A. CORMIER

12. Landscapes of the Past, Footprints of the Future:
Historical Ecology and the Study of
Contemporary Land-Use Change in the Amazon   365
EDUARDO S. BRONDÍZIO

Index   *407*

PREFACE

WILLIAM BALÉE AND CLARK ERICKSON

HISTORICAL ECOLOGY REPRESENTS a new perspective on understanding the complex historical relationship between human beings and the biosphere of earth. Contributors to this volume explore detailed interactions between local peoples and their associated landscapes through time in various regions within the New World Tropics (Neotropics). The twelve substantive chapters represent wide geographic coverage, including chapters on Mesoamerica, the western forested flank of the Ecuadorian Andes, the desert coast of Peru, northern South America east of the Andes, and the Amazon basin, with the temporal frame of reference ranging from early prehistory at the beginning of the Holocene period to the present time and future prospects. The authors cover such topics as modifications of genetics of plant and animal species, including domesticates, semidomesticates, and wild species; the geographical distribution and availability of domesticates; biodiversity and its origins in local and regional contexts; agrodiversity; origins of linguistic terms, narratives, oral histories, and memories relating to the environment; fire histories; material culture and its origins and development through time; the definition of archaeological sites, settlement patterns, and cultural landscapes; the development and expansion of prehistoric, complex polities; the origins and development of anthropogenic soils; documentation of pre-Columbian landscape engineering; archaeological and agronomic experimentation in prehistory; and relations between humans and domesticated as well as not-so-domesticated plants and animals through time.

The product of the collision between nature and culture, wherever it has occurred, is a landscape. These chapters view historical ecology as an interdisciplinary research program that takes the landscape to be the central unit of analysis and human beings as the principal mechanism for change in the environment. In the research program of historical ecology, natural environments, once modified by human action, may never regenerate themselves as such. The research contained in this volume shows how human societies of the neotropical lowlands—rather than adapt their subsistence activities, seasonal rounds, population sizes, settlement sizes, and so on to preexisting constraints of the environment—have transformed most of those constraints into negligible phenomena as concern species diversity, landscape heterogeneity, archaeological site formation, and the development of the built environment more generally.

Books like this usually materialize within a sustaining institutional context and a supportive community of individuals. The initial idea for this book originated at the Symposium on Neotropical Historical Ecology, organized by William Balée and hosted by the Neotropical Ecology Institute of Tulane University in October 2002. Most of the chapter authors presented original papers at that symposium. We are indebted to the Neotropical Ecology Institute for hosting the event within its symposium series. We also thank the Department of Anthropology and the Stone Center for Latin American Studies of Tulane University for supplying auxiliary support to the symposium. The Center for Bioenvironmental Research of Tulane and Xavier Universities underwrote the finances of the symposium by means of a generous COYPU Foundation grant awarded to Thomas Sherry, professor of ecology and evolutionary biology at Tulane University and then director of the Neotropical Ecology Institute. Other persons who contributed to the symposium and ultimately to the appearance of this volume were Thomas Reese, Janna Rose, and James Welch, whose support and services we gratefully acknowledge. Annie Barva offered excellent advice in revising the manuscript for publication. We thank science editor Robin Smith at Columbia University Press for encouragement, enthusiasm, and patience.

October 2004
New Orleans and Philadelphia

CONTRIBUTORS

William Balée
Department of Anthropology
Tulane University
1021 Audubon Street
New Orleans, LA 70118-5698

Michael Balick
Institute for Economic Botany
New York Botanical Garden
Bronx, NY 10458

Daniel Brinkmeier (artwork)
Program Developer, Community Outreach
Environment and Conservation Programs
The Field Museum
1400 Lake Shore Drive
Chicago, IL 60605-2496

Eduardo S. Brondízio
Department of Anthropology
Anthropological Center for Training and
Research on Global Environmental Change
(ACT)
Center for the Study of Institutions,
Population, and Environmental Change
(CIPEC)
Indiana University
701 Kirkwood Ave., Student Building 130
Bloomington, IN 47405

David Campbell
Department of Biology
Grinnell College
Grinnell, IA 50112

Charles R. Clement
Coordenação de Pesquisas em Ciências
Agronômicas (Department of Agronomic
Sciences)
Instituto Nacional de Pesquisas da Amazônia
(National Research Institute for Amazonia)
Caixa Postal 478, 69.011-970
Manaus, AM, Brazil

Loretta Cormier
Department of Anthropology
338 Ullman Building
University of Alabama at Birmingham
Birmingham, AL 35294-3350.
Web site: http://main.uab.edu/show.
asp?durki=40354

William M. Denevan
Department of Geography
University of Wisconsin–Madison (Emeritus)
P.O. Box 853
Gualala, CA 95445

Clark L. Erickson
Department of Anthropology
University of Pennsylvania
33rd and Spruce Streets
Philadelphia, PA 19104-6398
Web site: www.sas.upenn.edu/˜cerickso/

Anabel Ford
MesoAmerican Research Center
University of California
Santa Barbara, CA 93106-2150

email: marc.ucsb.edu
Web site: http://espmaya.org/

Elizabeth Graham
Institute of Archaeology
University College London
31–34 Gordon Square
London WC1H 0PY, England
Web site: www.belizecubadigs.com

Christine Hastorf
Department of Anthropology
232 Kroeber Hall
University of California
Berkeley, CA 94720-3710

Michael Heckenberger
Department of Anthropology
University of Florida
Turlington Hall, Room 1112
P.O. Box 117305
Gainesville, FL 32611-7305

Jeffrey K. Lake
Department of Plant Biology
2502 Miller Plant Sciences
University of Georgia
Athens, GA 30602
Web site: www.plantbio.uga.edu/˜shubbell/
Webpages/Members/jeff_wp.htm

Karen S. Lowell
Department of Biology
Grinnell College
Grinnell, IA 50112

Eduardo G. Neves
Museu de Arqueologia e Etnologia
Universidade de São Paulo
Av. Prof. Almeida Prado, 1466, 05508-900
São Paulo SP, Brazil

Constanza Ocampo-Raeder
Department of Anthropological Sciences
Stanford University Main Quad
Building 360, Room 361D
Stanford, CA 94305-2117

James B. Petersen
Department of Anthropology
University of Vermont
Williams Hall
Burlington, VT 05405

Peter W. Stahl
Department of Anthropology
Binghamton University
P.O. Box 6000
Binghamton, NY 13902-6000

Andrew Townesmith
William L. Brown Center for Plant Genetic
Resources
Missouri Botanical Garden
Box 299
St. Louis, MO 63166

Jay Walker
Department of Botany
University of Wisconsin
430 Lincoln Drive
Madison, WI 53706

TIME AND COMPLEXITY IN HISTORICAL ECOLOGY

Rivers
Amazon River 1
Jama River 2
Mamoré River 3
Negro River 4
Solimões River 5
Tapajós River 6
Tocantins River 7
Xingu River 8

Archaeological Sites
Açutuba Site 9
Ambergris Caye Site 10
Comunidade Terra Preta Site 11
El Pilar Site 12
Hatahara Site 9
Ibibate Mound Complex 13
Iviato Mound Site 13
Lago Grande Site 9
Lamanai Site 12
Los Buchillones Site 14
Marajó Island Sites 15
Osvaldo Site 9
Peña Roja Site 16
Santarém Site 17

Towns and Cities
Belém 18
Belterra 17
Brasília 19
Fordlândia 17
Manaus 9
Santarém 17

Case Study Locations
Amazon Estuary 20
Baures Region 21
Belém 18
Belize Cayes Islands 10
Belterra 17
Cayo District 12
Central Amazon Project 9
Coastal Peru 22
El Pilar 12
Fordlândia 17
Guajá (Maranhão) 23
Ibibate Mound Complex 13
Ix Chel 12
Jama River Valley 3
Llanos de Mojos 24
Marajó Island 15
Santarém 17
Stann Creek District 10
Terra Nova 12
Tukanoan Region 25
Upper Xingu Region 26

Location map showing archaeological sites, rivers, regions, and modern cities mentioned in the book.

TIME, COMPLEXITY, AND HISTORICAL ECOLOGY

WILLIAM BALÉE AND CLARK L. ERICKSON

THE PERSPECTIVE OF HISTORICAL ECOLOGY

HISTORICAL ECOLOGY IS a powerful perspective for understanding the complex historical relationship between human beings and the biosphere. The present volume proceeds from the axiom that humanity in its historic paths across earth has interceded in material and measurable ways in a biotic world that evolved previously by natural selection and other evolutionary forces, and that the changes thus imposed on nature have in turn been reflected in human cultures, societies, and languages through time. In effect, historical ecology encompasses the view that wherever humans have trodden, the natural environment is somehow different, sometimes in barely perceptible ways, sometimes in dramatic ways. The authors in this volume have been trained in various disciplines, including anthropology (especially the subdisciplines of archaeology and sociocultural anthropology), geography, plant genetics, integrative biology, and general ecology, and they recognize the interdependence of these fields in attempting to comprehend the effects and countereffects of human behavior in the lowlands of the New World Tropics (Neotropics). The Neotropics are the torrid zone of the New World, and the lowlands within them are tropical in climate, moist, usually heavily forested, and at altitudes below approximately 500 meters. As shown in this volume's case studies, the neotropical lowlands exhibit classic anthropogenic or cultural landscapes formed over thousands of years.

Historical ecology is an interdisciplinary approach. It focuses on the historical *landscape,* a multidimensional physical entity that has both spatial and temporal characteristics and has been modified by human activity such that human intentions and actions can be inferred, if not read as material culture, from it.

The landscape is like a text, but not one that is readily accessible to historians' and epigraphers' methods because it is not written in a decipherable script, but rather is inscribed in a subtle, physical sense by learned, patterned behavior and action—what anthropologists traditionally refer to as *culture*. Culture is physically embedded and inscribed in the landscape as nonrandom patterning, often a palimpsest of continuous and discontinuous inhabitation by past and present peoples. In contrast to text-based approaches, the historical perspective taken by practitioners of historical ecology also includes prehistory. This version of historical ecology is explicitly people centered or *anthropocentric*, in contrast to other human-environmental approaches that tend to reify extra-human and noncultural phenomena, such as natural selection, kin selection, self-organization, climate change in prehistory, ecosystemic change in prehistory, and ongoing randomness of pattern and event in the environment (Botkin 1990; Egan and Howell 2001a, 2001b; Gunn 1994; Kohler and Gumerman 2000; Winterhalder 1994). Our historical ecology also stands in sharp contrast to the neoenvironmental determinism popular in archaeology today (deMenocal 2001; Fagan 1999, 2000; Kolata 1996, 2002; McIntosh, Tainter, and McIntosh 2000).

As such, landscape ecology, which has been practiced almost exclusively by population ecologists, biologists, and conservationists, is not the same as historical ecology because landscape ecology has distinguished between landscapes without human influence (a modern version of the allegedly pristine environment, or what William Denevan [1992] aptly describes as the "pristine myth") and landscapes with human influence, usually assumed to be degraded or simplified (Alvard 1995; Alvard and Kuznar 2001; Chew 2001; Krech 1999; Redford 1991, Redford and Stearman 1993; Redman 1999; Soulé and Lease 1995; Stearman and Redford 1992). Historical ecology does not treat humans as simply another animal in a complex web of organisms, or as one species among many in an ecosystem understood within a system based on equilibrium and process. Rather, the human species can be understood as a "keystone" species (Mann 2002) and as a mechanism of environmental dynamics principally through disturbance (Balée 1998b), which sometimes enhances species diversity and landscape richness (Botkin 1990; Connell 1978).

In the perspective of historical ecology, natural environments, once modified by humans, may never regenerate themselves as such. The product of the collision between nature and culture, wherever it has occurred, is a landscape, the central object of analysis in historical ecology. Archaeologist and historical ecologist Carole Crumley points out that "historical ecology traces the ongoing dialectical relations between human acts and acts of nature, made manifest in the *landscape*. Practices are maintained or modified, decisions are made, and ideas are given shape; a landscape retains the physical evidence of these mental activities" (1994a:9, emphasis in original). The landscape is where people and the environment can be seen as a totality—that is, as a multiscalar, diachronic, and holistic unit of study and analysis. In historical ecology, the anthropogenic landscape is a form of the built environment, often having been intentionally designed as architecture or as some other symbolic appropriation of nature that has patterned, physical underpinnings.

In this sense, human agency is expressed as intentionality in resource management (Balée 2003; Posey 2002); sophisticated strategies of land use (Erickson 2000b, 2003), and structured productive activities within the landscape (Heckenberger et al. 2003). The physical record of intentionality is key to understanding interrelationships between human society and its biotic environs over multiple temporal and spatial scales. The authors of the case studies in this volume and of other works in historical ecology and allied viewpoints (Balée 1998a; Cormier 2003; Crumley 1994b, 1998; Egan and Howells 2001; Ellen, Parkes, and Bicker 2000; Fairhead and Leach 1996; Lentz 2000; Li 1999; Zimmerer and Young 1998) present the evidence for the contemporary, historical, and archaeological centrality of these concepts.

Historical ecology is probably not a paradigm in the sense provided by Thomas Kuhn (1970), who doubted that such paradigms occur at all in the social sciences. Paradigms require overwhelming consensus in the scientific community, and all essential problems in the field (in this case, research problems concerning long-term relations between humans and the environment) need to have their own models of explication and deduction generated from the paradigm in order to have validity. Such consensus does not yet exist with regard to historical ecology, nor has historical ecology yet developed a wide range of models. Various authors have employed the term *historical ecology* to emphasize climatic change, geomorphological processes, environmental history, value of historical documents, and human ecology (Biersack 1999; Egan and Howells 2001; Gunn 1994; Moran 2000; Rival 2002; Sugden and Stone 2001).

Some of this confusion regarding the meaning of *historical ecology* seems to be an initial reaction to what we consider to be a radically new idea—namely, that humans can and have at different times and places increased the richness and equitability of nature by enhancing biodiversity (especially *alpha diversity*, or diversity in a restricted locale), soil fertility, and landform heterogeneity (in this volume, see chapters 1, 5, 7, 9, and 10). Humans can also decrease richness and equitability, but that is not a new observation (see Kirch and Hunt 1997; Orlove and Brush 1996). Scholars who subscribe to historical ecology as we define it in this book have tended to reject the assumptions of earlier approaches—such as cultural ecology,[1] human ecology, systems theory, and systems ecology—in proposing this perspective on human relationships with the environment over time. Historical ecologists disclaim the adaptationist assumptions of cultural ecology (and its congeneric modeling systems, such as behavioral ecology, systems ecology, self-organizing systems, sociobiology, and cultural materialism) (Diamond 1997; Harris 1979; Kohler and Gumerman 2000; Lansing 2003;

Meggers 1996, 2001; Smith and Winterhalder 1992). Adams lumps these various approaches, which for him are ultimately deriving from the cultural ecology of Julian Steward, under the term *etic rationalism:* specifically, one axiomatic part of cultural ecology that is repudiated in historical ecology concerns the concept of adaptation, whereby cultures "must first and foremost adapt themselves to the resources and opportunities of their particular environments, and this is the main explanation … for conspicuous differences between one culture and another" (1998:66). In the Amazon region, the adaptationist model has been referred to as the "standard" model (Stahl 2002; Viveiros de Castro 1996), and it still has its defenders (Headland 1997; Meggers 2001; Moran 1993). Likewise, systems ecology considers the environment and its physical constraints on organisms, their food supplies, and their populations to be hegemonic, self-sustaining, self-organizing entities. Ecosystem ecologists do not envision the ideal environment as intrinsically subject to long-term, sometimes profound change by individual organisms, particularly through the associated technologies and environmental know-how of human societies, except where those changes produce significant degradation and biological simplification of the previously existing environment (Moran 1990; Rappaport 2000). Conservation biology likewise corresponds to these sets of theoretical understandings with the added proviso that human activity in the environment is destructive (Pullin 2002; Soulé and Orians 2001). The concepts of the ecosystem, systems ecology, and cultural ecology ultimately tend to deny human agency in positively shaping the environment over time (Kohler and Gumerman 2000; Lansing 2003; Moran 1990, 2000).

Research in historical ecology instead focuses on how human societies, instead of adapting their subsistence activities, seasonal schedules, population size, settlements, and so on to preexisting constraints in the environment (Meggers 1996, 2001; Gross 1975; Harris 1979; see also critiques in Heckenberger, Petersen, and Neves 1999; Heckenberger et al. 2003; Stahl 1996, 2002; and in Clements, chapter 6, and Erickson, chapter 8, this volume), begin at once to transform most of those constraints into negligible analytic phenomena as concern suites of species, their alpha diversity, and other significant environmental features, as well as the availability of these resources for human utilization and modification within what demonstrably have become constructed and managed landscapes. In other words, environments are in a sense adapted to the sociocultural and political systems (or to humans' needs and desires) that have coexisted with them, sometimes for long periods of historical time. Historical ecology is not the same as landscape ecology (cf. Moran 2000:69). That is, historical ecologists disavow the view that humans are essentially automatons in terms of their exploitative and acquisitive activities in their physical environs (Kirch and Hunt 1997); they understand this view to be a fallacy implicit in models deriving from sociobiology, behavioral ecology, evolutionary psychology, cultural ecology, and systems ecology. In observing human behavior within such a framework, ethnographers need not a priori ask natives specific questions about environmental phenomena because natives' discourse on their intentionality and their behavior vis-à-vis the environment is typically seen by ethnographers as *emic,* or nonscientific. At the same time, their scientifically observable, or *etic,* behaviors are assumed to be already selected for, either by a cultural or naturalistic mechanism (Durham 1991; Harris 1979; Rindos 1984) and are seen as economically rational and environmentally "sound" (see Adams 1998:338).

Historical ecologists seek to liberate scientific inquiry into human/nature relationships from these assumptions not only by incorporating the observable effects of human activity and resource management into the very definition of the landscape, but also by admitting that the central species in this ongoing relationship is endowed with unique and formidable cognitive, intellectual, and aesthetic ability as well as with inimitable agency in terms of environmental resources and productive strategies. Popular print and film media have recently picked up on this idea (Mann 2002; Sington 2002). Historical ecologists support a version of cultural determinism, at least for more extreme cases, of long-term creation and maintenance of engineered landscapes in the Americas (Balée 1989; Denevan 2001; Doolittle 2002; Erickson 2000b, forthcoming; Raffles 2002; Stahl 1996, 2002; Viveiros de Castro 1996; Whitmore and Turner 2002).

Perhaps a better philosophical guideline is to consider historical ecology as a research program (Lakatos 1980). The natural sciences have mechanisms for comprehending change in the environment, such as the laws of thermodynamics, relativity, and natural selection. Evolutionary ecology (also known as behavioral ecology) contains proposals of an interdependence of human genes and environmental conditions and constraints (e.g., Smith and Winterhalder 1992), whereas coevolution (Rindos 1984) exhibits a focus on an assumed interdependence of human genes and specific cultural phenomena. In contrast to historical ecologists, supporters of both approaches tend to deny human agency in the environmental milieus that encompass known societies. There is no need for consciousness of action or intentionality, moreover, in these models. Natural selection explains the evolution of species, whereas the social sciences only approximate such a mechanism by focusing on historical events, their chronology, and retrodiction (not prediction) of the motivating forces of history.

What historical ecology harbors as an explicit proposal is that *the human species is itself a principal mechanism of change in the natural world, a mechanism qualitatively as significant as natural selection.* In addition, the human species is not just a product of natural selection (though it is partly that) because it too makes histories and specific landscapes that bear its inscriptions. The cumulative effects of these undertakings influence the development and form of the exact cultural qualities of contemporary landscapes and are manifested in them.

Each major environment of the earth has a unique and often complex human history embedded in the local and regional landscape. Understanding the human role in the creation and maintenance of this uniqueness is a central goal of historical ecology. This approach involves the study of human effects on other life-forms, wherever they exist; historic changes in cultures due to these effects; and continuing (i.e., ethnographically documented) human effects on nature, sometimes in ways that increase the complexity and heterogeneity of the landscape through phenomena such as enhanced soils (Hecht 2003; Hecht and Posey 1989; Lehmann et al. 2003; McCann, Woods, and Meyer 2001; WinklerPrins 2001; Woods and McCann 1999), hydrology (Erickson, chapter 8, this volume; Raffles 2002), and species composition (Balée 1998b; see also Stahl, chapter 4, and Erickson and Balée, chapter 7, this volume).

Historical ecology is associated with some of the tenets of the new ecology (Botkin 1990; Little 1999; Scoones 1999; Zimmerer 1994; Zimmerer and Young 1999) such as "non-equilibrium dynamics, spatial and temporal variation, complexity, and uncertainty" (Scoones 1999:479). It does not brandish the *ecosystem concept* (cf. Moran 1990, 2000; Rappaport 2000) because that term has historically corresponded to synchronic views of arbitrarily defined spatial units that lack historical contingency (that are, in other words, in a supposed state of equilibrium). Practitioners of the new ecology also reject the ecosystem concept's equilibrium assumption (Begon, Harper, and Townsend [1990] and Botkin [1990] refer to landscapes as "culturalized ecosystems"; see also Worster 1994:390–391; cf. Egan and Howell 2001b:2). In fact, landscapes represent histories that unfold in a biotic and cultural domain in which inscriptions of an array of human activities across the temporal spectrum may be discerned by research. Historical ecology undertakes to present a historical (human and cultural) accounting of seemingly naturalistic events and processes, as with other contingency-based approaches to human-environmental dynamics (Prigogine and Stengers 1984). But it is not environmental history (Balée 1998b; Moran 2000; Worster 1993) because environmental history, like human ecology or ecological anthropology, is a subject field, whereas historical ecology actually instantiates a distinctive perspective on such fields.

Intentional and unintentional human activities can create—in addition to documented cases of environmental degradation—sustained levels of environmental disturbance considered important for ensuring resilience of biotas and landscapes (Connell 1978; Scoones 1999; Stahl 1996, 2000; Zimmerer 1994; Zimmerer and Young 1998). Nonequilibrium ecology is actually part of historical ecology. Historical ecology does not ignore catastrophic, chaotic disturbances that destroy (rather than merely alter) landscapes (Kirch and Hunt 1997). It emphasizes human activities in the environment over long periods of time that ultimately contribute to understanding the heterogeneity of landscapes across world regions, and it assesses the historical relationship among cultural, linguistic, and biological diversity (Maffi 2001).

OVERVIEW

Patterns of residues, anomalies, and cultural imprints (as palimpsest) of humans on the landscape are the primary data of historical ecology. In this book, these data include the genetics of plants and animals, especially those of semidomesticates within domesticated species; the geographical distribution of domesticates; biodiversity; agrodiversity; linguistic terms, narrative, oral history, and memory relating to the environment; agroforestry; fire histories; material culture; archaeological sites and settlement patterns; agricultural fields; anthropogenic soils; hydraulic engineering; archaeological and agronomic experimentation; and, finally, relations with domesticated animals.

Historical ecology recognizes two kinds of selection: one historical and the other properly evolutionary. One is not simply a variety of the other, yet in particular cases both are intertwined and analytically inseparable. In the case of the three sites in the Petén forest of lowland Guatemala studied by David Campbell and his collaborators in chapter 1, the diversity and patterning of vegetation cannot be understood apart from activities of the Maya people and their predecessors dating back at least 4,000 years. These people actively selected for economic species, and this suite of economically important plants can still be discerned in the present landscape as oligarchic forests, which by definition are dominated by just a few species and are often the result of human activity (Peters 2000). Indeed, the Maya landscape is incomprehensible without knowing this complex history and prehistory, in which humans are and have been the principal actors. The Maya landscape studied by Campbell and colleagues is highly patterned and cannot be described or understood without consideration of the human imprint inscribed on it.

As shown in the case studies of various chapters in the volume, history and prehistory are necessary to understand present-day landscapes. One can identify domestication of plants and animals, the introduction of these species into exotic habitats, and the effects such introductions have or have not had on local cultures, as Christine Hastorf examines (chapter 3). Elizabeth Graham (chapter 2) suggests that prehistoric peoples altered texture and chemical composition of natural soils, wittingly or not, not only in Amazonia, but in other neotropical regions; such human interventions in the ground had enhancing effects on soil fertility, which improved the results of agriculture. Graham also argues that local historical context and processes must be considered in order to understand the phenomena of dark earths recognized in many parts of the Neotropics. One can indicate how landscapes in eastern Bolivia have in effect been domesticated through engineering by rearranging soils, altering drainage, constructing massive earthworks, and enhancing effects on local diversity, as Erickson and Balée (chapter 7) and Erickson (chapter 8) demonstrate. Peter Stahl (chapter 4) documents in lowland Ecuador the heterogeneity of fauna in a local habitat thanks to human agricultural activity over time. Charles Clement (chapter 6) demonstrates how the long-term domestication of fruit trees from the beginning of the Holocene period onward appears to be direct evidence for how and when people in the Amazon became early managers, as opposed to merely foragers of the forest. Michael Heckenberger (chapter 10) highlights the continuities and disjunctures in the ethnographic, historical, and archaeological record in south-central Amazonia regarding a demonstrably complex social and political organization of society in what has traditionally been considered an unpromising environment for human development. Loretta Cormier (chapter 11) examines the trajectory of a foraging society, the Guajá of eastern Amazonian Brazil, and discusses how their subsistence in recent times—as hunter-gatherers, that is, people without agriculture—can be explained only through consideration of a historical dimension that in turn incorporates a notion of variably weighted disruptions of contact (including disease, depopulation, and slavery) and of temporal vagaries in the landscapes their forebears inhabited. Eduardo Brondízio (chapter 12) explores how conceptual models that focus either on negative or positive effects of urbanization in Amazonian environments are inadequate for understanding the intrinsic complexity of the interrelationships among biophysical, sociocultural, economic, and historical factors actively influencing contemporary land use.

Merely listing the effects that indigenous peoples have had on nature over time fails to capture the diverse forms of manipulation and transformation of lowland neotropical environments documented to a noticeable extent within the chapters of this volume. As the case studies presented in this volume and in others demonstrate, some neotropical landscapes were created by native people organized as "complex" hierarchical societies (the states of the Maya and Olmec; the chiefdoms of the eastern Bolivian Amazon and upper Xingu River; and the major polities along the Amazon River in late prehistory [Carneiro 1995; Heckenberger et al. 2003; Neves and Peterson, chapter 9, this volume]). Countless societies historically considered to be "simple" in terms of sociopolitical organization (egalitarian bands and autonomous villages such as the Sirionó, Ka'apor, Guajá, modern Xinguanos, and other peoples discussed in this volume and elsewhere) have also had measurable effects on their environments (Balée 1989; Heckenberger et al. 2003; Posey 2002). All of these societies and others like them contributed to the complex and long history of how the contemporary environment came to be through their activities in the living landscape, measurable by material evidence. These activities were driven, moreover, at least partly by human intentions.

Intentionality with regard to living resources is conditioned by time and the complexity of the landscape. It is a facet of knowledge relating to the biosphere or some part of it. Historical ecology of knowledge reveals the means by which changes in the environment induced by humans actually condition subsequent generations in terms of language, technology, and culture. Patterns of folk classification and the social constructs of nature, whereby some of the visible biota and landscape features of an environment have more psychological saliency than do others for a given group of people participating in shared knowledge of that environment, are molded by landscape transformation over time. Each such repertoire of landscape knowledge instantiates an ecological *epistéme* (cf. Descola 1996:93), a distinctive and historically defined way of knowing the environment that has its origins in the particular relationship it has had over time to local landscapes and to their metamorphosis at human hands. In other words, environmental knowledge is contingent on interactions people experience over time with their landscape (Ellen, Parkes, and Bicker 2000), and such an observed contingency is clearly not unique to the Neotropics (Ellen 1999; Fairhead and Leach 1996; Li 1999). That knowledge is not the result either of environmental (or biological) determinism or of cultural determinism alone (a point also made by Ingold 2000), but rather ensues from the conjunction of time and complexity in what is essentially a reciprocal dynamic between society and the environment.

Although human activities are assumed to have shaped the major environments of the earth, proponents of historical ecology are cautious about uncritically assigning the value-laden terms such as *beneficial, enhancing, sustainable, destructive,* and *degrading* to human activities past and present. These terms are often applied as black-box assumptions without clearly defining or considering the appropriate temporal or geographical scale of the case study. As Erickson stresses in chapter 8, these terms and their associated concepts imply an extant benchmark for a pristine, natural environment to which anthropogenic landscapes can be compared. As highlighted in the various case studies of this volume, however, pristine environments must be first proved, rather than assumed, in the Neotropics.

Conservation biologists have pointed to human-caused degradation of the environment such as predation (overhunting) leading to trophic cascades, anthropogenic eutrophication, air and water pollution, introduction of exotic species into new habitats, devastation by fire, habitat destruction and fragmentation, and extinctions 100 to 1,000 times the background rate (Pullin 2002; Soulé and Orians 2001; Wilson 1992). Historical ecologists maintain that human nature per se is not the culprit in these calamities; rather, causality can be addressed to historically defined configurations of interrelationships over time between specific societies and their economies, on the one hand, and given environments, on the other (Balée 1998b; Egan and Howell 2001a, 2001b). They maintain this view because in other cases of the human-environmental relationship, as documented in the Neotropics, local biodiversity (biological diversity as indicated by numbers and distribution of species of animals and plants, including agrodiversity) has increased thanks to human modifications and management of resources and the landscape (Balée 1994; Berkes 1999; Brookfield et al. 2002; Denevan and Padoch 1988; Posey 2002; Posey and Balée 1989; various chapters in this volume).

Likewise, under certain agricultural and agroeconomic regimes, soils have become organically and chemically impoverished (such as loss of topsoil in the North American Midwest due to industrial agriculture, or salinization of the Euphrates River due to ancient Mesopotamian irrigation), whereas under other regimes, soils have actually become highly fertile in terms of their nutrient content and physicochemical properties. The organic black and brown earths of upland Amazonia (Amazonian Dark Earths), typically the result of prehistoric agriculture and settlement, are actually much more fertile than surrounding soils not so utilized and subjected to management over time (Erickson 2003; Hecht 2003; Hecht and Posey 1989; Lehmann et al. 2003; McCann, Woods, and Meyer 2001; WinklerPrins 2001; Woods and McCann 1999; see also Denevan, chapter 5, Erickson and Balée, chapter 7, and Heckenberger, chapter 10, this volume).

Indeed, the chapters in this volume taken as a whole constitute powerful evidence that *Homo sapiens,* as an agent of landscape creation, modification, and artificial selection over the long term, is synonymous neither with the ecologically noble savage (*Homo ecologicus,* the idealized human species that is inherently custodial and nurturing of nonhuman nature) nor with the ecologically ignoble savage (*Homo devastans,* the idealized human species that is biologically programmed to destroy nonhuman nature). The authors agree that indigenous societies in the Neotropics have permanently and significantly transformed, built, and maintained environments to such a scale that they have determined local and regional species diversity, environmental richness in general, soil quality, and other palpably natural features that are often the object of modern conservationist efforts. In the specific areas studied by William Denevan (central and lower Amazon regions, chapter 5) and by Eduardo Góes Neves and James Peterson (the central Amazon, chapter 9), the black earths point unmistakably to humans' intentional, long-term, custodial influence on the environment, even under regimes of intensive agriculture that would have been feeding and supporting dense populations. The topographically diverse raised field and fish weir landscapes in the Bolivian Amazon described by Erickson (chapter 8) enhanced ecological heterogeneity and created conditions for a higher standard of living for the prehistoric human inhabitants.

Conservation biologists and historical ecologists are concerned with habitat degradation and species extinctions. Regarding the human capacity for both landscape degradation and enhancement, we lean more toward the "enhancement" side and have an admittedly anthropocentric bias. Historical ecology demonstrates numerous cases of human activities that by conservation standards actually have benefited biological richness and diversity. Forests are typically more species rich than adjacent savannas and grasslands per unit area. Fire has certainly been involved in destructive deforestation in Amazonia and other tropical regions worldwide (Pullin 2002:55), where savanna has expanded at the expense of forest and in some cases desertification has occurred. But Stephen Pyne (1998) has shown how North American Indians prehistorically used fire to manage forested and savanna landscapes actively. One outcome of such management by fire was to lower the risk of destructive wildfires of the sort that occurred frequently in the late twentieth century and are occurring in the early twenty-first century in the western United States and southeastern Australia. In other words, fire can certainly be damaging to a landscape and its attendant biota, and conservation biologists tend to focus exclusively on this damage, but fire can also be harnessed and used to enhance the diversity of the same.

Forest islands in the savannas of Guinea (West Africa) are now understood not to be relics of Pleistocene events or the remnants of once vast pristine forests, but rather direct and inescapable outgrowths of multiple generations of human settlement and intense resource management (Fairhead and Leach 1996; Leach and Mearns 1996). Forest islands in the upland savannas of central Brazil are likewise seen as anthropogenic, thanks to the activities of the Kayapó Indians (Anderson and Posey 1989; Posey 2002), although this view is still controversial (Balée 2003; Parker 1992; Posey 1992). Many if not most of the forest islands on the wet savanna of the Bolivian Amazon are now understood to be the result of settlement, farming, and mound building by its pre-Columbian inhabitants (Erickson 1995; Mann 2000; Walker 2003; see also Erickson and Balée, chapter 7, and Erickson, chapter 8, this volume). The savannas of the same region, which account for at least two-thirds of the total area, as Erickson reports in chapter 8, can be comprehended only as effects of intense human landscape management in the past. In cases of forest expansion and diversification directed by humans—that is, cultural forests (chapters 1, 5, 6, 9, and 10)—local biodiversity cannot be fully accounted for by using only a model of natural selection, but rather should be seen as artificially established by cultural conventions acting in tandem with given genotypes. In other words, through the study of traditional resource management and environmental knowledge in the past and present, we can begin to grapple with the implications of such knowledge for conservation and management of biodiversity and landscape diversity. The human activity that built earthworks, engineered soils and water, and constructed forests and savannas where there were none was more a product of human history than a result of evolutionary forces, such as natural selection (see Graham, chapter 2). The various species present on the forest islands of eastern Bolivia, West Africa, and central Brazil, which are biologically richer than the surrounding savannas, are likewise products both of natural and artificial selection acting in tandem, not in isolation. The formation of forest islands by human activity is one of the most dramatic examples of landscape research in historical ecology; many other less dramatic but equally intriguing examples of the dialogue between humans and nature can be noted.

Historical ecology represents a range of studies that permit comparison among diverse sociopolitical entities in relationship to local landscapes, larger phenomena such as regions, and ultimately the biosphere itself. In this volume, we present a range of studies as they relate specifically to the lowland Neotropics, an arbitrary geographical designation to be sure, but one with intrinsically well-documented cases of extensive resource and landscape management by humans over many millennia and across a tremendous array of habitats, environments, and distribution patterns of flora and fauna. Each lowland neotropical landscape presents us with a rich history of human activities, the effects of which in principle can be evaluated on their merits and not a priori presumed to be either conservationist or anticonservationist in character. Historical ecology applies a multiscalar geographical (local place to regional landscape) and temporal (short- to long-term) perspective for a historical understanding of human activities in the environment and how the environment itself came to be. As a consequence, historical ecology may provide practical strategies for managing landscapes in the present and future.

ACKNOWLEDGMENTS

We thank all the contributors to this volume for stimulating discussion of the ideas in this introduction, given both at the Symposium on Neotropical Historical Ecology at Tulane in 2002 and during the course of compiling the various chapters. We especially thank William Denevan and Peter W. Stahl for insightful suggestions.

NOTES

1. The term *cultural ecology* has long been used somewhat differently in geography; it does not refer so much to a point of view as to the subject matter of interactions between humans and the environment (Wagner and Mikesell 1962). We use the term here in the original sense of Julian Steward to indicate a perspective that assigns determinism of social and ideological phenomena to technology and the environment.

REFERENCES

Adams, W. Y. 1998. *The Philosophical Roots of Anthropology.* Stanford, Calif.: Center for the Study of Language and Information.

Alvard, M. S. 1995. Infraspecific prey choice by Amazonian hunters. *Current Anthropology* 36 (5): 789–818.

Alvard, M. S., and L. Kuznar. 2001. Deferred harvests: The transition from hunting to animal husbandry. *American Anthropologist* 103 (2): 295–311.

Anderson, A. B., and D. A. Posey. 1989. Management of a tropical scrub savanna by the Gorotire Kayapó of Brazil. In D. A. Posey and W. Balée, eds., *Resource Management in Amazonia: Indigenous and Folk Strategies,* 159–173. Advances in Economic Botany no. 7. Bronx: New York Botanical Garden.

Balée, W. 1989. The culture of Amazonian forests. In D. A. Posey and W. Balée, eds., *Resource Management in Amazonia: Indigenous and Folk Strategies,* 1–21. Advances in Economic Botany no. 7. Bronx: New York Botanical Garden.

——. 1994. *Footprints of the Forest: Ka'apor Ethnobotany—the Historical Ecology of Plant Utilization by an Amazonian People.* New York: Columbia University Press.

——, ed. 1998a. *Advances in Historical Ecology.* New York: Columbia University Press.

——. 1998b. History ecology: Premises and postulates. In W. Balée, ed., *Advances in Historical Ecology,* 13–29. New York: Columbia University Press.

——. 2003. Native views of the environment in Amazonia. In H. Selin, ed., *Nature Across Cultures: Views of Nature and the Environment in Non-Western Cultures,* 277–288. Manchester, U.K.: Kluwer Academic.

Begon, M., J. L. Harper, and C. R. Townsend, eds. 1990. *Ecology: Individuals, Populations, and Communities.* 2d. ed. Boston: Blackwell Scientific.

Berkes, F. 1999. *Sacred Ecology: Traditional Ecological Knowledge and Resource Management.* Philadelphia: Taylor and Francis.

Biersack, A. 1999. Introduction: From the "new ecology" to the new ecologies. *American Anthropologist* 101 (1): 5–18.

Botkin, D. 1990. *Discordant Harmonies: A New Ecology for the Twenty-First Century.* New York: Oxford University Press.

Brookfield, H., C. Padoch, H. Parsons, and M. Stocking, eds. 2002. *Cultivating Biodiversity: Understanding, Analyzing, and Using Agricultural Diversity.* New York: United Nations University/United Nations Environmental Program, Columbia University Press.

Carneiro, R. L. 1995. The history of ecological interpretations of Amazonia: Does Roosevelt have it right? In L. Sponsel, ed., *Indigenous Peoples and the Future of Amazonia: An Ecological Anthropology of an Endangered World,* 45–70. Tucson: University of Arizona Press.

Chew, S. C. 2001. *World Ecological Degradation: Accumulation, Urbanization, and Deforestation 3000 B.C.–A.D. 2000.* Walnut Creek, Calif.: Altamira Press.

Connell, J. H. 1978. Diversity in tropical rain forests and coral reefs. *Science* 199: 1302–1310.

Cormier, L. A. 2003. *Kinship with Monkeys.* New York: Columbia University Press.

Crumley, C. L. 1994a. Historical ecology: A multidimensional ecological orientation. In C. L. Crumley, ed., *Historical Ecology: Cultural Knowledge and Changing Landscapes,* 1–13. Santa Fe, N.Mex.: School of American Research Press.

——, ed. 1994b. *Historical Ecology: Cultural Knowledge and Changing Landscapes.* Santa Fe, N.M.: School of American Research Press.

——. 1998. Foreword. In W. Balée, ed., *Advances in Historical Ecology,* ix–xiv. New York: Columbia University Press.

deMenocal, P. B. 2001. Cultural responses to climate change during the Late Holocene. *Science* 292: 667–673.

Denevan, W. M. 1992. The pristine myth. *Annals of the Association of American Geographers* 82 (3): 369–385.

——. 2001. *Cultivated Landscapes of Native Amazonia and the Andes.* New York: Oxford University Press.

Denevan, W. M., and C. Padoch, eds. 1988. *Swidden-Fallow Agroforestry in the Peruvian Amazon.* Advances in Economic Botany no. 5. Bronx: New York Botanical Garden.

Descola, P. 1996. Constructing natures: Symbolic ecology and social practice. In P. Descola and G. Pálsson, eds., *Nature and Society: Anthropological Perspectives*, 82–102. New York: Routledge.

Diamond, J. 1997. *Guns, Germs, and Steel: The Fates of Human Societies*. New York: W. W. Norton.

Doolittle, W. E. 2002. *Cultivated Landscapes of Native North America*. New York: Oxford University Press.

Durham, W. H. 1991. *Coevolution: Genes, Culture, and Human Diversity*. Stanford, Calif.: Stanford University Press.

Egan, D., and E. Howell, eds. 2001a. *The Historical Ecology Handbook: A Restorationist's Guide to Reference Ecosystems*, 1–23. Washington, D.C.: Island Press.

——. 2001b. Introduction. In D. Egan and E. Howell, eds., *The Historical Ecology Handbook: A Restorationist's Guide to Reference Ecosystems*, 1–23. Washington, D.C.: Island Press.

Ellen, R. 1999. Forest knowledge, forest transformation: Political contingency, historical ecology, and the renegotiation of nature in central Seram. In T. M. Li, ed., *Transforming the Indonesian Uplands*, 131–157. Amsterdam: Harwood.

Ellen, R., P. Parkes, and A. Bicker, eds. 2000. *Indigenous Environmental Knowledge and Its Transformations: Critical Anthropological Perspectives*. Amsterdam: Harwood.

Erickson, C. L. 1995. Archaeological perspectives on ancient landscapes of the Llanos de Mojos in the Bolivian Amazon. In P. Stahl, ed., *Archaeology in the American Tropics: Current Analytical Methods and Applications*, 66–95. Cambridge, U.K.: Cambridge University Press.

——. 2000a. An artificial landscape-scale fishery in the Bolivian Amazon. *Nature* 408: 190–193.

——. 2000b. The Lake Titicaca basin: A pre-Columbian built landscape. In D. Lentz. ed., *Imperfect Balance: Landscape Transformations in the Precolumbian Americas*, 311–356. New York: Columbia University Press.

——. 2003. Historical ecology and future explorations. In J. Lehmann, D. C. Kern, B.Glaser, and W. I. Woods, eds., *Amazonian Dark Earths: Origin, Properties, Management*, 455–500. Dordrecht, Netherlands: Kluwer Academic.

——. Forthcoming. Intensification, political economy, and the farming community: In defense of a bottom-up perspective of the past. In C. Stanish and J. Marcus, eds., *Agricultural Practices and Strategies*. Los Angeles: Cotsen Archaeological Institute, University of California, Los Angeles.

Fagan, B. 1999. *Floods, Famines, and Emperors: El Nino and the Fate of Civilizations*. New York: HarperCollins.

——. 2000. *The Little Ice Age: The Prelude to Global Warming, 1300–1850*. New York: Basic.

Fairhead, J., and M. Leach. 1996. *Misreading the African Landscape: Society and Ecology in the Forest-Savanna Mosaic*. Cambridge, U.K.: Cambridge University Press.

Gross, D. 1975. Protein capture and cultural development in the Amazon basin. *American Anthropologist* 77 (3): 526–549.

Gunn, J. S. 1994. Global climate and regional biocultural diversity. In C. Crumley, ed., *Historical Ecology: Cultural Knowledge and Changing Landscapes*, 67–97. Santa Fe, N.Mex.: School of American Research Press.

Harris, M. 1979. *Cultural Materialism: The Struggle for a Science of Culture*. New York: Random House.

Headland, T. 1997. Revisionism in ecological anthropology. *Current Anthropology* 38(4): 605–630.

Hecht, S. B. 2003. Indigenous soil management and the creation of Amazonian Dark Earths: Implications of Kayapó practices. In J. Lehmann, D. C. Kern, B. Glaser, and W. I. Woods, eds., *Amazonian Dark Earths: Origin, Properties, Management,* 355–371. Dordrecht, Netherlands: Kluwer Academic.

Hecht, S. B., and D. A. Posey. 1989. Preliminary findings on soil management of the Kayapó Indians. In D. A. Posey and W. Balée, eds., *Resource Management in Amazonia: Indigenous and Folk Strategies,* 174–188. Advances in Economic Botany no. 7. Bronx: New York Botanical Garden.

Heckenberger, M. J., A. Kuikuruo, U. T. Kuikuro, J. C. Russell, M. Schmidt, C. Fausto, and B. Franchetto. 2003. Amazonia 1492: Pristine forest or cultural parkland? *Science* 301:1710–1713.

Heckenberger, M. J., J. B. Petersen, and E. G. Neves. 1999. Village size and permanence in Amazonia: Two archaeological examples from Brazil. *Latin American Antiquity* 10:353–376.

Ingold, T. 2000. *The Perception of the Environment.* London: Routledge.

Kirch, P. V., and T. L. Hunt, eds. 1997. *Historical Ecology in the Pacific Islands: Prehistoric Environmental and Landscape Change.* New Haven, Conn.: Yale University Press.

Kohler, T., and G. Gumerman, eds. 2000. *Dynamics in Human and Primate Societies: Agent-Based Modeling of Social and Spatial Processes.* New York: Oxford University Press.

Kolata, A. L., ed. 1996. *Agroecology.* Vol. 1 of *Tiwanaku and Its Hinterland: Archaeological and Paleoecological Investigations of an Andean Civilization.* Washington, D.C.: Smithsonian Institution Press.

———. 2002. *Urban and Rural Archaeology.* Vol. 2 of *Tiwanaku and Its Hinterland: Archaeological and Paleoecological Investigations of an Andean Civilization.* Washington, D.C.: Smithsonian Institution Press.

Krech, S. 1999. *The Ecological Indian: Myth and History.* New York: W. W. Norton.

Kuhn, T. 1970. *The Structure of Scientific Revolutions.* 2d ed. Chicago: University of Chicago Press.

Lakatos, I. 1980. *The Methodology of Scientific Research Programmes.* Philosophical Papers, vol. 1. Cambridge, U.K.: Cambridge University Press.

Lansing, J. S. 2003. Complex adaptive systems. *Annual Review of Anthropology* 32:183–204.

Leach, M., and R. Mearns, eds. 1996. *The Lie of the Land: Challenging Environmental Orthodoxies in Africa.* London: James Currey.

Lehmann, J., D. C. Kern, B. Glaser, and W. I. Woods, eds. 2003. *Amazonian Dark Earths: Origin, Properties, Management.* Dordrecht, Netherlands: Kluwer Academic.

Lentz, D., ed. 2000. *Imperfect Balance: Landscape Transformations in the Precolumbian Americas.* New York: Columbia University Press.

Li, T. M., ed. 1999. *Transforming the Indonesian Uplands.* Amsterdam: Harwood.

Little, M. 1999. Environments and environmentalisms. *Annual Review of Anthropology* 29:253–284.

Maffi, L., ed. 2001. *On Biocultural Diversity: Linking Language, Knowledge, and the Environment.* Washington, D.C.: Smithsonian Institution Press.

Mann, C. 2000. Earthmovers of the Amazon. *Science* 287:786–789.

———. 2002. 1491. *Atlantic Monthly* 289 (3): 41–53.

McCann, J. M., W. I. Woods, and D. W. Meyer. 2001. Organic matter and anthrosols in Amazonia: Interpreting the Amerindian legacy. In R. M. Rees, B. Bau, C. Watson, and C. Campbell, eds., *Sustainable Management of Soil Organic Matter,* 180–189. New York: CABI.

McIntosh, R. J., J. A. Tainter, and S. K. McIntosh. eds. 2000. *The Way the Wind Blows: Climate, History, and Human Action.* New York: Columbia University Press.

Meggers, B. J. 1996. *Amazonia: Man and Culture in a Counterfeit Paradise.* 2d ed. Washington, D.C.: Smithsonian Institution Press.

——. 2001. The continuing quest for El Dorado: Round two. *Latin American Antiquity* 12 (3): 304–325.

Moran, E. F., ed. 1990. *The Ecosystem Approach in Anthropology: From Concept to Practice.* Ann Arbor: University of Michigan Press.

——. 1993. *Through Amazonian Eyes: The Human Ecology of Amazonian Populations.* Iowa City: University of Iowa Press.

——. 2000. *Human Adaptability: An Introduction to Ecological Anthropology.* 2d ed. Boulder, Colo.: Westview Press.

Orlove, B., and S. Brush. 1996. Anthropology and the conservation of biodiversity. *Annual Review of Anthropology* 25:329–352.

Parker, E. 1992. Forest islands and Kayapó resource management in Amazonia: A reappraisal of the *apêtê*. *American Anthropologist* 94 (2): 406–428.

Peters, Charles. 2000. Precolumbian silviculture and indigenous management of neotropical forests. In D. Lentz, ed., *Imperfect Balance: Landscape Transformations in the Precolumbian Americas,* 203–223. New York: Columbia University Press.

Posey, D. A. 1992. Reply to Parker. *American Anthropologist* 94 (2): 441–443.

——. 2002. *Kayapó Ethnoecology and Culture.* New York: Oxford University Press.

Posey, D. A., and W. Balée, eds. 1989. *Resource Management in Amazonia: Indigenous and Folk Strategies.* Advances in Economic Botany no. 7. Bronx: New York Botanical Garden.

Prigogine, I., and I. Stengers. 1984. *Order Out of Chaos: Man's New Dialogue with Nature.* Toronto: Bantam.

Pullin, A. S. 2002. *Conservation Biology.* Cambridge, U.K.: Cambridge University Press.

Pyne, S. J. 1998. Forged in fire: History, land, and anthropogenic fire. In W. Balée, ed., *Advances in Historical Ecology,* 64–103. New York: Columbia University Press.

Raffles, H. 2002. *In Amazonia: A Natural History.* Princeton, N.J.: Princeton University Press.

Rappaport, R. A. 2000. *Pigs for the Ancestors: Ritual in the Ecology of a New Guinea People.* 2d ed. Prospect Heights, Ill.: Waveland Press.

Redford, K. H. 1991. The ecologically noble savage. *Cultural Survival Quarterly* 15(1): 46–48 (Reprinted from *Orion Nature Quarterly* 9 [3] [1990]: 24–29.)

Redford, K. H.,. and A. M. Stearman. 1993. Forest-dwelling native Amazonians and the conservation of biodiversity. *Conservation Biology* 7: 248–255.

Redman, C. 1999. *Human Impact on Ancient Environments.* Tucson: University of Arizona Press.

Rindos, D. 1984. *The Origins of Agriculture.* New York: Academic Press.

Rival, L. M. 2002. *Trekking Through History.* New York: Columbia University Press.

Scoones, I. 1999. New ecology and the social sciences: What prospects for a fruitful engagement? *Annual Review of Anthropology* 28:479–507.

Sington, D., director. 2002. *The Secret of El Dorado* (videotape). London: BBC Horizon Series.

Smith, E. A., and B. P. Winterhalder, eds. 1992. *Evolutionary Ecology and Human Behavior.* Hawthorne, N.Y.: Aldine de Gruyter.

Soulé, M. E., and G. Lease, eds. 1995. *Reinventing Nature? Responses to Postmodern Deconstruction.* Washington, D.C.: Island Press.

Soulé, M. E., and G. H. Orians, eds. 2001. *Conservation Biology: Research Priorities for the Next Decade*. Washington, D.C.: Island Press.

Stahl, P. W. 1996. Holocene biodiversity: An archaeological perspective from the Americas. *Annual Review of Anthropology* 25:105–126.

———. 2000. Archaeofaunal accumulation, fragmented forests, and anthropogenic landscape mosaics in the tropical lowlands of prehispanic Ecuador. *Latin American Antiquity* 11 (3): 241–257.

———. 2002. Paradigms in paradise: Revising standard Amazonian prehistory. *Review of Archaeology* 23 (2): 39–51.

Stearman, A. M., and K. H. Redford. 1992. Commercial hunting by subsistence hunters: Sirionó Indians and Paraguayan caiman in lowland Bolivia. *Human Organization* 51 (3): 235–244.

Sugden, A., and R. Stone, eds. 2001. Ecology through time. *Science* (special issue) 293:623–660.

Viveiros de Castro, E. B. 1996. Images of nature and society in Amazonian ethnology. *Annual Review of Anthropology* 25:179–200.

Wagner, P. L., and M. W. Mikesell, eds. 1962. *Readings in Cultural Geography*. Chicago: University of Chicago Press.

Walker, J. H. 2003. *Agricultural Change in the Bolivian Amazon*. Latin American Archaeology Reports. Pittsburgh: University of Pittsburgh.

Whitmore, T. M., and B. M. Turner II. 2002. *Cultivated Landscapes of Middle America on the Eve of Conquest*. New York: Oxford University Press.

Wilson, E. O. 1992. *The Diversity of Life*. Cambridge, Mass.: Harvard University Press.

WinklerPrins, A. M. G. A. 2001. Why context matters: Local soil knowledge and management among an indigenous peasantry on the lower Amazon floodplain, Brazil. *Etnoecológica* 5 (7): 6–20.

Winterhalder, B. P. 1994. Concepts in historical ecology: The view from evolutionary theory. In C. L. Crumley, ed., *Historical Ecology: Cultural Knowledge and Changing Landscapes*, 17–41. Santa Fe, N.Mex.: School of American Research Press.

Woods, W. I., and J. M. McCann. 1999. The anthropogenic origin and persistence of Amazonian Dark Earths. *Yearbook of the Conference of Latin Americanist Geographers* 25:7–14.

Worster, D. 1993. *The Wealth of Nature: Environmental History and the Ecological Imagination*. New York: Oxford University Press.

———. 1994. *Nature's Economy: A History of Ecological Ideas*. 2d ed. Cambridge, U.K.: Cambridge University Press.

Zimmerer, K. S. 1994. Human geography and the "new ecology": The prospect and promise of integration. *Annals of the Association of American Geographers* 84 (1): 108–125.

Zimmerer, K., and K. Young, eds. 1998. *Nature's Geography: New Lessons for Conservation in Developing Countries*. Madison: University of Wisconsin Press.

PART 1

Las Piedras no son mudas.
Ellas solamente
guardan el silencio.

1

THE FERAL FORESTS OF THE EASTERN PETÉN

DAVID G. CAMPBELL, ANABEL FORD, KAREN S. LOWELL, JAY WALKER,
JEFFREY K. LAKE, CONSTANZA OCAMPO-RAEDER, ANDREW TOWNESMITH,
AND MICHAEL BALICK

THE NEW DISCIPLINE of historical ecology recognizes that human culture and the environment mutually influence each other (Balée 1998), rather than following the conventional one-way paradigm in which humans are ever adapting to their environments. We examine the contemporary Maya forest of the eastern Petén from this realistic vantage point, suggesting that its species composition and phytosociology are human artifacts dating from the Late Classic period. Wiseman (1978) was the first to use the term *man-made* to describe the Maya forest. About the same time, Turner (1978) and Hammond (1978) wrote the epithet for the Maya swidden and milpa, suggesting instead a much more sophisticated and biologically diverse system of mixed farming and silviculture. In an evolving series of papers, Edwards (1986), Fedick (1996a, 1996b; Fedick and Ford 1990), Ford (1986, 1991, 1998; Ford and Wernecke 2002), and Gómez-Pompa (1987) suggested that the contemporary forests of the Yucatán Peninsula were largely anthropogenic due to the ancient Maya's manipulation of its species composition. Moreover, Gómez-Pompa, Flores, and Sosa (1987) provided a modern-day mechanism to explain this transformation—the Yucatec Maya forest garden, known as the *pet kot*—and they described how it was created and managed. Very simply, the Maya actively selected certain species in their gardens and discouraged others. Moreover, the boundary of pet kot and forest was not distinct. Each was derived from the other. Maya folk ecology reveals an intimate knowledge of myriad forest utilities (Atran 1993), and the Yucatecan Mayan language itself is evocative of the forest as garden: *kannan k'aax,* translated as "well-cared-for forest," implies a human curatorial relationship to the forest.

We agree that the contemporary forests of the southeastern Petén—that is, eastern Guatemala and western Belize—are anthropogenic, the result of the active enrichment, encouragement, and culling of various woody plant species by the Maya and of disturbance by periodic fires, whether intentional or unintentional, that swept the area beginning with the introduction of silviculture and agriculture in the Preclassic period. In other words, the contemporary Maya forest may be one huge feral forest garden, the origins of which date back to at least 4000 BP (Pohl et al. 1996).

We explore this historical ecological concept, however, from a vantage point different from history and archaeology alone: by using the analytical methods of phytosociology. Toward this end, we pose and test three hypotheses: (1) the alpha diversities of these forests are low compared to the alpha diversities of areas of similar latitude and climate that have not been submitted to such pervasive disturbance; (2) the beta diversity among widely spaced samples of these forests is small (in other words, the forests of the eastern Petén have a uniform species composition); and (3) the oligarchies of these forests are composed principally of species that were—and are—of economic value to the Maya.

RESEARCH BACKGROUND

We conducted botanical inventories of three forest sites in distinct settings within 30 kilometers of each other, located in the Cayo District of western Belize (the extreme eastern Petén): El Pilar, Terra Nova, and Ix Chel (figure 1.1). All three sites were abandoned around 1000 BP and have not been recolonized or slashed and burned in any substantive way since. Our forest inventories were initially conducted in support of the Belize Ethnobotany Project (Balick 1991; Balick and Mendelsohn 1992; Balick et al. 2000), designed to provide data as to the diversity, quantity, and patterns of distribution of as many economically valuable plants as possible in Cayo District. For this reason, we chose the sites for their edaphic heterogeneity in terms of inclination, terrain, and drainage. El Pilar, the site of a major Classic period Maya city, is in the undulating, well-drained uplands on an escarpment north of the Belize River valley. Terra Nova is in the nearly level, poorly drained lowlands north of the Belize River and is the site of the world's first ethnomedicinal plant reserve (Balick, Arvigo, and Romero 1994). Ix Chel is on a steep, rocky slope above the Macal River, the site of a medicinal plants trail (Arvigo 1992, 1994). The settlement densities of the Late Classic period on the sites reflect these patterns: El Pilar, Terra Nova, and Ix Chel represent a gradient of structure density during the Late Classic period of 200, 50, and 2 per square kilometer, respectively (Fedick and Ford 1990).

All woody stems (including lianas) greater than or equal to 1.5 centimeters DBH were sampled on each site (appendix 1.1). This criterion gave a far greater species number than the conventional 10 centimeters DBH threshold of inclusion (figure 1.3) used in the majority of phytosociological studies (Campbell 1989). We continued sampling until asymptote was approximated on a species-area curve,

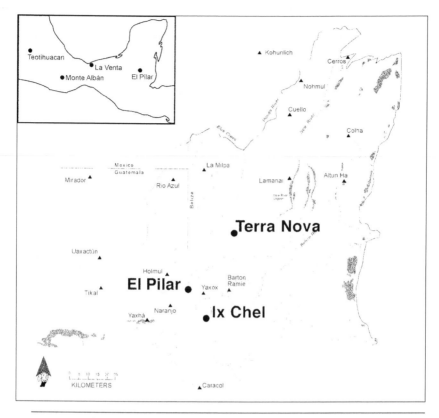

FIGURE 1.1 Three Maya forests.

giving us a reasonable certainty that we had sampled most of the species in the area (figure 1.2). We measured the DBH of each stem, from which we calculated its basal area (for individuals with multiple stems or lianas, the sum of the basal areas), and the relative dominance[1] for each taxon (table 1.3).

We collected voucher specimens for each individual plant until we were confident of our ability to recognize discrete species (or morphocategories) in the field. Representative voucher specimens of every taxon were identified by comparing them with reference specimens in the herbarium of the New York Botanical Garden. For various reasons (such as damage or loss), we were unable to identify 4 percent of the plants to either species or morphocategory, so we classified these plants as "unknown." We determined the Maya's ethnobotanical uses of the species in our samples from the literature (for all linguistic subgroups of the Maya, not just the Yucateca Maya), as well as from interviews with contemporary Yucatecan Maya in Cayo District (appendix 1.2).

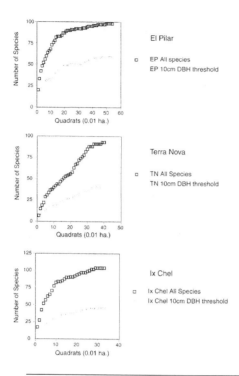

FIGURE 1.2 Species/area curves for three Maya forests.

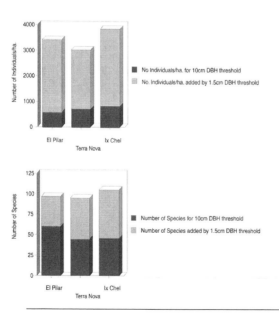

FIGURE 1.3 The attenuation of information using the 10-centimeter DBH threshold of inclusion versus the 1.5-centimeter DBH threshold.

Since first being applied to the phytosociology of tropical forests (Peters et al. 1989), the word *oligarchy* has undergone various definitions (Campbell 1994; Pitman et al. 2001). In a general sense, it means that a small number of species usurp a disproportionate share of the resources—in terms of space and light, for example—while the majority of species scrap for the remainder. For this chapter, we define *oligarchy* as the subsets of 10 (and 20) species in each sample with the highest relative dominances. Having identified the species that comprised our oligarchies, we tested the hypothesis that the oligarchies were anthropogenic. The test involved: (1) determining whether the mean relative dominance of the oligarchic species was significantly higher than that of nonoligarchic species or than that of the forest as a whole and (2) whether the percentage utilization of the oligarchic species was significantly higher than that of the nonoligarchic species or than that of the forest as a whole.

ANALYSIS OF A FERAL FOREST

We identified 179 species in the three inventories of the El Pilar, Terra Nova, and Ix Chel forests. (They and the number and relative dominance of each are listed in appendix 1.1.) Table 1.1 summarizes the general parameters of the inventories: the area of the sample, the number of stems, basal area (extrapolated to one hectare), number of species, Shannon's index of diversity, and Shannon's index of equitability.[2] The three sites showed a striking uniformity of species richness (93, 88, and 103 species, respectively) at approximate asymptote on the species/area curve. Each site also approached asymptote rapidly (figure 1.2), suggesting that a small sample—less than a hectare (and in the case of Ix Chel, a third of a hectare)—would be representative of the forest as a whole. Not surprisingly, therefore, the similarities (as measured by Sorenson's index)[3] among the various paired sites were also high, ranging from 0.46 to 0.61 for all species in the three forests (table 1.2).

Each site had a relatively low index of equitability (0.80, 0.77, and 0.90, respectively)—a hint of the presence of oligarchies described later, although the indices of similarity among the oligarchies of the three sites were more variable. Note that the oligarchy of Ix Chel appears to be the outlier among the three in terms of its species composition; we address this issue later in this chapter.

Among the 179 species found on the three sites, 76 (42 percent) were found to have been, or to be today, of economic value to the Maya (appendix 1.2). This value would appear to be low compared to the 80–90 percent utilization found in some other neotropical forests inhabited by Native Americans (Boom 1985); however, we did not count species that were used for no purpose other than "firewood" or "construction" in our analysis. When these species were added to our analyses, our sites had rates of utilization comparable to those of other neotropical forests.

TABLE 1.1 Summary of Three Maya Forests

EXTRAPOLATED

Site	Area	Minimum DBH	No. Stems ha^{-1}	Basal Area (m^{2}ha^{-1})	No. Spp.	H[a]	J[b]
El Pilar	0.54 ha.	$\geq$1.5 cm	3,036	25.83	93	3.65	0.80
Terra Nova	0.40 ha.	$\geq$1.5 cm	3,021	33.83	88	3.45	0.77
Ix Chel	0.33 ha.	$\geq$1.5 cm	4,087	35.96	103	4.13	0.90

[a] Shannon diversity index, $(H) = -\Sigma_i P_i ln P_i$, where P_i = no. individuals of sp i/total no. individuals all spp. (Greig-Smith, 1983).

[b] Shannon equitability index, $(J) = H(lnS)^{-1}$, where S = total no. spp. (Greig-Smith 1983).

TABLE 1.2 Indices of Similarity of Three Maya Forests

SORENSON'S INDEX[a]

Site	All Spp.	20 Most Dominant Spp.	10 Most Dominant Spp.
El Pilar and Terra Nova	0.57	0.62	0.50
El Pilar and Ix Chel	0.46	0.40	0.30
Ix Chel and Terra Nova	0.61	0.35	0.30

[a] $H = 2a(2a + b + c)^{-1}$, where a = number of species common to both plots; b = number of species unique to area 1; c = number of species unique to area 2 (Greig-Smith 1983).

The species that composed the oligarchies (defined as both the top 10 and 20 species in terms of their relative dominances) of the three forests are listed in tables 1.3 through 1.5. Each of the three samples was highly oligarchic: the top 10 most dominant species in each forest usurped a minimum of 57 percent (and in the case of Terra Nova an astonishing 73 percent) of the forests' footprint in terms of basal area. Species that are of economic importance to the Maya (appendix 1.2) are in boldface type in tables 1.3 through 1.5. Note that they comprise a majority of the species in the oligarchies, leading to the principal question of this chapter: Are the oligarchies anthropogenic? One way to address this question is cultural. At all three sites, the percentage utilization of the oligarchic species was significantly greater (P < 0.001) than that of the nonoligarchic species (table 1.6).

Another way to address this question is ecological and structural, by comparing the mean relative dominance of species of economic value to the Maya with those of no value (table 1.7). Again, at all three sites, the differences were

TABLE 1.3 Twenty Most Dominant Tree Species ≥ 1.5-centimeter DBH at El Pilar

SPECIES	RELATIVE DOMINANCE[a]
Pouteria reticulata (Engl.) Eyma	9.57
***Cryosophila stauracantha* (Heynh.) R. Evans**[b]	7.55
***Sabal morrisiana* H. H. Bartlett & L. H. Bailey**	7.35
***Alseis yucatanensis* Standl.**	6.00
***Swietenia macrophylla* King**	5.89
***Spondias radlkoferi* Donn. Sm.**	5.17
***Licania platypus* (Hemsl.) Fritsch**	4.51
***Tabebuia rosea* (Bertol.) DC**	4.11
***Attalea cohune* Mart.**	3.76
***Aspidosperma cruentum* Woodson**	2.96
Total Relative Dominance Top 10 Species	56.87
Simarouba glauca DC.	2.68
***Pseudolmedia spuria* (Swartz) Griseb.**	2.68
***Simira salvadorensis* (Standl.) Steyermark**	2.35
***Pouteria campechiana* (HBK) Baehni**	2.19
***Lonchocarpus castilloi* Standl.**	2.07
***Bursera simaruba* (L.) Sarg.**	2.02
***Protium copal* (Schltdl. & Cham.) Engl.**	1.79
Lippa myriocephala Schltdl. & Cham.	1.61
Croton guatemalensis Lotsy	1.54
***Terminalia amazonia* (J. F. Gmel.) Exell**	1.48
Total Relative Dominance Top 20 Species	77.28
Total Relative Dominance Remaining 73 Species	22.72

[a] Relative dominance = total basal area for species A/total basal area for all species.
[b] **Bold face** indicates species that are of economic importance to the Maya.

statistically significant (ranging from P < 0.05 to P < 0.001). Moreover, a comparison of the mean relative dominance of those species not of utility to the Maya and the mean relative dominance of all of the species in each sample in every case proves not to be significant. These results imply that the oligarchy may indeed be the result of human selection. In other words, the species that have fared best in this forest over the past several thousand years have been those encouraged by the Maya.

TABLE 1.4 Twenty Most Dominant Tree Species ≥ 1.5-Centimeter DBH at Terra Nova

SPECIES	RELATIVE DOMINANCE
Attalea cohune **Mart.**[a]	19.27
Alseis yucatanensis **Standl.**	14.44
Vitex gaumeri **Greenm.**	9.30
Cryosophila stauracantha **(Heynh.) R. Evans**	5.47
Pouteria reticulata (Engl.) Eyma	5.18
Sabal mauritiiformis **(H. Karst.) Griseb. & H. Wendl ex Griseb.**	4.05
Pouteria campechiana **(H.B.K.) Baehni**	3.94
Manilkara zapota **(L.) P. Royen**	3.87
Unknown TN 4	3.55
Simira salvadorensis **(Standl.) Steyerm.**	3.49
Total Relative Dominance Top 10 Species	72.56
Bursera simaruba **(L.) Sarg.**	2.80
Bunchosia lindeniana **A. Juss.**	2.27
Tabebuia rosea **(Bertol.) DC**	2.13
Brosimum alicastrum **Sw.**	1.94
Talisia oliviformis **(H.B.K.) Radlk.**	1.46
Terminalia amazonia **(J. F. Gmel.) Exell**	1.39
Guettarda combsii **Urban**	1.25
Protium copal **(Schltdl. & Cham.) Engl.**	1.10
Aspidosperma cruentum **R. E. Woodson**	0.92
Dendropanax arboreus **(L.) Decne. & Planch.**	0.76
Total Relative Dominance Top 20 Species	88.58
Total Relative Dominance Remaining 68 Species	11.42

[a] **Bold face** indicates species that are of economic importance to the Maya.

TABLE 1.5 Twenty Most Dominant Tree Species ≥ 1.5-Centimeter DBH at Ix Chel

SPECIES	RELATIVE DOMINANCE
Attalea cohune **Mart.**[a]	**29.12**
Bursera simarouba **(L.) Sarg.**	**12.75**
Vitex gaumeri **Greenm.**	**7.76**
Cryosophila stauracantha **(Heynh.) R. Evans**	**5.86**
Simira salvadorensis **Standl.**	**4.89**
Lonchocarpus castilloi (Standl.) Steyerm.	3.56
Zuelania guidonia **(Sw.) Britton & Millsp.**	**2.00**
Brosimum alicastrum **Sw.**	**1.76**
Piscidia piscipula **(L.) Sarg.**	**1.65**
Spondias radlkoferi **Donn. Sm.**	**1.63**
Total Relative Dominance Top 10 Species	70.98
Protium copal **(Schltdl. & Cham.) Engl.**	**1.50**
Nectandra hihua (Ruiz & Pav) Rohmer	1.50
Pouteria campechiana **(H.B.K.) Baehni**	**1.43**
Thouinia canescens **v. paucidentata Radlk.**	**1.41**
Simarouba glauca **DC.**	**1.39**
Piper neesianum **C. DC.**	**1.34**
Hyperbaena mexicana Miers	1.27
Swartzia cubensis **(Britton & P. Wilson) Standley**	**1.26**
Metopium brownei (Jacq.) Urb.	1.03
Annona reticulata **L.**	**0.93**
Total Relative Dominance Top 10 Species	84.04
Total Relative Dominance Remaining 83 Species	**15.96**

[a] **Bold face** indicates species that are of economic importance to the Maya.

TABLE 1.6 Percentage Utilization by the Maya

SITE	EL PILAR	TERRA NOVA	IX CHEL	MAYA MTN. 1[2]	MAYA MTN. 2[2]
% Utilization of Oligarchic Spp.[1]	90	90	80	60	60
vs.					
% Utilization of Nonoligarchic Spp.[3]	35***	40***	42***	40 (NS)	48 (NS)

[1] Defined as the top 20 species in terms of relative dominance.
[2] Analysis did not include one morphocategory and six species not listed in Balick et al. 2000.
[3] Does not include unidentified species that could not be assigned to morphocategories.
*** P < 0.001 (Chi-square Test performed on actual numbers, not on percentages).

TABLE 1.7 Mean Relative Dominance of Species of Economic Value vs. No Economic Value

MEAN RELATIVE DOMINANCE	EL PILAR	TERRA NOVA	IX CHEL
Of Spp. of Economic Value to Maya	1.68 (2.12)	1.77 (3.68)	1.52 (4.22)
vs.			
Of Spp. Not of Economic Value to Maya[1]	0.43 (1.34)*	0.38 (0.94)**	0.19 (0.24)***

[1] Does not include unidentified species that could not be assigned to morphocategories.
* P < 0.05 (Mann-Whitney U Test).
** P < 0.02 (Mann-Whitney U Test).
*** P < 0.001 (Mann-Whitney U Test).

DISCUSSION

One might argue that the high rates of utility of the oligarchic species versus the rates of utility of the nonoligarchic species are simply a function of the Maya having experimented first (and most often) on the species that they would be most likely to encounter in the wild—in other words, in the oligarchy. This argument might be valid if the rare species were unnamed in Yucatecan Mayan. But the facts show the contrary: the contemporary Yucatecan Maya recognize and name most species in the samples—common as well as rare. The Maya forest is a well-understood biota, one of cognizance (Atran 1999).

The individual alpha diversities of the three samples of Maya forest, as well as the beta diversities[4] among them, are low. This is surprising given that we selected the three sites for their edaphic heterogeneity. Data sets from other seasonal subtropical New and Old World forests are rare, making it difficult to make rigorous comparisons. Consider, for example, inventories of seasonal lowland Amazonian forests (Campbell 1994; Campbell et al. 1986), which do not approach asymptote even after 3 hectares (using a threshold of inclusion of 10 centimeters DBH, no less). Inventories (using a comparable DBH to our Belizean studies) of seasonal subtopical forests in southern China, close to the same latitude as Belize, yield approximately two times as many species at asymptote (An et al. 1999; Wang et al. 1999), and surveys of the Atlantic coastal forests of Brazil, which have a comparable (but southern) latitude to that of Belize, have revealed some of the highest alpha and beta diversities ever measured (Mori et al. 1983; Thomas et al. 1998).

Likewise, the extreme oligarchies of these Maya forests are comparable to what one would expect in a naturally disturbed tropical forest, such as occurs on the *várzea* floodplains of Amazonia. In the *várzea*, there is a distinct positive correlation between the magnitude of disturbance and the magnitude of the oligarchy (Campbell 1994). Clearly, the large oligarchies of the three Maya forest sites indicate that the sites have been submitted to a pervasive disturbance regime. We submit that this disturbance is to a significant extent anthropogenic, consisting of enrichment with species of economic value to humans and, of course, culling by fire. However, it is important to note that the oligarchies of our three sites are not composed of fire-adapted or early-succession postswidden species (with the exception of *Attalea cohune*). Yet fire may nevertheless have been and may still be an important agent for the creation of these oligarchies. Throughout the eastern Petén, we have observed the Maya using stone walls and cleared breaks to protect desirable forest saplings or gardens from the wildfires that periodically range over the land. The fires do the culling, but the Maya do the selection.

These conclusions are supported by our affirmation that the oligarchic species of the three forests have a significantly greater rate of utilization than the nonoligarchic species and that the mean relative dominance of the oligarchic species is significantly greater than that of the nonoligarchic species. Our results suggest an explicit intentionality and human agency (see Erickson and Balée, chapter 7, this volume) in the creation of the contemporary Maya forest. The most parsimonious explanation for these patterns of impoverishment and concomitant high utility in our samples of the Maya forests is that they are the descendants of an ancient kannan k'aax that once covered vast areas of the greater Petén.

The Ix Chel forest is distinct from the others in terms of its oligarchic species composition—perhaps because it is located on steep, rocky terrain unsuitable for agriculture and, not surprisingly, had a low settlement density during the Classic period. For this reason, it simply may not have been as extensively manipulated over time as the other sites. For example, it abounds with poisonwood (*Metopium brownei*), by any measure an undesirable species (in the Anacardiaceae, a family infamous for its toxic principles) that is the first to be culled from contemporary Maya forest gardens and that, although present, is rare on the other two sites. Therefore, of the three sites, Ix Chel may most closely resemble the original pre-Maya forest. Regardless, in spite of its edaphic and oligarchic dissimilarity with the other sites, it is part of the same pervasive community as the two other sites, especially Terra Nova. We suspect that we may never be able to locate a pre-Maya (control) forest in Cayo District—one that would be representative of the Cayo forest before human perturbation— a frustrating concession. Certainly, access to such a forest would quickly validate or invalidate our hypotheses. However, it may simply be the circumstance that human perturbations and modifications of the Maya forest over the past 4,000 years have been so pervasive that a control forest in the lowland Maya forest does not exist.

Forested upland areas within 100 kilometers of our study sites are known, however, to have been sparsely occupied by the ancient Maya, and thus are areas where the forest may not be as anthropogenic. One is the southern foothills of the Maya Mountains, where Brewer and Webb (2002) conducted two 1-hectare inventories of all stems greater than or equal to 5.0 centimeters DBH on the Bladen Nature Reserve (BNR). It is a region of complex geology: a conjunction of granite, shale, and limestone (Wright et al. 1959), dissected by streams and concomitant alluvial deposits (Durham 1996). Although the settlement patterns of the region are not well studied, the evidence points to low settlement densities for most of the Maya occupation (Graham 1987; Hammond 1981). In the BNR region, there are sites of small-scale mineral extraction and, in the Late Classic period, a few raised terraces in pockets of alluvial soil.

Even at a hefty threshold of inclusion of 5.0 centimeters DBH, the two BNR forest plots, having 114 and 104 species, are more diverse than those of Cayo District. At one hectare, the BNR plots do not approach asymptote on a species-area curve. These high diversities are as one would expect in a region of limited human agency. Like the Cayo forests, the BNR forests are highly oligarchic. However, unlike Cayo forests, the BNR oligarchies are not significantly richer in species of economic value (utilities derived from the same references as given earlier) than the nonoligarchies (table 1.6). Moreover, there is no significant difference in the mean relative dominance of the species of economic importance versus those of no value (table 1.7). These results are opposite to what we found in Cayo District and precisely what one would expect from a forest that was not heavily manipulated by humans.

The Maya signature on the Cayo forests, last penned more than a millennium ago, is discernible not just in their phytosociologies, however. It is physical as well, illustrated by the inverse functions of total forest basal area and the density of Maya structures. Where blocky limestone structures and rubble occupy much of the forest floor, there is less room for the establishment of massive tree trunks and roots. Liana frequency, by contrast, is a direct function of the structure density. In several parts of the Neotropics, high numbers of lianas have been correlated with human perturbation and, in Amazonia, with anthropogenic dark earth (Balée and Campbell 1990; Graham, chapter 2, and Neves and Peterson, chapter 9, this volume).

In conclusion, the contemporary Maya forests of Cayo District may be regarded from various perspectives. A preservationist would argue that they are the survivors of a much richer biotic community that was demolished by humans and that the forests we see today are the residue of that impoverishment. Remarkably, these forests, more than a millennium after the Maya civilization collapsed and abandoned them, still bear the scars of human perturbation, in both physical and phytosociological terms—casting an ominous perspective on the current rampant deforestation in the tropics and subtropics, including the Neotropics in general.

Anthropologists might be more optimistic. They might argue that the forests are a human-engineered wonder that once supported a large Maya population, yet still maintained a moderate biological diversity and forest cover. Certainly, the high degree of utility of these forests reveals the Maya's mastery of their biota and a concomitant knowledge of its virtues—a virtuosity that has been threatened since the European occupation of region.

A conservationist might employ a hybrid argument: that the patterns of Maya land use (both ancient and contemporary) may show us the path to a more rational way of making a living from the Maya forest without destroying it—certainly a better alternative to contemporary cattle ranching and agrarian monocultures.

ACKNOWLEDGMENTS

We dedicate this paper to Leopoldo Romero, Yucatecan Maya bushmaster, parataxonomist, and teacher of Santa Elena, Belize, without whom this research would have been impossible. This research was supported by funds from the Henry Luce Foundation, U.S. AID, Grinnell College, and the late Ann Smeltzer. We thank the New York Botanical Garden (NYBG) for making its Belizean collections available to us in order to identify our voucher specimens, Daniel Atha of the NYBG for reviewing our botanical nomenclature, and William Balée and Clark Erickson for numerous helpful suggestions.

APPENDIX 1.1 Species in Three Maya Forests

FAMILY, GENUS, SPECIES	VULGAR NAME	NUMBER HA^{-1} (RELATIVE DOMINANCE)		
		EL PILAR	TERRA NOVA	IX CHEL
Acanthaceae				
Aphelandra scabra (Vahl) Sm.	anal; anal-chae			
	palo verde; wild ax canan			60(0.05)
Agavaceae				
Dracaena americana Donn. Sm.	wild isote	56(1.00)	33(0.10)	
Anacardiaceae				
Astronium graveolens Jacq.	jabin; jobillo	27(0.25)	3(0.21)	
Comocladia guatemalensis Donn. Sm.				7(0.01)
Metopium brownei (Jacq.) Urb.	che-chen;			
	black poisonwood	9(0.75)	13(0.66)	97(1.03)
Spondias radlkoferi Donn. Sm.	rum-p'ok; hog plum	63(5.17)	20(0.63)	47(1.63)
Annonaceae				
Annona glabra L.	cork wood	2(<0.005)		
A. reticulata L.	wild custard apple		23(0.23)	113(0.93)
Malmea depressa (Baill.) R. E. Fr.	sufrekaya	19(0.35)	23(0.10)	160(0.60)
Apocynaceae				
Aspidosperma cruentum Woodson	my lady	65(2.96)	28(0.92)	
Tabernaemontana arborea Rose	juevo de caballo	7(0.05)	8(0.30)	30(0.68)
Thevetia ahouai (L.) A. DC.	juevo de chucho	15(0.03)	3(<0.005)	27(0.03)
Araliaceae				
Dendropanax arboreus (L.) Decne. & Planch.	lion hand; lion heart; mano e león tree	13(0.34)	10(0.76)	10(0.07)

FAMILY, GENUS, SPECIES	VULGAR NAME	NUMBER HA^{-1} (RELATIVE DOMINANCE)		
		EL PILAR	**TERRA NOVA**	**IX CHEL**
Arecaceae				
Bactris mexicana Mart.	warrie cahoon	15(0.01)		
Chamaedorea TN1			3(<0.005)	
Cryosophila stauracantha (Heynh.) R. J. Evans	give-and-take	433(7.55)	558(5.47)	90(5.86)
Desmoncus orthacanthos Mart.	basket tietie			17(0.03)
Attalea cohune Mart.	corozo; cahoon	17(3.76)	110(19.27)	130(29.12)
Sabal mauritiiforme Griseb. & H. Wendl.	botán; bay leaf	63(7.35)	63(4.05)	
Schippia concolor Burret	silver thatch; silver palmetto			17(0.24)
Asteraceae				
Critonia morifolia (Mill.) R. M. King & H. Rob.	palo verde	202(0.80)		3(0.01)
Bignoniaceae				
Callichlamys latifolia (Rich.) K. Schum.	shna'corts; bejuco negro	30(0.14)		
Godmania aesculifolia (HBK.) Standl.				23(0.05)
Tabebuia rosea (Bertol.) DC.	roble	22(4.11)	33(2.13)	37(0.09)
Arrabidaea pubescens (L.) A. H. Gentry	allspice tie tie	44(0.20)	65(0.23)	
Bombacaceae				
Pachira aquatica Aubl.	uacut; provision tree			3(0.63)
Burseraceae				
Bursera simaruba (L.) Sarg.	cha-ca; gumbo limbo	13(2.02)	13(2.80)	230(12.75)
Protium copal (Schltdl. & Cham.) Engl.	pomte; copal	111(1.79)	35(1.10)	47(1.50)
Capparidaceae				
Unknown IX 1	fiddle wood			3(0.01)
Cecropiaceae				
Cecropia peltata L.	trumpet leaf	4(0.40)	3(0.09)	

FAMILY, GENUS, SPECIES	VULGAR NAME	NUMBER HA^{-1} (RELATIVE DOMINANCE)		
		EL PILAR	TERRA NOVA	IX CHEL
Pourouma bicolor Mart.	bahb; big leaf	7(0.13)		
Chrysobalanaceae				
Hirtella americana L.			8(0.11)	
Licania platypus (Hemsl.) Fritsch	succótz; monkey apple	41(4.51)		
Combretaceae				
Bucida buceras L.	puk te; bullet tree	13(0.78)		
Terminalia amazonia (J. F. Gmel.) Exell	nargusta	22(1.48)	18(1.39)	
Dioscoreaceae				
Dioscorea bartlettii C. V. Morton	ya-ya-chil cocomecca blanca; red China root	11(0.01)	8(<0.005)	7(<0.005)
Euphorbiaceae				
Croton guatemalensis Lotsy	black pepper; Sta. María	11(1.54)		70(0.46)
Sebastiania tuerckheimiana Pax & K. Hoffm.) Lundell	white poison wood		10(0.02)	40(0.69)
Flacourtiaceae				
Casearia sylvestris Sw.				13(0.10)
Laetia procera (Poepp.) Eichler	drunken bayman wood	24(0.59)		
L. thamnia L.	bullyhob			60(0.60)
Unknown IX 1			35(0.29)	3(<0.005)
Zuelania guidonia (Sw.) Britton & Millsp.	tamai		10(0.30)	107(2.00)
Lauraceae				
Licaria 1	tabaquillo		12(0.04)	7(0.01)
Licaria IX 1				50(0.55)
Licaria peckii (I. M. Johnston) Kosterm.	bahon		5(0.02)	10(0.04)
Nectandra hihua (Ruiz & Pav.) Rohwer	laurél	15(0.07)	5(0.15)	236(1.50)
N. nitida Mez	cot tree		3(0.16)	
Nectandra Unknown 1			18(0.15)	27(0.27)
Phoebe 1	granadillo		48(0.48)	3(0.03)
Lauraceae Unknown EP 1	rosewood	4(1.07)		

FAMILY, GENUS, SPECIES	VULGAR NAME	NUMBER HA^{-1} (RELATIVE DOMINANCE)		
		EL PILAR	TERRA NOVA	IX CHEL
Lauraceae Unknown EP 2	false rosewood	7(0.11)		
Lauraceae Unknown TN 1			28(0.41)	
Legum. Caesalpiniaceae				
Bauhinia calderonii (Rose) Lundell	bull hoof	20(0.49)	3(0.04)	27(0.09)
B. divaricata L.	bull hoof			7(0.02)
Legum. Fabaceae				
Erythrina EP 1	red seeds	2(0.70)		
Inga EP 1	inga	7(0.04)		
Lonchocarpus amarus Standl.	bitterwood; barbasco	15(0.49)		
L. castilloi Standl.	white/black cabbage bark	13(2.07)	8(0.07)	100(3.56)
Piscidia piscipula (L.) Sarg.	jabim; palo de gusano	4(0.03)	3(0.05)	110(1.65)
Swartzia cubensis (Britton & P. Wilson) Standl. (UNK EP 6)	llora sangre/blood vine	40(0.46)	10(0.01)	87(1.26)
Legum. Mimosaceae				
Acacia cornigera (L.) Willd.	white cockspur	43(0.58)	3(0.01)	
Acacia cf. *dolichostachya*	John Crow (unknown)			3(0.24)
Acacia gentlei Standl.	red cockspur		13(0.17)	176(0.81)
Acacia globulifera Saff.	white cockspur	9(0.05)		50(0.50)
Zygia 1	deerhorn		5(0.02)	97(0.48)
Mimosaceae IX 1				3(0.15)
Leguminosae				
Unknown EP 1	pole wood	7(0.21)		
Unknown IX 1				13(0.59)
Loganiaceae				
Strychnos panamensis Seem.	chicoloro	19(0.08)	18(0.07)	7(0.01)
Malphigiaceae				
Bunchosia swartziana Griseb.	mourín			3(0.027)
Bunchosia lindeniana A. Juss.	luín	37(1.22)	100(2.27)	20(0.029)

FAMILY, GENUS, SPECIES	VULGAR NAME	NUMBER HA^{-1} (RELATIVE DOMINANCE)		
		EL PILAR	TERRA NOVA	IX CHEL
Malvaceae				
Hampea trilobata Standl.	mahow; mahawa	7(0.65)		30(0.10)
Melastomataceae				
Miconia EP 1		31(0.44)		
Meliaceae				
Swietenia macrophylla King	mahogany	15(5.89)		
Trichilia havanensis Jacq.	spoon tree		5(0.03)	33(0.06)
T. hirta L.			3(<0.005)	
Trichilia cf. *pallida* Sw.	brown berries		10(0.02)	27(0.40)
Menispermaceae				
Hyperbaena mexicana Miers				50(1.27)
Monimiaceae				
Mollinedia guatemalensis Perkins	wild coffee	9(0.04)	38(0.15)	
Moraceae				
Brosimum alicastrum Sw.	ramon	20(1.27)	23(1.94)	7(1.76)
Ficus radula Willd.	higo de la montaña	9(0.64)		
F. schippii Standl.	mata palo	13(0.82)	13(0.61)	
Pseudolmedia spuria Griseb.	wild cherry	154(2.68)	85(0.63)	3(0.02)
Myrtaceae				
Eugenia cf. *axillaris* (sw.) Willd.	little green berries		3(<0.005)	23(0.03)
E. buxifolia (Sw.) Willd.	semillón; guaya-billo; bastard gumbo limbo	137(1.12)	5(0.11)	
E. oerstediana O. Berg	little green berries			3(0.03)
Myrcianthes fragrans (Sw.) McVaugh	little green berries		5(0.15)	90(0.25)
Pimenta dioica (L.) Merr.	allspice	4(0.50)	3(0.02)	37(0.15)
Nyctaginaceae				
Neea psychotrioides Donn. Sm.			5(0.01)	37(0.25)
Pisonia aculeata L.	una de gato vine	13(0.19)	5(0.03)	7(0.05)

FAMILY, GENUS, SPECIES	VULGAR NAME	NUMBER HA^{-1} (RELATIVE DOMINANCE)		
		EL PILAR	TERRA NOVA	IX CHEL
Ochnaceae				
Ouratea nitida (Swartz) Engl.			10(0.01)	3(<0.005)
Opiliaceae				
Agonandra EP 1	man vine	9(0.04)		
Piperacae				
Piper amalgo L.	red cordonsillo	92(0.42)	163(0.23)	33(0.05)
Piper neesianum C. DC.	black cordoncillo	18(0.20)	18(0.03)	193(1.34)
Polygonaceae				
Coccoloba hondurensis Lundell	wild grape		10(0.71)	
Coccoloba schiedeana Lindau	wild grape	2(<0.005)		10(0.33)
Rhamnaceae				
Krugiodendron ferreum Urb.	ax master	9(0.04)	3(0.03)	20(0.38)
Rubiaceae				
Alseis yucatanensis Standl.	lión; dzón	124(6.00)	275(14.44)	13(0.58)
Chiococca alba (L.) Hitchc.	kibish			3(0.01)
Chiococca 1	skunk vine	2(0.01)	13(0.05)	7(0.03)
Guettarda combsii Urb.	verde lucero; xtez-tab; glassy wood; casacb	9(0.18)	78(1.25)	20(0.10)
Guettarda IX 1	wild grape			10(0.23)
Psychotria chiapensis Standl.			3(<0.005)	
P. nervosa Swartz	wild ix canon		5(0.02)	13(0.01)
Psychotria EP 1	bastard wild coffee	216(0.86)		
Psychotria IX 1				10(0.01)
Randia cf. *pleiomeris* Standl.	wild okra		3(0.01)	17(0.10)
Randia xalapensis M. Martens & Galeotti	wild okra			7(0.01)
Simira salvadorensis (Standl.) Steyerm.	redwood; John Crow	50(2.35)	88(3.49)	120(4.89)
Rutaceae				
Zanthoxylum riedelianum Engl.	prickly yellow tree	55(0.14)		

FAMILY, GENUS, SPECIES	VULGAR NAME	NUMBER HA^{-1} (RELATIVE DOMINANCE)		
		EL PILAR	TERRA NOVA	IX CHEL
Z. juniperinum Poepp	sinan ché; lemoncillo	6(0.4)	8(0.25)	
Sapindaceae				
Allophylus cominia (L.) Sw.	tabaquillo		3(0.03)	
Cupania belizensis Standl.	grande Betty	52(0.87)	45(0.45)	153(0.10)
Sapindus saponaria L.	kinep (guinep)			40(0.17)
Talisia olivaeformis (H.B.K.) Radlk.			13(1.46)	77(0.22)
Thouinia paucidentata Radlk.	trifolia	56(0.28)		120(1.41)
Sapotaceae				
Chrysophyllum cf. *mexicanum* Brandegee ex Standl.				3(0.01)
Dipholis salicifolia A. DC.	sapote			3(<0.005)
Manilkara zapota (L.) P. Royen	chico sapote; chic ibúl	13(0.16)	3(3.87)	30(0.26)
Mastichodendron foetidissimum (Jacq.) Cronquist	mastic		10(0.09)	
Pouteria IX 1				7(0.10)
Pouteria campechiana (H.B.K.) Baehni	sapotillo rojo	56(2.19)	148(3.94)	47(1.43)
Pouteria reticulata (Engl.) Eyma	sapotillo blanco	111(9.57)	243(5.18)	3(<0.005)
Sapotaceae Unknown TN 1	mamay		3(0.28)	
Sapotaceae Unknown IX 1				3(0.05)
Simaroubaceae				
Simarouba glauca DC.	negrito; dysentery bark	41(2.68)	13(0.04)	83(1.39)
Smilacaceae				
Smilax 1	red China root vine	2(0.01)	6(0.01)	
Solanaceae				
Solanum EP 1	lava plato vine	24(0.09)		
Solanum EP 2	velvet leaf	2(<0.005)		

FAMILY, GENUS, SPECIES	VULGAR NAME	NUMBER HA^{-1} (RELATIVE DOMINANCE)		
		EL PILAR	TERRA NOVA	IX CHEL
Theophrastaceae				
Jacquinia aurantiaca Ait.			3(0.01)	
Deherainia smaragdina Decne.				10(0.02)
Unknown				
Unknown 2	yax coch; red ramon	6(0.20)		
Unknown 16	waco/ wild ocro vine	4(0.01)	3(<0.005)	
Unknown EP 1	(white latex)	2(0.02)		
Unknown EP 5 (EP 3)		2(0.07)		
Unknown EP 7	wairú	2(0.01)		
Unknown EP 8	Wild star apple; caimito	2(0.06)		
Unknown EP 9	conch root vine	2(0.01)		
Unknown EP 10	cuello de sapo	2(0.52)		
Unknown EP 11	fluffy flowers/sm. lvs.	22(0.28)		
Unknown EP 13	fuzzy stem	2(<0.005)		
Unknown EP 14	habanero vine	13(0.03)		
Unknown EP 17	false grande Betty	2(<0.005)		
Unknown EP 18		2(<0.005)		
Unknown EP 19		2(0.37)		
Unknown EP 20	bejuco blanco	80(0.07)		
Unknown TN 1	kibish		5(0.02)	
Unknown TN 2	(vine)		3(0.01)	
Unknown TN 3	(vine)		3(0.02)	
Unknown TN 4			10(3.55)	
Unknown TN 5	guinego		3(0.01)	
Unknown TN 6	puntero		3(0.01)	
Unknown IX 1				3(0.02)
Unknown IX 4	trifolia			27(0.38)
Unknown IX 5	kinep			3(0.01)
Unknown IX 6				3(0.04)
Unknown IX 7				27(0.30)

FAMILY, GENUS, SPECIES	VULGAR NAME	NUMBER HA^{-1} (RELATIVE DOMINANCE)		
		EL PILAR	TERRA NOVA	IX CHEL
Unknown IX 8	jobillo			3(0.03)
Unknown IX 9				3(0.14)
Unknown IX 10				10(0.24)
Unknown IX 11	guinego			13(0.04)
Unknown IX 12	yellow manchich			3(0.01)
Unknown IX 13	bohuna			3(<0.005)
Unknown IX 14	jobillo			3(<0.005)
Unknown IX 15	kibish			3(0.01)
Unknown vines		24(0.08)	70(0.35)	53(0.18)
No Voucher		9(0.16)	10(1.38)	57(0.30)
Verbenaceae				
Lippia myriocephala Schltdl. & Cham.	sakuché	24(1.61)		17(0.43)
Vitex gaumeri Greenm.	fiddle wood; yax-nik	63(0.43)	53(9.30)	63(7.76)
Violaceae				
Rinorea 1	bahon		100(0.40)	3(0.05)
Vitaceae				
Vitis tilliaefolia Humb. & Bonpl.	bejuco de agua	9(0.02)	5(0.02)	
Total		3,230(100.45)[1]	3,056(100.98)	3,889(98.16)
Total Species 179		93	88	103

Source: Vouchers were identified using specimens at the New York Botanical Garden, and in Balick, Nee, and Atha 2000; Nash and Dieterle 1976; Nash and Williams 1976; Standley and Record 1936; Standley and Steyermark 1946a, 1946b, 1949a, 1949b, 1952; Standley and Williams 1961, 1962, 1963, 1967, 1970, 1973, 1975; Standley, Williams, and Gibson 1974; Swallen 1955.

SPECIES (FAMILY)	MAYA USE (REFERENCES)
Acacia cornigera (Mimosaceae)	*home garden* (Allison 1983:29) *medicinal* (Arvigo 1992:24; Arvigo and Balick 1993:47; Roys 1931:312)
A. dolichostachya	*construction* (Atran 1993:656; Mutchnick and McCarthy 1997:176) *medicinal* (Mutchnick and McCarthy 1997:176)
A. globulifera	*medicinal* (Roys 1931:312)
Allophylus cominia (Sapindaceae)	*food* (Standley and Steyermark 1949b:237)
Alseis yucatanensis (Rubiaceae)	*construction* (178, Atran 1993:667; Mutchnick and McCarthy 1997) *brooms* (Romero 1997–2001:1)
Annona reticulata (Annonaceae)	*fiber* (Atran 1993:647) *food* (Arvigo 1992:46; Atran 1993:647; Roys 1931:272; Standley and Steyermark 1946a:279) *medicinal* (Arvigo 1992:46; Arvigo and Balick 1993:187; Comerford 1996:332; Rico-Grey, Cheams, and Mandujano 1991:152; Roys 1931:272) dye (Arvigo 1992:46; Standley and Steyermark 1946a:279) *plant remains at Colha: charcoal, seeds* (Caldwell 1980:264)
Attalea cohune (Arecaceae)	*construction* (Arvigo 1992:6; Standley and Steyermark 1949a:276) *food* (Arvigo 1992:6; Mutchnick and McCarthy 1997:174; Romero 1997–2001:1; Standley and Steyermark 1949a:276) *medicinal* (Arvigo 1992:6; Romero 1997–2001:1)
Bauhinia divaricata (Caesalpiniaceae)	*construction* (Rico-Grey, Cheams, and Mandujano 1991:152) *fiber* (Rico-Grey, Cheams, and Mandujano 1991:152; Standley and Steyermark 1946b:91) *fodder* (Rico-Grey, Cheams, and Mandujano 1991:152) *medicinal* (Arvigo 1992:48; Comerford 1996:333; Rico-Grey, Cheams, and Mandujano 1991:152; Roys 1931:308, 315)

SPECIES (FAMILY)	MAYA USE (REFERENCES)
Brosimum alicastrum (Moraceae)	*construction* (Standley and Steyermark 1946a:13; Atran 1993:663) *fodder* (Arvigo 1992:56; Atran 1993:663; Rico-Grey, Cheams, and Mandujano 1991:152; Roys 1931:272; Standley and Steyermark 1946a:13) *food* (Arvigo 1992:56; Atran 1993:663; Mutchnick and McCarthy 1997:178; Rico-Grey, Cheams, and Mandujano 1991:152; Romero 1997–2001:1; Roys 1931:272; Standley and Steyermark 1946a:13) *home garden* (Allison 1983) *magico-religious* (Roys 1931:272) *medicinal* (Orellana 1987:184; Rico-Grey, Cheams, and Mandujano 1991:152; Roys 19316:272) *plant remains at Colha: charcoal, seeds, phytoliths* (Caldwell 1980:266) *tools* (Atran 1993:663 Rico-Grey, Cheams, and Mandujano 1991:152)
Bucida buceras (Combretaceae)	*plant remains at Colha: charcoal, seeds, pollen, phytoliths* (Caldwell 1980:265) *seeds at Cerros* (Dunham, Jamison, and Leventhal 1989:309)
Bunchosia lindeniana (Malpighiaceae)	*medicinal* (Romero 1997–2001:8)
B. swartziana	*construction* (Rico-Grey, Cheams, and Mandujano 1991:152) *magico-religious* (Rico-Grey, Cheams, and Mandujano 1991:152) medicinal (Rico-Grey, Cheams, and Mandujano 1991:152)
Bursera simaruba (Burseraceae)	*construction* (Mutchnick and McCarthy 1997:174; Rico-Grey, Cheams, and Mandujano 1991:152) *fodder* (Rico-Grey, Cheams, and Mandujano 1991:152) *food* (Mutchnick and McCarthy 1997:174; Rico-Grey, Cheams, and Mandujano 1991:152) *home garden* (Allison 1983:28) *living Fences* (Arvigo 1992:40, Roys 1931:228; Standley and Steyermark 1946b:440) *magico-religious* (Arnason et al. 1980:359; Rico-Grey, Cheams, and Mandujano 1991:152; Standley and Steyermark 1946b:440)

SPECIES (FAMILY)	MAYA USE (REFERENCES)
	medicinal (Allison 1983, Arnason et al. 1980:347, 348, 355; Arvigo 1992:40; Arvigo and Balick 1993:89; Atran1993:653; Comerford 1996:333; Mutchnick and McCarthy 1997:174; Orellana 1987:184; Rico-Grey, Cheams, and Mandujano 1991:152; Romero 1997–2001:3; Roys 1931:228; Standley and Steyermark 1946b:440) *plant remains at Colha: charcoal, seeds, pollen, phytoliths* (Caldwell 1980:264) *tools* (Rico-Grey, Cheams, and Mandujano 1991:152)
Casearia sylvestris (Flacourtiaceae)	*fuel* (Alcorn 1984:582) *medicinal* (Alcorn 1984:582)
Chiococca alba (Rubiaceae)	*medicinal* (Arvigo 1992:22; Arvigo and Balick 1993:16; Comerford 1996:334; Roys 1931:223)
Cryosophila stauracantha (Arecaceae)	*construction* (Arvigo 1992:30; Mutchnick and McCarthy 1997:174) *food* (Arvigo 1992:30) *medicinal* (Arvigo 1992:30; Arvigo and Balick 1993:79; Romero 1997–2001:2) *tools* (Arvigo 1992:30; Arvigo and Balick 1993:79; Romero 1997–2001:2)
Chrysophyllum mexicanum (Sapotaceae)	*construction* (Mutchnick and McCarthy 1997:180) *food* (Mutchnick and McCarthy 1997:180) *fuel* (Atran 1993:670) *tools* (Standley and Williams 1967:219)
Chrysophyllum sp.	*pollen at Cerros* (Dunham, Jamison, and Leventhal 1989:311)
Coccoloba acapulcensis (Polygonaceae)	*artisanal, making bastones* (Ortiz 1994:16) *construction* (Rico-Grey, Cheams, and Mandujano 1991:153) *fodder* (Rico-Grey, Cheams, and Mandujano 1991:153) *food* (Rico-Grey, Cheams, and Mandujano 1991:153)
Critonia morifolia (Asteraceae)	*medicinal* (Romero 1997–2001:4 and 5)
Croton guatemalensis (Euphorbiaceae)	*medicinal* (Arvigo and Balick 1993:151; Standley and Steyermark 1949b:72)
Cupania belizensis (Sapindaceae)	*fuel* (Arvigo 1992:54; Arvigo and Balick 1993:83) *medicinal* (Arvigo 1992:54; Arvigo and Balick 1993:83)

SPECIES (FAMILY)	MAYA USE (REFERENCES)
Dendropanax arboreus (Araliaceae)	*construction* (Atran 1993:649; Mutchnick and McCarthy 1997:174) *medicinal* (Atran 1993:649)
Desmoncus orthacanthos (Arecaceae)	*fiber* (Arvigo 1992:10; Standley and Steyermark 1949a:258)
Dioscorea bartlettii (Dioscoreaceae)	*food* (Nations and Nigh 1980:16) medicinal (Arvigo 1992:58; Romero 1997–2001:15)
Diphysa carthagenensis (Fabaceae)	*fodder* (Rico-Grey, Cheams, and Mandujano 1991:154) *medicinal* (Arnason et al. 1980:355, 358; Rico-Grey, Cheams, and Mandujano 1991:154; Roys 1931:316)
Dracaena americana (Liliaceae)	*fiber* (Rico-Grey, Cheams, and Mandujano 1991:154) *ornamental* (Rico-Grey, Cheams, and Mandujano 1991:154)
Eugenia axillaris (Myrtaceae)	*medicinal* (Roys 1931:248)
Eugenia oestediana (Myrtaceae)	*food* (Alcorn 1984:643) *medicinal* (Alcorn 1984:643)
Eupatorium morifolium (Asteraceae)	*construction* (Nash and Williams 1976:82) *medicinal* (Arvigo and Balick 1993:85)
Ficus maxima and **F. schippi** (Moraceae)	*bark paper* (Standley and Steyermark 1946a:31) *construction* (Standley and Steyermark 1946a:31) *fodder* (Standley and Steyermark 1946a:31) *magico-religious* (Standley and Steyermark 1946a:31)
Guazuma ulmifolia (Sterculiaceae)	*construction* (Rico-Grey, Cheams, and Mandujano 1991:155; Roys 1931:276) *fiber* (Atran 1993:671; Roys 1931:276; Standley and Steyermark 1949b:412) *fodder* (Arvigo 1992:16, Standley and Steyermark 1949b:411) *food* (Arvigo 1992:16; Arvigo and Balick 1993:25; Atran 1993:671; Rico-Grey, Cheams, and Mandujano 1991:155; Standley and Steyermark 1949b:411) *home gardens* (Allison 1980:64) *medicinal* (Allison 1980:70; Arvigo 1992:16; Arvigo and Balick 1993:25; Atran 1993:671; Comerford 1996:335; Mutchnick and McCarthy 1997:180; Rico-Grey, Cheams, and Mandujano 1991:155; Roys 1931:276; Standley and Steyermark 1949b:412)

SPECIES (FAMILY)	MAYA USE (REFERENCES)
	tools (Rico-Grey, Cheams, and Mandujano 1991:155; Standley and Steyermark 1949b:412)
Guettarda combsii (Rubiaceae)	*construction* (Atran 1993:667; Mutchnick and McCarthy 1997:178)
Hampea trilobata (Malvaceae)	*fiber* (Atran 1993:651; Rico-Grey, Cheams, and Mandujano 1991:155; Standley and Steyermark 1949b:396) *home garden* (Allison 1983:29)[1]
Hirtella americana (Chrysobalanaceae)	*food* (Lundell 1938:46)
Jacquinia aurantiaca (Theophrastaceae)	*barbasco* (Roys 1931:267; Standley and Williams 1967:132) *medicinal* (Standley and Williams 1967:132; Roys 1931:267) *ornamental* (Standley and Williams 1967:132) *plant remains at Colha: charcoal, phytoliths* (Caldwell 1980:268)
Laetia thamnia (Flacourtiaceae)	*construction* (Atran 1993:659) *medicinal* (Comerford 1996:333)
Licania platypus (Chrysobalanaceae)	*food* (Lundell 1938:46; Mutchnik and McCarthy 1997:176–177)
Licaria sp. (Lauraceae)	*construction* (Atran 1993:659) *tools* (Atran 1993:659)
Lonchocarpus castilloi (Fabaceae)	*alcoholic beverage* (Atran 1993:666) *construction* (Arvigo 1992:44; Mutchnick and McCarthy 1997:176) *medicinal* (Mutchnick and McCarthy 1997:176) *tools* (Mutchnick and McCarthy 1997:176)
Malmea depressa (Annonaceae)	*construction* (Mutchnick and McCarthy 1997:174; Rico-Grey, Cheams, and Mandujano 1991:156) *food* (Rico-Grey, Cheams, and Mandujano 1991:156; Standley and Steyermark 4:288) *medicinal* (Comerford 1996:332; Rico-Grey, Cheams, and Mandujano 1991:156) *tools* (Mutchnick and McCarthy 1997:174)
Manilkara zapota (Sapotaceae)	*chicle* (Arvigo 1992:62; Atran 1993:669; Rico-Grey, Cheams, and Mandujano 1991:156; Roys 1931:297; Standley and Williams 1967:224) *construction* (Arvigo 1992:62; Atran 1993:669; Rico-Grey, Cheams, and Mandujano 1991:156) *food* (Arvigo 1992:62; Rico-Grey, Cheams, and Mandujano 1991:156; Roys 1931:297; Standley and Williams 1967:224) *medicinal* (Arnason et al. 1980:357; Rico-Grey, Cheams, and Mandujano 1991:156)

SPECIES (FAMILY)	MAYA USE (REFERENCES)
	plant remains at Colha: seeds (Caldwell 1980:267) *Manilkara* sp. *pollen at Cerros* (Dunham, Jamison, and Leventhal 1989:311)
Neea psychotrioides (Nyctaginaceae)	*ink* (Rico-Grey, Cheams, and Mandujano 1991:156)
Pachira aquatica (Bombacaceae)	*food* (Mutchnick and McCarthy 1997:174) *medicinal* (Arnason et al. 1980:351; Arvigo and Balick 1993:137; Atran1993:651; Mutchnick and McCarthy 1997:174; Orellana 1987:224) *plant remains at Colha: charcoal, seeds, pollen, phytoliths* (Caldwell 1980:264)
Pimenta dioica (Myrtaceae)	*construction* (Atran1993:664; Mutchnick and McCarthy 1997:178) *embalming* (Arvigo 1992:26; McVaugh 1963:385) *home garden* (Allison 1983) *medicinal* (Arnason et al. 1980:357; Arvigo 1992:26; Arvigo and Balick 1993:3; Atran1993:664; Comerford 1996:334; McVaugh 1963:385; Mutchnick and McCarthy 1997:178; Orellana 1987:227; Romero 1997–2001:5) *plant remains at Colha: charcoal, pollen, phytoliths* (Caldwell 1980:267) *Spice* (Arvigo 1992:26; McVayh 1963:385; Romero 1997–2001:4) *tools* (Mutchnick and McCarthy 1997:178)
Piper amalago (Piperaceae)	*medicinal* (Arnason et al. 1980:347, 356; Arvigo 1992:38; Arvigo and Balick 1993:31; Comerford 1996:334; Romero 1997–2001:13–14)
P. neesianum	*medicinal* (Comerford 1996:334)
Piscidia piscipula (Fabaceae)	*barbasco* (Rico-Grey, Cheams, and Mandujano 1991:156; Standley and Steyermark 1946b:337) *construction* (Atran1993:665; Mutchnick and McCarthy 1997:176; Rico-Grey, Cheams, and Mandujano 1991:156; Standley and Steyermark 1946b:337) *handcrafts* (Rico-Grey, Cheams, and Mandujano 1991:156) *medicinal* (Arvigo and Balick 1993:97; Atran1993:665; Orellana 1987:228; Rico-Grey, Cheams, and Mandujano 1991:156; Standley and Steyermark 1946b:337) *timber* (Rico-Grey, Cheams, and Mandujano 1991:156)
Pisonia aculeate (Nyctaginaceae)	*medicinal* (Rico-Grey, Cheams, and Mandujano 1991:156; Roys 1931:217) *seeds at Cerros* (Dunham, Jamison, and Leventhal 1989:309)

SPECIES (FAMILY)	MAYA USE (REFERENCES)
Pouteria campechiana (Sapotaceae)	*chicle adulterant* (Standley and Williams 1967:236) construction (Standley ad Williams 1967:236) food (Standley and Williams 1967:236)
Protium copal (Burseraceae)	*fencing* (Atran1993:653) *magico-religious* (Arnason et al. 1980:358; Arvigo 1992:36; Arvigo and Balick 1993:57; Atran1993:653; Romero 1997–2001:7; Roys 1931:277; Standley and Steyermark 1946b:442) *medicinal* (Arnason et al. 1980:347; Arvigo 1992:36; Arvigo and Balick 1993:57; Atran1993:653; Comerford 1996:333; Romero 1997–2001:7; Roys 1931:278) *varnish* (Arvigo 1992:36; Standley and Steyermark 1946b:442; Romero 1997–2001:7) *resin at Cerros* (Dunham, Jamison, and Leventhal 1989:309)
Pseudolmedia oxyphyllaria (Moraceae)	*construction* (Atran1993:663; Mutchnick and McCarthy 1997:178) *food* (Atran1993:663; Mutchnick and McCarthy 1997:178; Romero 1997–2001:14) *medicinal* (Atran1993:663, Romero 1997–2001:14)
Pseudolmedia spuria (Moraceae)	*food* (Standley and Steyermark 1946a:55) *medicinal* (Mutchnick and McCarthy 1997:178)
Psychotria nervosa (Rubiaceae)	*magico-religious* (Alcorn 1984:768) *medicinal* (Alcorn 1984:768)
Sabal mauritiiforme (Arecaceae)	*construction* (Mutchnick and McCarthy 1997:174, Romero 1997–2001:3 and 15; Standley and Steyermark 1949a:289) *food* (Mutchnick and McCarthy 1997:174, Romero 1997–2001:3) *medicinal* (Romero 1997–2001:3) *seeds at Cerros* (Dunham, Jamison, and Leventhal 1989:309)
Sapindus saponaria (Sapindaceae)	*barbasco* (Standley and Steyermark 1949b:256) *construction* (Rico-Grey, Cheams, and Mandujano 1991:157) *home garden* (Allison 1983) *magico-religious* (Roys 1931:309, Standley and Steyermark 1949b:256) *soap* (Roys 1931:309, Standley and Steyermark 1949b:256) *tanning* (Rico-Grey, Cheams, and Mandujano 1991:157)
Sebastiana tuerckheimiana (Euphorbiaceae)	*barbasco* (Atran1993:646)

SPECIES (FAMILY)	MAYA USE (REFERENCES)
	construction (Mutchnick and McCarthy 1997:176) *fuel* (Mutchnick and McCarthy 1997:176) *medicinal* (Mutchnick and McCarthy 1997:176)
Sideroxylon foetidissimum (Sapotaceae)	*chicle adulterant* (Atran 1993:669; Standley and Williams 1967:221) *construction* (Atran 1993:669; Standley and Williams 1967:221)
Simarouba glauca (Simaroubaceae)	*construction* (Arvigo and Balick 1993:121; Atran 1993:670; Mutchnick and McCarthy 1997:180) *food* (Atran 1993:670; Standley and Steyermark 1946b:433) *medicinal* (Arnason et al. 1980:357; Arvigo 1992:28; Arvigo and Balick 1993:121; Comerford 1996:335; Mutchnick and McCarthy 1997:180, Standley and Steyermark 1946b:434; Romero 1997–2001:11) *tools* (Arvigo and Balick 1993:121; Mutchnick and McCarthy 1997:180)
Simira salvadorensis (Rubiaceae)	*dye* (Standley and Williams 1975:199)
Spondias mombin (Anacardiaceae)	*construction* (Mutchnick and McCarthy 1997:174) *food* (Arvigo 1992:8; Atran 1993:647; Cano et al. 2000:33; Mutchnick and McCarthy 1997:174; Romero 1997–2001:7) *medicinal* (Arvigo 1992:8, Romero 1997–2001:7, Orellana 1987:243) *tools* (Mutchnick and McCarthy 1997:174) *plant remains at Colha: charcoal, seeds, pollen, phytoliths* (Caldwell 1980:264) *remnant of forest garden* (Kelley 1988:149)
Spondias radlkoferi (Anacardiaceae)	*medicinal* (Arvigo and Balick 1993:95)
Swartzia cubensis (Fabaceae)	*construction* (Atran 1993:654)
Swietenia macrophylla (Meliaceae)	*home gardens* (Allison 1980:64)
Tabebuia rosea (Bignoniaceae)	*home gardens* (Allison 1980:64)
Tabernaemontana alba (Apocynaceae)	*glue* (Alcorn 1984:803) *medicinal* (Alcorn 1984:803)
Tabernaemontana arborea (Apocynaceae)	*latex for "rubber" balls* (Romero 1997–2001:5)
Talisia olivaeformis (Sapindaceae)	*fodder* (Rico-Grey, Cheams, and Mandujano 1991:157) *food* (Atran 1993:669; Mutchnick and McCarthy 1997:180; Rico-Grey, Cheams, and Mandujano 1991:157; Standley and Steyermark 1946b:269)

SPECIES (FAMILY)	MAYA USE (REFERENCES)
	medicinal (Rico-Grey, Cheams, and Mandujano 1991:157) *tools* (Mutchnick and McCarthy 1997:180)
Terminalia amazonia (Combretaceae)	*construction* (Mutchnick and McCarthy 1997:176, Atran1993:656)
Thouinia paucidentata (Sapindaceae)	*construction* (Atran1993:669, Rico-Grey, Cheams, and Mandujano 1991:157) *fuel* (Atran1993:669) *medicinal* (Rico-Grey, Cheams, and Mandujano 1991:157)
Trichilia havanensis (Meliaceae)	*medicinal* (Standley and Steyermark 1946b:463; Orellana 1987:249) *tools* (Standley and Steyermark 1946b:463; Romero 1997–2001:5)
Trichilia hirta (Meliaceae)	*cosmetic* (Standley and Steyermark 1946b:464)
Vitex gaumeri (Verbenaceae)	*construction* (Arvigo 1992:50; Arvigo and Balick 1993:73; Atran1993:671; Mutchnick and McCarthy 1997:180; Rico-Grey, Cheams, and Mandujano 1991:158) *medicinal* (Arnason et al. 1980:348; Arvigo 1992:50; Arvigo and Balick 1993:73; Atran1993:671; Mutchnick and McCarthy 1997:180; Rico-Grey, Cheams, and Mandujano 1991:158; Romero 1997–2001:15; Roys 1931:300) *tools* (Atran1993:671; Mutchnick and McCarthy 1997:180; Rico-Grey, Cheams, and Mandujano 1991:158) *fodder* (Rico-Grey, Cheams, and Mandujano 1991:158) *musical instruments* (Arvigo 1992:50; Arvigo and Balick 1993:73; Romero 1997–2001:15)
Vitis tiliafolia (Vitaceae)	*fiber* (Standley and Steyermark 1949b:302) *medicinal* (Orellana 1987:252) *water source* (Nations and Nigh 1980:22; Standley and Steyermark 1949b:302)
Zuelania guidonia (Flacourtiaceae)	*construction* (Rico-Grey, Cheams, and Mandujano 1991:158) *medicinal* (Rico-Grey, Cheams, and Mandujano 1991:158)

Note: The ethnobotanical uses of "construction" and "firewood" were considered insufficient for inclusion in this appendix. However, "construction" was listed as a use if the species was already represented as being utilized in another manner.

NOTES

1. *Relative dominance* is defined in table 1.3.
2. Shannon's index of diversity is a numerical measurement of species richness and the spatial distribution of those species (equitability) based on information theory. The index of equitability is one of the components of Shannon's index. Both indexes are defined mathematically in table 1.1.
3. Sorenson's index is a numerical measure of the similarity of species compositions between two sites, ranging from 0.0 (no species in common) to 1.0 (complete similarity). Sorenson's index is defined mathematically in table 1.2.
4. *Alpha diversity* is a measure of species richness in one sample, in one place. *Beta diversity* is the turnover in species compositions among one or more samples that occur in different places. Therefore, the inverse of Sorenson's index of similarity is a good numerical measurement of beta diversity.

REFERENCES

Alcorn, J. B. 1984. *Huastec Mayan Ethnobotany.* Austin: University of Texas Press.

Allison, J. L. 1983. An ecological analysis of home gardens *(huertos familiares)* in two Mexican villages. Master's thesis, University of California, Santa Cruz.

An, S., X. Zhu, Z. Wang, D. G. Campbell, G. Li, and X. Chen. 1999. Plant species diversity in a tropical montane forest on Wuzhi Mountain, Hainan. *Acta Ecologica Sinica* 19 (6): 803–809.

Arnason, T., F. Uck, J. Lambert, and R. Hebda. 1980. Maya medicinal plants of San Jose Succotz, Belize. *Journal of Ethnopharmacology* 2: 345–364.

Arvigo, R. 1992. *Panti Maya Medicine Trail Field Guide.* San Ignacio, Belize: Ix Chel Tropical Research Foundation.

Arvigo, R., and M. J. Balick. 1993. *Rainforest Remedies: One Hundred Healing Herbs of Belize.* Twin Lakes, Wisc.: Lotus Press.

Atran, S. 1993. Itza Maya tropical agro-forestry. *Current Anthropology* 34:633–700.

———. 1999. Itzaj Maya folk-biological taxonomy: Cognitive universals and cultural particulars. In D. Medin and S. Atran, eds., *Folk Biology,* 119–204. Cambridge, Mass.: MIT Press.

Balée, W. 1998. Introduction. In W. Balée, ed., *Advances in Historical Ecology,* 1–10. New York: Columbia University Press.

Balée, W., and D. G. Campbell. 1990. Ecological aspects of liana forest, Xingu River, Amazonian Brazil. *Biotropica* 22 (1): 36–47.

Balick, M. J. 1991. The Belize ethnobotany project: Discovering the resources of the tropical rain forest. *Fairchild Tropical Garden Bulletin* 46 (2): 16–24.

Balick, M. J., R. Arvigo, and L. Romero. 1994. The development of an ethnomedicinal forest reserve in Belize: Its role in the preservation of biological and cultural diversity. *Conservation Biology* 8 (1): 316–317.

Balick, M. J., R. Arvigo, G. Shropshire, J. Walker, D. G. Campbell, and L. Romero. 2000. The Belize ethnobotany project: Safeguarding medicinal plants and traditional knowledge in Belize. In: M. Iwu and J. Wootten, eds., *Ethnomedicine and Drug Discovery,* 267–281. Advances in Phytomedicine no. 1. New York: Elsevier.

Balick, M. J., and R. Mendelsohn. 1992. Assessing the economic value of traditional medicines from tropical rain forests. *Conservation Biology* 6 (1): 128–130.

Balick, M. J., M. H. Nee, and D. E. Atha. 2000. Checklist of the vascular plants of Belize. Common names and uses. *Memoirs of the New York Botanical Garden* 85:vii–246.

Boom, B. 1985. Amazonian Indians and the forest environment. *Nature* 314:324.

Brewer, S. W., and M. A. H. Webb. 2002. A seasonal evergreen forest in Belize: unusually high tree species richness for northern Central America. *Botonical Journal of the Linnean Society* 138:275–296.

Cadwell, J. R. 1980. Archaeological aspects of the 1980 field season, Colha, Belize. In T. R. Hester, J. D. Eaton, and H. J. Shafer, eds., *The Colha Project, Second Season, 1980 Interim Report,* 257–268. San Antonio, Tex., and Venizia, Italy: Center for Archaeological Research, University of Texas, and Centro Studi e Ricerche Ligabue.

Campbell, D. G. 1989. Quantitative inventory of tropical forests. In D. G. Campbell and H. D. Hammond, eds., *Floristic Inventory of Tropical Countries,* 523–533. New York: New York Botanical Gardens.

——. 1994. Scale and patterns of community structure in Amazonian forests. In: P. J. Edwards, R. May, and N. R. Webb, eds., *Large-Scale Ecology and Conservation Biology,* 179–197. Oxford: Blackwell.

Campbell, D. G., D. C. Daly, G. T. Prance, and U. N. Maciel. 1986. Quantitative ecological inventory of *terra firme* and *várzea* tropical forest on the Rio Xingu, Brazilian Amazon. *Brittonia* 38 (4): 369–393.

Cano, E., F. Cabrera, C. Salazar, and J. S. Flores. 2000. *Anacardiaceae.* Etnoflora Yucatanense no. 15. Mérida: Universidad Autónoma de Yucatán.

Comerford, S. C. 1996. Medicinal plants of two Mayan healers from San Andrés, Petén, Guatemala. *Economic Botany* 50 (3): 327–336.

Dunham, P. S. 1996. Resource exploitation and exchange among the Classic Maya: Some initial findings of the Maya Mountains Archaeological Project. In S. L. Fedick, ed., *The Managed Mosaic: Ancient Maya Agriculture and Resource Use,* 315–334. Salt Lake City: University of Utah Press.

Dunham, P. S., T. R. Jamison, and R. M. Leventhal. 1989. Secondary development and settlement economics: The Classic Maya of southern Belize. *Research in Economic Anthropology,* Supplement 4:255–292.

Edwards, C. R. 1986. The human impact on the forest in Quintana Roo, Mexico. *Journal of Forest History* 30:120–127.

Fedick, S. L. 1996a. An interpretive kaleidoscope: Alternative perspectives on ancient agricultural landscapes of the Maya lowlands. In S. L. Fedick, ed., *The Managed Mosaic: Ancient Maya Agriculture and Resource Use,* 107–131. Salt Lake City : University of Utah Press.

——. 1996b. Introduction: New perspectives on ancient Maya agriculture and resource use. In: S. L. Fedick, ed., *The Managed Mosaic: Ancient Maya Agriculture and Resource Use,* 1–14. Salt Lake City: University of Utah Press.

Fedick, S. L., and A. Ford. 1990. The prehistoric agricultural landscape of the central Maya lowlands: An examination of local variability in a regional context. *World Archaeology* 22 (1): 18–33.

Ford, A. 1986. *Population Growth and Social Complexity: An Examination of Settlement and Environment in the Central Maya Lowlands.* Tempe: Arizona State University.

——. 1991. Economic variation of ancient Maya residential settlement in the upper Belize River area. *Ancient Mesoamerica* 2:35–45.

——. 1998. *The Future of El Pilar: The Integrated Research and Development Plan for the El Pilar Archaeological Reserve for Maya Flora and Fauna, Belize-Guatemala.* Washington, D.C.: U.S. Man and the Biosphere Program, Bureau of Oceans and International Environmental and Scientific Affairs.

Ford, A., and C. Wernecke. 2002. *Trails of El Pilar: A Comprehensive Guide to the El Pilar Archaeological Reserve for Maya Flora and Fauna. Exploring Solutions Past.* Santa Barbara, Calif.: Maya Forest Alliance.

Gómez-Pompa, A. 1987. Tropical deforestation and Maya silviculture: An ecological paradox. *Tulane Studies in Zoology and Botany* 26:19–37.

Gómez-Pompa, A., J. S. Flores, and V. Sosa. 1987. The "pet kot": A man-made tropical forest of the Maya. *Interciencia* 12 (1): 10–15.

Graham, E. A. 1987. Resource diversity in Belize and its implications for models of lowland trade. *American Antiquity* 52 (4): 753–767.

Greig-Smith, P. 1983. *Quantitative Plant Ecology.* 3rd ed. Oxford: Blackwell.

Hammond, N. 1978. The myth of the milpa: Agricultural expansion in the Maya lowlands. In: P. D. Harrison and B. L. Turner, eds., *Pre-Hispanic Maya Agriculture,* 23–34. Albuquerque: University of New Mexico Press.

———. 1981. Settlement patterns in Belize. In: W. Ashmore, ed., *Lowland Maya Settlement Patterns,* 157–186. Albuquerque. University of New Mexico Press.

Kelley, J. H. 1988. *Cihuatán, El Salvador: A Study in Intrasite Variability.* Publications in Anthopology no. 35. Nashville: Vanderbilt University Press.

Lundell, C. L. 1938. Plants probably utilized by the old empire Maya of Petén and adjacent lowlands. *Papers of the Michigan Academy of Science, Arts, and Letters* 24:37–56.

McVaugh, R. 1963. Flora of Guatemala part VII. *Fieldlandia* 24, part 7, no. 3.

Mori, S. A., B. M. Boom, A. M. de Carvalho, and T. S. dos Santos. 1983. Southern Bahian moist forests. *Botanical Review* 49 (2): 155–232.

Mutchnick, P. A., and B. C. McCarthy. 1997. An ethnobotanical analysis of the tree species common to the subtropical moist forests of the Petén, Guatemala. *Economic Botany* 51 (2): 158–183.

Nash, D. L., and J. V. A. Dieterle. 1976. Flora of Guatemala part XI. *Fieldlandia* 24, part 11, no. 4.

Nash, D. L., and L. O. Williams. 1976. Flora of Guatemala part XII. *Fieldlandia* 24, part 12.

Nations, J. D., and R. B. Nigh. 1980. The evolutionary potential of Lacandon Maya sustained-yield tropical forest agriculture. *Journal of Anthropological Research* 36 (1): 1–30.

Orellana, S. L. 1987. *Indian Medicine in Highland Guatemala.* Albuquerque: University of New Mexico Press.

Ortiz, J. J. 1994. *Polygonaceae.* Etnoflora Yucatense no. 10. Mérida: Universidad Autónoma de Yucatán.

Peters, C., M. J. Balick, F. Kahn, and A. Anderson. 1989. Oligarchic forests of economic plants in Amazonia: Utilization and conservation of an important tropical resource. *Conservation Biology* 3 (4): 341–349.

Pitman, N., J. W. Terborgh, M. R. Silman, P. Nuñez, D. A. Neill, C. E. Cerón, W. A. Palacios, and M. Aulestia. 2001. Dominance and distribution of tree species in upper Amazonian *terra firme* forests. *Ecology* 82 (8): 2101–2117.

Pohl, M. D., K. O. Pope, J. G. Jones, J. S. Jacob, D. R. Piperno, S. D. DeFrance, D. L. Lentz, J. A. Gifford, M. E. Danforth, and J. K. Josserand. 1996. Early agriculture in the Maya lowlands. *Latin American Antiquity* 7 (4): 335–372.

Rico-Grey, V., A. Cheams, and S. Mandujano. 1991. Uses of tropical deciduous forest species by the Yucatecan Maya. *Agroforestry Systems* 14:149–161.

Romero, L. 1997–2001. Personal communication. Page numbers are from notes taken by Andrew Townesmith.

Roys, R. L. 1931. *The Ethnobotany of the Maya.* New Orleans: Department of Middle American Research, Tulane University.

Standley, P. C., and S. J. Record. 1936. *The Forests and Flora of British Honduras.* Publication 350, Botanical Series, vol. 12. Chicago: Field Museum of Natural History.

Standley, P. C., and J. A. Steyermark. 1946a. Flora of Guatemala part IV. *Fieldlandia* 24, part 4.

———. 1946b. Flora of Guatemala part V. *Fieldlandia* 24, part 5.

———. 1949a. Flora of Guatemala part I. *Fieldlandia: Botany* 24, part 1.

———. 1949b. Flora of Guatemala part VI. *Fieldlandia* 24, part 6.

———. 1952. Flora of Guatemala part III. *Fieldlandia* 24, part 3.

Standley, P. C., and L. O. Williams. 1961. Flora of Guatemala part VII. *Fieldlandia* 24, part 7, no. 1.

———. 1962. Flora of Guatemala part VII. *Fieldlandia* 24, part 7, no. 2.

———. 1963. Flora of Guatemala part VII. *Fieldlandia* 24, part 7, no. 4.

———. 1967. Flora of Guatemala part VIII. *Fieldlandia* 24, part 8, no. 3.

———. 1970. Flora of Guatemala part IX. *Fieldlandia* 24, part 9, nos. 1–2.

———. 1973. Flora of Guatemala part IX. *Fieldlandia* 24, part 9, nos. 3–4.

———. 1975. Flora of Guatemala part XI. *Fieldlandia* 24, part 11, nos. 1–3.

Standley, P. C., L. O. Williams, and D. N. Gibson. 1974. Flora of Guatemala part X. *Fieldlandia* 24, part 10, nos. 3–4.

Swallen, J. R. 1955. Flora of Guatemala part II: Grasses of Guatemala. *Fieldlandia* 24, part 2.

Thomas, W. W., A. M. de Carvalho, A. M. Amorim, J. Garrison, and A. L. Arbeláez. 1998. Plant endemism in two forests in southern Bahia, Brazil. *Biodiversity and Conservation* 7 (3): 311–322.

Turner, B. L. 1978. The development and demise of the swidden thesis of Maya agriculture. In: P. D. Harrison and B. L. Turner II, eds., *Pre-Hispanic Maya Agriculture,* 13–22. Albuquerque: University of New Mexico Press.

Wang, Z., S. An, D. G. Campbell, X. Yang, and X. Zhu. 1999. Biodiversity of the montane rain forest in Diaoluo Mountain, Hainan. *Acta Ecologica Sinica* 19 (1): 61–67.

Wiseman, F. M. 1978. Agricultural and historical ecology of the Maya lowlands. In: P. D. Harrison and B. L. Turner II, eds., *Pre-Hispanic Maya Agriculture,* 63–115. Albuquerque: University of New Mexico Press.

Wright, A. C. S., R. H. Arbuckle, D. H. Romney, and V. E. Vial. 1959. *Land in British Honduras: Report of the British Honduras Land Use Survey Team.* Colonial Research Publications no. 24. London: Her Majesty's Stationery Office.

2

A NEOTROPICAL FRAMEWORK FOR *TERRA PRETA*

ELIZABETH GRAHAM

IN THIS CHAPTER, I take as my starting point the view that Amazonian Dark Earth (ADE) studies (Glaser and Woods 2004; Lehmann, Kern et al. 2003), originally known as research into *terra preta* or *terra preta do índio* (Kern and Kämpf 1989; Sombroek 1966; Woods 2003:3; Zech, Pabst, and Bechtold 1979), comprise the most important scientific contribution thus far to our understanding of human impact on the environment over the long term. An interest in the earth as transformed by human action is by no means in a fledgling state, but the consequences of human impact that attract intensive study are mostly negative, such as degradation and deforestation (see, e.g., papers in Turner et al. 1990), and the doctrine of *Homo devastans* (Balée 1998:16–19) is a view widely held. As observed by Greenland (1994:4), soil degradation has been mapped (Oldeman, Hakkeling, and Sombroek 1990), but there is no comparable map of world soil improvement.

Studies of ADE (Woods 2003:3) contribute powerfully and uniquely to the idea that long-term human impact can be measured positively rather than negatively and viewed constructively rather than destructively. That intentional additives can enrich soils has been known since ancient times, but also integral to ADE research is the idea that unintentional consequences of human depositional activity can result in soil enrichment (Kern, Lima da Costa, and Lima Frazão 2004; Woods 2003:4), a connection not well recognized and, outside of ADE research, not well studied.

Although there is no doubt in my mind regarding the importance of ADE research, I take the opportunity in this chapter to suggest that ADE as a conceptual term or tool will have reached its limits heuristically if its recognition as an anthrosol[1] remains confined to Amazonian studies and linked so closely to the Amazonian experience. To understand the complexities of soil-formation processes influenced by a generative anthropogenic component—that is, a component that contributes to the transformation of the character of soils by increasing fertility—Amazonian studies of the environmental effects of human activities should be expanded to include other places in the humid Neotropics, such as lowland Mesoamerica or the Caribbean, where dark earths also exist but have not been productively problematized in the same way. As noted by McCann, Woods, and Meyer with regard to ADE, "important basic questions regarding origin, formation, persistence, variation, distribution and use remain unresolved" (2001:180).

With a view to encouraging the expansion of dark-earth research in the Neotropics beyond the Amazonian region, I discuss here research trends in ADE studies that can be applied productively within a broader neotropical framework and note other trends that should be followed with caution. Four points provide the framework for my discussion of these trends:

- The first and to some extent the most important point is that the working hypothesis for studying the relationship between human activity and dark earth should be that the relationship is associational and not causal.
- The assumption that there is a range of established depositional characteristics that is indicative of human activity, whereas absence of these characteristics reflects natural activity, is based on ideas about the interaction among humans, nonhumans, vegetation, and the abiotic environment that may not be adequate to the task of clarifying the complexities of the ADE phenomenon.
- If we wish to apply to today's world what we learn about human-environmental interaction from the experience of past Amazonians and the world they left us—and in my case from the Maya and the Taino and the world they left us—then the use of the culture/nature dichotomy loses its utility. Especially in the humid Neotropics, where domesticated grazing animals were not present to simplify the landscape, and where incidental and specialized domestication seem to have played large roles (Pearsall 1995), what was cultural and what was natural are not always clear or easy to define.
- One might argue that the concept of an ADE has outlived its utility because the close association of dark earths with the Amazon may be as much a product of the history of Amazonian research as it is a product of the history or prehistory of the Amazonian people. The way soils have been conceptualized is related to the way the pre-Columbian experience has been conceptualized through archaeology and ethnohistory. This is as true for lowland Mesoamerica or the Caribbean as it is for the Amazonian region.

AN ASSOCIATIONAL, NOT CAUSAL, RELATIONSHIP

In northern Belize, which is limestone country (King et al. 1992:26), preservation of artifacts is variable, but ceramics abound. Many ruins lie in bush, so that trees, vines, and other forest vegetation form part of an archaeologist's daily commute. At the time I began fieldwork in 1972, Puleston (1968) had introduced the hypothesis that ramon trees, *Brosimum alicastrum,* were so numerous around Maya ruins that their modern concentrations might be due to ancient planting by the Maya. I found this proposal intriguing. When I began my own research for my thesis, I moved south, to the Stann Creek District in Belize (figure 2.1), and one of the first things I noticed after I became accustomed to the local forests was that there were no ramon trees. Given that ramon was thought to have been in widespread use among the Maya, even if only as a famine food (Hammond 1982:156), why were there no ramon trees here?

It turned out that Stann Creek District soils are acidic and the parent materials of the district—outliers of the Maya Mountains—include no limestone (Wright et al. 1959:148). The absence of soils derived from limestone along with the absence of ramon adds support to Lambert and Arnason's (1982) proposal that ramon is a lime-loving tree, and its connection with the ruins of limestone buildings reflects its requirements for growth and reproduction, and thus not necessarily deliberate planting by the ancient Maya. Peters's (1983) autoecological study of the growth, reproduction, and population dynamics of ramon in

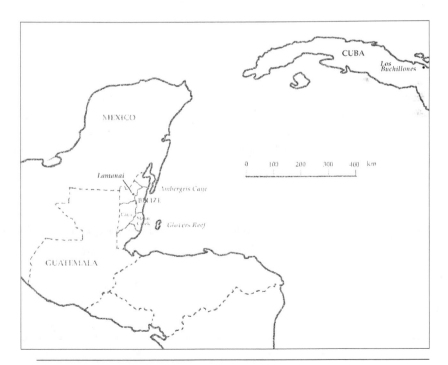

FIGURE 2.1 Map showing sites discussed in the text.

southern Mexico suggests further that a wide range of factors is involved in the ramon-ruin association, among them the food and habitat preferences of bats as well as factors that affect *B. alicastrum* breeding and productivity, with the possibility remaining that modern ramon characteristics and distribution may in part reflect artificial selection by the ancient Maya.

The implications of the ramon story lie at the methodological heart of any future dark-earth studies. In the Maya area, Puleston conflated association with cause, as did Sanders (1977) in his assumption that the fertile soils that exist around the present-day site of Tikal were the stimulus for settlement in the past (Graham 2002:24–25). Amazonian researchers are happily not guilty of this level of conflation, and several have explicitly emphasized the complexity of dark-earth formation and have advised the exercise of caution in assessing the nature of the human contribution (McCann 1994; Woods, McCann, and Meyer 2000). Nonetheless, conflating association with causation is an ever-present danger in ADE research and should be recognized explicitly as problematic.

For example, dark earth is discussed as a discrete Amazonian "technology," which is proposed to parallel the function of irrigation and terracing practices elsewhere in the world (Myers et al. 2003:15,17). Sociocultural practices are said to generate terra preta (Neves et al. 2003:29), whereas a more cautious statement would be that sociocultural practices generate processes that in some circumstances become what we call terra preta. In other words, sociocultural practices are implicated, but neither has a causal connection been established, nor is terra preta a consistent effect (Erickson 2003:464). The most that can be said is that humans, sometimes through sociocultural practices, create conditions for change (Woods and McCann 1999:9; Woods, McCann, and Meyer 2000). Under particular, yet to be fully defined environmental or ecological conditions, both synchronic and diachronic, these changes result in what we call ADE.

On the one hand, the formation processes proposed for ADE are assumed to affect soils but not to determine them, which thereby reinforces an associational rather than a causal connection between humans and ADEs (Kämpf et al. 2003:79). On the other hand, the described processes are so unambiguously linked to discrete, identifiable human activities that the picture of the human contribution is almost too good to be true: (1) long-term habitation with additions of domestic refuse, known as the midden model; (2) intensive cultivation practices, such as swidden or patch cultivation; and (3) activities involving the shifting of earth, such as mound building or raised field construction (Kämpf et al. 2003:79 for a summary). The midden model is said to be responsible for terra preta and the agricultural model for *terra mulata* (Andrade 1986:53,54; Sombroek 1966). Terra preta is described as dark in color, rich in phosphorous, calcium, and other elements and with abundant artifact content. Terra mulata is dark grayish brown, has phosphorous and calcium levels no higher than non-anthropogenic soils, has few artifacts, but a higher content of charcoal—all of which is interpreted as evidence of long-term soil-management practices, especially mulching and burning under intensive agriculture (Kämpf et al. 2003:79; Woods and McCann 1999).

The archaeological remains of refuse disposal, cultivation, and earth used as core in construction are, based on present evidence, strongly associated with ADEs; and middens may well prove to influence soils in ways distinctive from cultivation practices. Indeed, the definition of terra preta as the product of midden deposition has been called into question because, although household refuse is implicated chemically in the formation of terra preta, the terra preta itself is not simply (very) old household waste (McCann, Woods, and Meyer 2001:181,185).

Other factors need to be considered in analyses of the human connection. The first is that the categories of archaeological activities identified, such as midden or refuse disposal or cultivation (summarized in Kämpf et al. 2003:79), are simplistic if causal factors are ever to be isolated. Is there only one kind of refuse or only one kind of disposal? Is refuse never mixed or reused in various types of construction or in garden cultivation? Is the chemistry of refuse always the same? Is land never used for more than one purpose? Do people never live near cultivation plots for part of the year? The proposition that what was a midden when it accumulated in the past might consistently equate with a named soil we can recognize and characterize scientifically today assumes an ahistorical outlook, as if human activities never changed or were carried out in the same locale in the same way for all of Amazonian prehistory. This view would be convenient and would facilitate the identification of causal factors, but such stability or systematic behavior has been shown not to exist even within a single settlement or settlement zone (Mora et al. 1991:47–61, 75–82; Silva 2003). In addition, it is widely recognized in ADE research (e.g., Kämpf et al. 2003:32–83) that certain nonhuman factors are as critical as human input both in the soil-formation process and in processes that can obscure the record (Woods 1995). Factors indirectly related to input by humans, such as microbial activity and the presence of residues from incomplete combustion of plant materials, have been identified as critical in the persistence of the organic content of dark earths (Glaser 1999; Glaser et al. 2000; Sombroek et al. 2002; Woods and McCann 1999; Zech, Haumaier, and Hempfling 1990). Yet the relationships assumed to have existed between the archaeological remains and the human activities implicated by the remains are—compared to the sophistication and range of the soil studies—lacking in complexity and thus do not represent the potentially causally linked human dynamics that are so critical in understanding dark-earth formation.

MATERIAL, CULTURE, AND HUMAN ACTIVITY

THE STANN CREEK SITES AND PRESERVATION HISTORY

The presence of artifacts can be said to be a reflection of human activity. What are sometimes called "ecofacts" (Schiffer 1987:290–91)—material not directly altered or modified by humans but associated with human exploitation, such as faunal remains, carbon from burning, or stone transported to home bases as the raw material for tool manufacture—can also be indicators of human activity, as in the case of terra mulata, which sometimes has no artifacts but always contains elevated levels of soil organic matter in comparison with undisturbed forest soils (Sombroek et al. 2002). The question remains, however, whether the absence of artifacts is always a reflection of the absence of human activity, as in the observation by Sombroek and colleagues (2002) and by Kern and colleagues (2003:71) that both terra preta and terra mulata soils are delimited by natural soils under primary or unaltered forest. It can be argued that a key ecofact present in both terra preta and terra mulata—the elevated level of soil organic matter—is missing from forest soils, but might there be other ecofacts, such as soil micromorphological attributes or particular palms or entire vegetational communities or even landscapes (Erickson 2003:461) that reflect past human influence on the forest? Or is it possible that some types of artifacts are not preserved in soils under forest, but might potentially be detected by chemical imprints?

When I began the archaeological survey that was part of my dissertation research in Belize's Stann Creek District, I discovered that despite the publications that had reported tombs in which spectacular jade artifacts were recovered at both Pomona on the North Stann Creek River and Kendal on the Sittee River (Kidder and Ekholm 1951; Price 1899), ceramics were elusive, both on the surface and in the excavations. When we could locate mounds at all—which itself was difficult in some areas owing to hilly terrain, in others owing to the thickness of alluvium, in others owing to widespread destruction—we dug test pits, excavated test trenches, and in some cases cleared cut stone faces and exposed core faces. When ceramics manifested themselves, however, they were largely in the form of worn reddish lumps. In one case, we cleared a deeply buried terraced platform and found virtually no pottery, bone, or shell. Burned clay floors and some stone tools made up the bulk of our assemblage, but even the presence of chert was the result of exceptional circumstances.

Because limestone is rare in the region, there are no chert deposits. Chert, like obsidian, had to be imported, and therefore the low occurrence of chert is probably the result of efficient use of a scarce material (Graham 1994:263–288). In the case of pottery, bone, and shell, however, soil pH tests revealed high levels of acidity (Graham 1994:appendix II). Bone and shell, of course, contain high levels of calcium, and the percolation of groundwater through acid soils seems to have had a devastating effect on these materials. Decay of ceramics was more difficult to explain until I noticed that the cruder, utilitarian pottery that had been tempered with sand fared better than the diagnostic wares tempered with calcite. The reaction of the calcite temper with acid groundwater had the same effect on the potsherds as the groundwater did on the bones and shell. As for the relative dearth of architecture, rivers had been at work, repeatedly overflowing their banks during rainy seasons and effectively covering structures, in many cases completely, with thick deposits of alluvium.

If Maya cities can "disappear" in terms of artifacts and structural features, what does this tell us about smaller-scale occupations or settlements characterized by wooden architecture? The Stann Creek site phenomenon teaches an important lesson about established characteristics considered indicative of human activity. The Stann Creek District landscape is almost devoid of portable artifacts, structures, and recognizable features, yet it clearly once supported extensive and intensive Maya settlement. If a situation like this can exist where populations had reached almost urban proportions (based on equivalent sites elsewhere and on coastal evidence that reflects inland occupation from at least 400 BC until the fifteenth century AD), then it is possible to conceive of situations elsewhere in the Maya lowlands and in the Amazon in which low artifact density is not necessarily a direct reflection of the presence of humans or even of the intensity of human activity, but instead reflects preservation history.

For example, two of the criteria that differentiate terra mulata from terra preta are the absence or minimal presence of artifacts and the lower levels of soil organic matter in terra mulata (see Kern et al. 2003:71 for a comprehensive review of how terra mulata has historically been differentiated from terra preta). The generally accepted interpretation (papers in Lehmann, Kern et al. 2003) is that both terra preta and terra mulata reflect primary but distinctive past human activities (household refuse deposition versus cultivation), and what distinguishes the soils from each other today reflects past distinctions in the primary human activity. An alternative interpretation is that the primary activities may have been the same or different or represent mixed use through time (Erickson 2003:474), or perhaps they are not strictly determinable, but the deposits have been differentially exposed to environmental degradation, and this differential exposure, possibly attributable to humans, has helped to create the modern pattern. In other words, there is no reason to expect that the differences in past activity should correspond with the gross-level terra preta versus terra mulata distinction (see also McCann et al. 2001:183).

In addition to the midden/cultivation dichotomy, even the concept of a "site" remains problematic (see critique in Erickson 2003:461) when the features recorded to reflect sites are artifacts (Graham 1994; Kern et al. 2003:67, table 1; Zeidler 1995:26). Certainly in the Maya area it is rare to look for sites where artifacts and architecture cannot be detected in surface survey, although there are notable exceptions, such as the microrelief patterns that indicate wetland agriculture (Pohl 1990:7), which have also been so effectively surveyed in lowland Bolivia (Erickson 1995), and the rigorous methods that take into account environmental processes and features that can bury or hide sites (Graham 2002:16–17; Johnston 2002, 2004). Patterned human activities that alter landscapes, such as management of forests and savannas, have been defined and described in Brazil (Anderson and Posey 1989; Balée 1989, 1992; Balée and Gély 1989), although the implications have yet to be applied widely in detecting or defining ancient human occupation archaeologically. Methods and sampling strategies to find buried sites or to identify activities not indicated by artifacts (Arrhenius 1955, 1963; Woods 2003:6) are widely used (Ball and Kelsay 1992; Erickson 1995; Roosevelt 1989, 1991; Terry et al. 2000; Woods 1977, 1984), although as noted by Woods and colleagues (1996), problems still exist in terms of sample-collection strategy, accuracy of provenience, and methods of analysis. The key seems to be that where we already expect to find human occupation, we are more apt to seek it out by any means possible, but if we do not know whether humans were ever present, or if our expectations exclude in advance the possibility of human activity, as in the assumption of an unaltered forest, then absence of artifacts looms large as a criterion used to infer a dearth of human activity.

THE LOS BUCHILLONES SITE, CUBA

There is another point to be made about the relationship between people, artifacts, and conditions of preservation. In Cuba, at the site of Los Buchillones (Pendergast et al. 2001, 2002), surface ceramics are uncommon. In fact, cultural material has rarely if ever been found on the surface near the lagoon and seashore where we now know a submerged archaeological site is located, although artifacts can be found on dry land some distance behind the lagoon. The buildings that we have begun to expose turned up by accident. Two fishermen recovered well-preserved wooden artifacts from the lagoon and shore over the years, but the assumption was that the wooden artifacts were secondarily deposited. Wet-site excavation has since revealed the remains of two structures in which preservation of wood and other vegetal matter is excellent. Ceramics preserve reasonably well, but they are not common. The Taino here clearly favored wood, which is rarely preserved archaeologically in humid tropical conditions. In places such as the Los Buchillones site, one might easily interpret ceramic scarcity and the relative scarcity of other nonperishables as evidence of low population. Or one might interpret the Taino as lacking an elaborate material culture. But no one who has seen the artifacts from the Los Buchillones lagoon can say that the Taino lacked an elaborate material culture (Deagan and Cruxent 2002:23–45; Graham et al. 2000; Pendergast et al. 2001:figure 7.3). Not only are appropriate techniques needed to understand a site, but expectations are important, too, akin to what motivated the "big models" attributed to Lathrap (Erickson 1995:68; Oliver 1992), which helped to change the course of investigations into Amazonian prehistory (Stahl 2002:42–44). It is not surprising that people who live in the Neotropics use a great deal of wood. Yet because the Taino did not produce large quantities of pottery, and because wood preservation is rare, our initial view of Taino material culture was restricted, which helped to structure our expectations.

REFRAMING EXPECTATIONS

In a setting with preservation conditions such as those exhibited in the Stann Creek District, the stimulus behind the rejection of an equation of no pottery with no occupation was that I had come into the district with a model of Maya settlements developed elsewhere over a long period. Thus, I persisted in the search for features that reflected human occupation because I *believed* that occupation was dense and that what appeared as a largely natural environment was an artifact of preservation.

Later, with added experience on the coast, I was able to extrapolate information about inland occupation from coastal data (Graham 1989, 1994:135–247). But to interpret the inland landscape, the pressure was on to read sediments and deposits in more detail than would be expected in an area where cultural material abounded. Sediment particle size, sorting, depositional history, soil type, and contexts of sediments that were natural in terms of their content, such as alluvium, but that could not have occurred naturally where I found them, all burgeoned in importance in the absence of artifacts. Vegetational associations that suggested anthropogenic input to soils took on added interest as well (Graham 1998). Expectations should not create evidence out of thin air; nonetheless, given humid tropical conditions and what we are coming to learn about the complexities of human use of the tropical landscape, absence of artifacts might best be expressed as only a hypothetical dearth of human activity. Ultimately, all possible avenues of evidence need to be explored—and the way expectations are framed is critical to the detection of all possible avenues.

It is well known in the Amazon, but perhaps less well acknowledged and practiced in Maya archaeology with its focus on urban populations, that a human imprint can be detected from more than the presence of artifacts (Stahl 1995; see also Stahl, chapter 4, this volume). Although it was the association of pottery with terra preta that tied terra preta originally to past human occupation (Myers et al. 2003:22–23; H. Smith 1879:26, 144–145), and although absence of artifacts is still considered a significant factor in inferring a nonhabitation source for dark earth—for example, in the case of terra mulata (Woods, McCann, and Myer 2000:114)—the chemical and physical properties of heightened fertility that characterize terra preta, its pronounced chemical enrichment markers

(Woods, McCann, and Myer 2000:114), are now accepted as a consequence of human activity (Kern and Kämpf 1989; Lehmann, Kern et al. 2003; N. Smith 1980; Woods 1995).

But can human impact exclusive of artifacts potentially be detected in the Amazon and elsewhere from more than the presence of phosphate, carbon, nitrogen, or calcium? Woods (2003:11) has reviewed the major anthropic indicators in detail and emphasizes that anthropogenic alterations should always be considered in relation to the local depositional and pedological system. This is surely the most critical factor in dark-earth analysis. Geologists working with me at the Maya archaeological site of Santa Cruz on Ambergris Caye in Belize were able to determine that a particular deposit, devoid of all artifact content, bore a mineralogic and crystallographic imprint related to lime processing and hence could only be the result of human production activity (Mazzullo, Teal, and Graham 1994). This kind of sleuthing was possible because the geology of the caye was, after many years of study, well known to Mazzullo and his colleagues, an illustration of Woods's point about local knowledge.

In addition to soil and sediment indicators, vegetation and land use also leave imprints that can potentially be linked to humans. For example, incomplete combustion of organic material has been linked to particular kinds of agricultural practices (see review in Kern et al. 2003). But might modern studies of vegetational succession provide us with ideas regarding further links to changes triggered by humans? In the Neotropics, trees and their presence are critical to the creation and spread of soil-enriched microsites (Kellman 1979). Settlements in the humid Neotropics, particularly in pre-Columbian times prior to the adoption of agropastoralism (Melville 1994), were characterized by suites of fruit and shade trees (Marcus 1982). These areas of common use may never have been zones of artifact deposition or discard, but they were certainly areas of habitation, and the presence of trees would have played a critical role in nutrient cycling and thereby in vegetational succession. Is it not possible that the trees selected by humans for planting in household compounds produced distinctive kinds of soil-enrichment zones, which in turn affected both vegetational succession and the character of subsequent animal exploitation (soil microfauna, birds, bats [Peters 1983])? What remains for us is the identification of a suite of vegetational, soil, and archaeological evidence that will facilitate the detection of such activity in space and in time.

CULTURE, NATURE, AND LEARNING FROM THE PAST

DARK-EARTH ORIGINS, ARCHAEOLOGY, AND APPLIED RESEARCH

Although two nineteenth-century observers assumed a human connection with the origins of ADE (Glaser, Zech, and Woods 2004; Woods 2003:3), later studies seem to have bent over backward to prove that terra preta was the product of solely natural processes. Examples are discussed in N. Smith (1980:554), such as fallout from volcanoes (Camargo n.d.), sedimentation in lakes (Falesi 1974), or, in some areas, small depressions that filled with water during the rainy seasons (Cunha Franco 1962). As noted earlier, however, ADE is now generally understood to be a consequence of human activities (Erickson 2003:477; Lehmann, Kern et al. 2003; N. Smith 1980:555).

Archaeologists working in the Amazon emphasize the value of ADE in drawing inferences about the lives of people in the past (Neves et al. 2003:39). Expanding dark-earth studies to the Maya lowlands is equally important in clarifying Maya prehistory, but my ultimate interest is to apply what can be learned from anthropogenic processes to modern land use and urban planning. This emphasis on an applied aspect of dark-earth studies has always been a major goal of ADE research, in particular among geographers and soil scientists (e.g., Lehmann, Kern et al. 2003:xv–xvii; Lehmann, Pereira da Silva et al. 2003; Woods, McCann, and Meyer 2000:115). However, the ways in which archaeology can contribute to applied knowledge, in addition to providing a context for the origins of inorganic and organic input, have yet to be fully explored, although archaeological research in association with ADE studies, as pioneered by Kern (1988), for example, is expanding rapidly (see summary in Erickson 2003; Kern, Lima da Costa, and Lima Frazão 2004; Mora 2003; Neves et al. 2003; Ruivo et al. 2003).

To develop a body of applied knowledge, archaeologists must be able to make a connection between an identifiable human activity or behavior, which would need to be placed in space and time, and its resultant effect, whether the effect turns out to be sediment accumulation (building platforms out of earth), soil protection (covering areas with buildings or shelters), chemical input (artifact or ecofact decay), or organic input (land clearing or kiln firing). For example, if a bounded stratum can be identified as refuse (broken artifacts mixed with remains from food preparation and cooking in a discrete deposit), and the evidence indicates that the midden is primary (via dated structural associations or artifact patterning—for example, sherds from reconstructable vessels in discrete pockets), the archaeologist can place this activity in time as well as space. Several additional steps can be taken; an important one would be to explore the chemistry and petrology of ceramic resource procurement and manufacture in order to list the potential minerals or organics that entered the soil subsequent to discard and degradation of pottery. Ceramic manufacturers can draw their tempering materials from a variety of sources, local and nonlocal, and if clays have a high shrinkage rate, the minerals or inorganic substances added to mitigate the shrinkage, in addition to naturally occurring minerals and added tempers, would be available for dissolution (as also observed by Erickson 2003:480).

Organics are used in the Maya area in clay that coats the wooden walls of buildings and in paints and slips. Pottery, stone, shell, bone, textiles, thatch, wood, ash, lime, food remains, furniture, skin, hair, feces, urine, corpses, transferred sediments, dyes, feathers, herbs, plants, gourds, stone and wood construction materials, clay floors, thatched roofs—not to mention the debris from manufacturing processes such as fermentation of plant juices, lime production, tobacco curing, and leather tanning—all decay and have the potential to affect soil chemistry and potentially soil-formation processes, as does constructing a house (seals deposits from weathering) or clearing land (alters vegetation and exposes surface soils) or penning deer (differential grazing and browsing, availability of manure) or keeping peccary (soil disturbance, feces) or managing floodwaters (sediment alteration).

Nonetheless, encountering potsherds or bones or daub or any of these materials or their decay products in a test pit or core will not provide archaeologists, or anyone else for that matter, with enough information to determine the *origins* of the depositional activity. Only extensive, open-area excavations (Erickson 2003:460) will provide answers on the dynamics of context, dating, and details of depositional origins, if such answers exist; if they do not, no amount of test pitting or coring will ever improve the picture. However, examining exposed profiles, test pitting, and coring are highly appropriate in extending our information on dark-earth distribution, variability, persistence, and modern use (Woods, McCann, and Meyer 2000:120). The field collection and sampling strategies described by Woods, McCann, and Meyer (2000) in a discussion of ADE analysis and directions for future research are geared to examine soil profiles in exposures (clay pits, privies, cut banks) without time-consuming hand excavation into archaeological sites (Woods, McCann, and Meyer 2000:116). Once profiles have been described and sampled, a coring program can be initiated. Although it is not stated explicitly, these procedures seem to be aimed at first locating and then discriminating among dark earths: terra preta from terra mulata from background soils. Further discussion of research concentrates on finer-grained analyses designed to answer questions concerning nutrient availability and to improve characterization of soil organic matter and understanding of the persistence of dark earths (Woods, McCann, and Meyer 2000:117).

With the wider application of the field and laboratory strategies discussed by Woods, McCann, and Meyer (2000), knowledge of the distribution, variability, persistence, and use of ADEs is certain to be expanded. The origins of dark earths, however, and the processes that create the conditions for ADE development will remain generalized and simplistic without large-scale archaeological studies that focus on defining human activities, activity areas, specific behaviors, and the material concomitants of behavior. With regard to the origins of dark earths, Woods, McCann, and Meyer themselves state: "Only replicate field studies and intensive micro-scale pedological investigations coupled with comprehensive archaeological studies of individual sites and surrounding use areas will provide the data necessary to attain fully an understanding of the processes of dark earth formation" (2000:119).

Although the common link among dark earths is the association with pyrolitic by-products connected to human habitation and cultivation, together with the association with the pre-Columbian period (Woods, McCann, and Meyer 2000:119), this association may not be enough (Erickson 2003:476–482). With expanded archaeological work both inside and outside of the Amazon, we have the potential to be able to link well-dated and well-described ancient depositional events with particular chemical or pedological microprocesses and ultimately with the effects of these processes on longer-term soil formation. There are implications not just for modern cultivation practices, but for disposal of refuse, approaches to landfill, and recycling in urban landscapes. Extending ADE research to other parts of the neotropical lowlands ought to facilitate making the detailed connections that need to be established between specific kinds of human activity and their chemical, morphological, vegetational, or even ecological imprints.

WITH THE BEST OF INTENTIONS...

In the preceding discussion, I have attempted to show the ways in which knowledge about the origins of dark earths is dependent on detailed archaeological investigations coupled with microscale pedologic investigations. I have emphasized that the links to be made are between specific ancient behaviors or activities and their chemical or pedological concomitants. The word *behavior* is critical here because it is used widely in archaeology. Although in recent years there has been a rise in interest in cognitive archaeology (Oliver 2000; Pearson 2001), which considers agency or intent in the past, the interpretation of the stratigraphy of archaeological sites is generally not discussed in terms of intent on the part of the people under study. Except for the distinction between primary deposits, such as burials or even middens that reflect original human placement or discard patterns, and nonprimary deposits (which comprise the bulk of archaeological remains), in which artifact locus and context do not reflect original placement or discard, stratigraphy and context are discussed in terms of the material concomitants of behavior. *Behavior* describes what can be observed rather than what in life can be communicated only by language, such as intent. In other words, whether ancient action was intentional or inadvertent is irrelevant to an analysis of the chemical or pedological profile of the deposits; either the deposits contain carbon or phosphorous, or they do not. Considering what we already know about the complexities of ADE origins and history (Mora et al. 1991; Neves et al. 2003), it seems to me that what may have been intended in the past is methodologically irrelevant to what was left behind for us to study.

I first became interested in terra preta as a vehicle for demonstrating the importance of anthropogenic activity in long-term soil-formation processes (Graham 1998), but I have always used the term *anthropogenic* to subsume both inadvertent and intentional activity and have not attempted to differentiate between the two. The terminology in any case seems not to be standardized. For example, Neves and colleagues distinguish *anthropic* as inadvertent and *anthropogenic* as intentional (2003:35), whereas soil science terminology employs *anthropic* to reflect the intentional effects of human cultivation and fertilization (*Glossary of Soil Science Terms* 1987:2–3). The literature often makes distinctions between intentionally and unintentionally formed ADE (Erickson 2003:483; Kämpf et al. 2003:79; Neves et al. 2003:35), but most ADE research seems operationally to subsume both. Ancient cultivation practices, perhaps because of their links to intentionality, seem more readily quantifiable and possibly more attractive to study than the idea that humans and their attendant litter and trash inadvertently transformed the earth in a positive way. Even Hecht's three large classes of human activities that have shaped Amazonian landscapes—large-scale earthworks, fluvial management and floodplain soil manipulation, and manipulation of vegetation through domestication (2003:355–356)—exclude the human behaviors of refuse or garbage accumulation, disposal of human waste, and burial, the remains of which could be said to be inadvertent potential contributions to long-term environmental change. At the level of analysis in which we explore the origins—but not the implications—of dark-earth deposits, I am uncomfortable with intentions, and I do not think we can disentangle the threads of the association between ADE and human activity if we privilege a dichotomy of intentionality versus inadvertency. This discomfort spills over onto my views of the culture/nature dichotomy.

HOW CULTURE AND NATURE GET MIXED UP

The ancient Maya use of alluvial deposits and *bajo* (swamp) clays as core of their massive terraced platforms in some areas resulted in naturally formed deposits making up artificial structures. Refuse or rubbish was always being shifted about; it was frequently used as core material in masonry platforms or as fill in low-lying, periodically inundated areas. As the refuse was exposed to natural processes of water percolation and erosion, the character of what had been a cultural or human-induced deposit changed.

On the coast and cayes in Belize, although artifacts and ecofacts abound in the form of what archaeologists call shell middens, the stratigraphy of the middens, if it was once representative of human depositional patterns, is now representative of land crab (*Cardisoma* sp.) activity. These industrious creatures dig burrows down to water level to replenish their oxygen supply; as a result, they have literally turned deposits upside down and inside out. To add to the land crab activity, hermit crabs (*Coenobita clypeatus*), seeking "houses" as they grow, have used and thereby shifted almost every conch shell discarded by the Maya in ancient times. These phenomena are good examples of how what we call cultural and natural activity can be inextricably intertwined.

PROBLEMS WITH C- AND N-TRANSFORMS

Schiffer (1972, 1987) is best known for his attempts to detail what he calls the formation processes of the archaeological record, in which he differentiates between *c-transforms* (cultural, in which the agency of transformation is human behavior) and *n-transforms* (noncultural, in which the agencies stem from processes of the natural environment) (Schiffer 1987:7, 22). His work has been important in getting archaeologists to think critically about how artifacts come to be deposited where they are found, but it has not guaranteed that archaeologists will act effectively in all cases in applying knowledge about c- and n-transforms. In recent years at Lamanai, building on Pendergast's research (e.g., Pendergast 1981), my colleagues and I have discovered evidence of continuous palace or elite residential construction from the Late Classic through the Terminal Classic to the Postclassic period (Graham 2004). Late Classic construction is manifested by substantial masonry buildings that comprise more than 2 meters of deposit. Terminal Classic and Postclassic construction phases that overlay Late Classic construction took us twice as long to find and expose, and they compose no more than 30 to 40 centimeters of dark earth that represents the decay of sizeable wooden structures. Such deposits are routinely cleared away as natural humic accumulation above structural remains, a trap into which we almost fell ourselves. Is Lamanai's record of noncollapse and continuity of occupation an aberrant phenomenon? Or is it possible that other sites were continuously occupied, but that construction in wood has been missed in excavation because the deposits appeared to be natural decay over large masonry buildings and were swept away? Similarly, has settlement unmarked by masonry buildings been missed (Johnston 2002, 2004)? If palace architecture can be made to "disappear" because it is mistaken for a phenomenon of natural decay, how much more caution should be exercised in areas where evidence may be even more ephemeral owing to rarity of masonry architecture or to former residents' complete dependence on perishable construction?

At one level, the classification of c- and n-transforms constitutes a rigorous approach to the interpretation of artifactual and depositional phenomena in archaeology. Yet the nomenclature can be problematic and is highlighted here because it is based on assumptions that occasionally surface in ADE research with regard to culture and its role in ADE formation. For example, Schiffer's categories imply that what is cultural will always involve a human agent; that what is environmental never masks human agency; that humans are always recognizably active agents of transformation; and that the contradiction between the conflicting forces of culture (artifacts) and nature (the environment) is the determining factor in their continuing interaction.

What is cultural archaeologically does not always have an exclusively human agent. Conch shells are brought from the reef and modified (sometimes) by humans to permit removal of the conch and/or to remove bits for decorative carving; the shells are discarded and later modified and moved around by soldier crabs and sometimes further modified and moved by birds trying to get at the crabs to eat them—all well within the perimeters of the archaeological site and sometimes within the limits of the original shell midden accumulation. From a Schifferian perspective, one would simply lump the crab and bird activities in with the natural environment and make them n-transforms. But from the point of view of ADE research, in which interest lies in the details of how humans have helped to transform environments positively, interpreting the presence of the crab and the bird as "natural" phenomena—as noncultural formation processes acting on a cultural material—is not productive. From the point of view of human-associated environmental transformation, it is important to make the ecological connection between human activity and the presence of the crab and the bird. Without the human, neither the crab nor the bird would be there, and *all* are conceivably critical in the transformation of an environment. In this scenario, where does culture stop and nature begin?

What is natural can also mask human agency. The first example that comes to mind is the work of Balée and others (e.g., Balée 1989, 1992; Balée and Gély 1989) on the emerging concept of cultural forests. If, as I suggested earlier, humans selected a range of trees that can be associated with particular soil nutrient–retention properties, then macrobotanical remains of such trees can conceivably be seen as both cultural and noncultural transformation agents. And does it matter whether humans are active or passive agents of transformation? What is the difference in environmental terms whether humans are active and add mulch to fields or practice defecation in fields without thinking about its effects? When a human defecates, is this a cultural or natural process?

CREATING CONDITIONS FOR CHANGE

There is no question that human beings initiated changes in the past that in many places—in the Maya lowlands, along the Amazon, and elsewhere in the Neotropics—led to dark-earth accumulation. As Woods notes, "The terra preta of Amazonia are not unique, but rather are part of a global process of accumulation of heterogeneous sediments at sites of human occupation" (1995:163). Methodologically, however, should dark earth be approached as the result of human activity, or should it be approached ecologically (Clement, McCann, and Smith 2003; German 2003; Hecht 2003; Hiraoka et al. 2003; Silva 2003; Woods, McCann, and Meyer 2000) as a particular set of relationships (which we have yet to define) that have important implications for global health?

We can legitimately start out with the idea that ADE is part of a cultural landscape (Hecht 2003:355). However, the term *cultural* covers a multitude of depositional environments, many of which are deeply affected by rainfall, temperature, insects, invertebrates, and other less-than-cultural phenomena. The same circumstances affected the coast and cayes of Belize, where humans started something by occupying seasonally inundated spits of land. They fished, collected shellfish, and concentrated various chemicals by tanning hides and processing salt; they built thatched houses, cooked food, swept away the remains of hearths, urinated, defecated, buried their dead, and dumped rubbish. As refuse from human occupation accreted both on and around the sites, inundated land rose and dried out. The area became inviting to invertebrates, birds came to eat the invertebrates, seeds from fruit trees were deposited by birds, snakes and rodents fed on the birds, mammals came to feed on the fruits and nuts and seeds. In a hundred years or so, the site no longer resembled the original camp. In a thousand years, it was completely transformed. But can labeling input as *cultural* or *natural* help us to understand the dynamics of this transformation? Is there a cultural dynamic that is discrete from the natural one? By using the term *culture*, we remain distinct from *nature*; but as Ingold has observed, how can we account for our existence in nature by taking ourselves out of it (2001:256)?

Any pre-excavation aerial photograph of Tikal, Guatemala, shows a richly forested area, and one that has been described as having high soil-nutrient capacity (Sanders 1977). This environment is partly a product of human activity and one from which modern Petén cultivators have benefited. Yet even in these conditions vegetation and soils are not strictly speaking culturally fashioned objects—that is, artifacts. They are subjected to rainfall; solar activity; insect activity; use by mammals, birds, and other animals; erosion; leaf litter deposition; and organic decay. Where you have dark earth, or terra preta, it is true that you have chemical markers distinctive to human activity (Kern and Kämpf 1989:224), but human activity does not always result in dark earth (Erickson 2003:464). As Sombroek and colleagues (2002) have shown with regard to their studies of soil organic matter, it is the *interaction* between what has been deposited by humans and the processes that go on outside of human control that results in terra preta and terra mulata. Clearly, dark earth cannot be studied outside of the framework of human activity because human behavior plays an important role in its genesis. As Silva states, "An interpretation of Amazonian Dark Earths can only be successful once the natural and cultural processes that played a role in their formation have been understood" (2003:384). But as I hope I have shown regarding c- and n-transforms, the details of interaction that must be isolated in the case of the origins of dark-earth formation are such that employing the terms *cultural* or *natural* is inadequate and in some cases misleading.

The working research question in a study of origins (but not so critical in studying distribution, variability, persistence, or use [Woods, McCann, and Meyer 2000:120]) might be better framed as: What human-animal-plant-abiotic interactions are associated with the genesis of these distinctive areas of cultivability? This question would focus efforts on an ecological or *relational* approach. In this frame, it does not matter whether humans inadvertently or purposefully brought about change, just as the question of *why* ants modify soil properties when they build their nests does not arise in investigations of the degree and persistency of changes of soil morphology and chemistry in abandoned anthills (Kristiansen and Amelung 2001:355). What matters is that soils under anthills are enriched in organic matter for considerable periods and influence the heterogeneity of the soil within the forest (Kristiansen and Amelung 2001). For ADEs, what matters in investigations of origins is *how* humans have created conditions for change and whether this relationship is measurable and predictable.

SHOULD ADE BECOME NDE (NEOTROPICAL DARK EARTH)?

Here I build on my fourth point, which is that dark-earth studies should be expanded to include other areas of the Neotropics, although obviously, given my own experience, I have an interest in convincing geographers, soil scientists, and other archaeologists that both Belize and Cuba provide productive locations for dark-earth studies. This interest is not entirely opportunistic in that I think I can show that the way terra preta has been conceptualized reflects the way pre-Columbian Amazonia has been viewed. Extending studies to other regions should broaden horizons and expand opportunities for testing relationships established exclusively within Amazonian research. I emphasize the Neotropics rather than a global expansion (Neotropical Dark Earths, NDEs, rather than Global Dark Earths, GDEs), at least at this juncture, because the absence of the human/grazing animal complex in the Neotropics was a critical factor in the intensity and character of pre-Columbian domestication and land management (Melville 1994). Even urban populations existed in Mesoamerica without dependence on grazing animals (Graham 1999), whose needs, as I have noted, required simplifying the landscape (Melville 1994), not managing existing diversity. Thus, forests became transformed into grasslands. The human/grazing animal complex was so important and ubiquitous in the Old World that comparisons of European or African or Asian dark earths with pre-Columbian America may not be instructive or valid.

DARK EARTH IN BELIZE AND CUBA: WHY COASTAL AND
NOT INLAND?

I made the point earlier that in coastal locations in Belize, land crabs and hermit crabs seem to have had as much influence on stratigraphy as humans have had. But there is more to the equation. When I moved to the Stann Creek District coast, I found that conditions contrasted strongly with the inland environment. Ceramics were well preserved at coastal sites. Indications were that the thousands of shells left by the Maya interacted with the acid river deposits to produce a pH that was close to neutral, a good matrix for ceramic preservation (Graham 1994:27,346), and apparently a fertile environment for the growth of lush vegetation. Ancient sites were detectable both in aerial photos and on the ground by the presence of broadleaf forest vegetation within a zone dominated by mangrove. Where the broadleaf trees had been cleared for planting, dense stands of coconut palms took their place.

On offshore islands such as Ambergris Caye (Graham 1989; Graham and Pendergast 1989), broadleaf forest vegetation, or in some circumstances mangrove forests with an unusually high canopy, distinguish zones of ancient use. The situation is somewhat different on Middle Caye, part of Glover's Reef Atoll, where I worked in 1999–2000. Here, the intensity of pre-Columbian and historic human use on such a small island—in recent times, mainly clearing for the planting of coconuts—may have limited the utility of vegetation as an indicator of human activity. Some island studies have shown that replacement of natural vegetation by coconuts results in erosion and beach retreat (e.g., Preu 1991:124). On Middle Caye, however, where some natural vegetation is still found, there is also evidence of refuse buildup by humans from pre-Columbian times through the British Colonial period to the present day, and preliminary data suggest that human activity has facilitated island accretion.

The soils found in association with the vegetation described along the coast and on the cayes and atolls is called "black earth" in Belize in the same way that terra preta has been distinguished in the Amazon. Fishermen and other locals whom I have interviewed believe that the black soil was brought to the coastal lagoons, coastal islands, and cayes in canoes by the ancient Maya. The same is said about the dark earth along the coast and cayes near the Los Buchillones site, in Cuba. Both stories are a version of the "fertility occurred first and the people came later" approach (Neves et al. 2003:35). The black soils of Belize, like those of the Amazon (Hiraoka et al. 2003), are highly sought after for gardens and for cultivation in general.

Why should a dark-earth phenomenon be distinguished or named on the coast and on cayes but not inland? The vegetational environment of the coast and cayes is distinctive: white mangrove (*Laguncularia racemosa*), red mangrove (*Rhizophora mangle*), black mangrove (*Avicennia nitida*), black and white poisonwood

(*Metopium brownei, Cameraria* sp.), beach grasses, and sedges. Both the broadleaf vegetation in black-earth zones and the distinctive black color of the soils stand out against the quartz sand deposits brought to the coast by the rivers of Stann Creek or against the white sands generated by the weathering of coral.

Extensive and long-term occupation occurred inland, but limestone soils, which dominate the heavily studied areas of northern Belize, are black. Cultural deposits are extensive, but no one (archaeologists, local Belizeans, geographers, land-use surveyors, soils scientists), as far as I am aware, has ever distinguished soils with cultural deposits from soils without artifacts, or named separate soil types on the basis of cultural content. In Stann Creek or in the Cayo District, where soils are red in color, there are zones of dark soils, but ancient Maya occupation is not seen to be limited to darker-colored soils; sherds and cultural features are found in soils of many colors and in many different contexts. In my experience, darker soils, which indeed are often richer in artifacts, exist because they have been protected from erosion or degradation in some way. They are often buried by platform construction or have been protected by vegetation. The platforms or the buildings they support need not be masonry—they just need to have been faced by material that prevented erosion for a time. All this is by way of saying that people have never used the presence of a dark earth as an indicator of long-term settlement. Long-term settlement by the Maya is simply assumed to have a wide range of manifestations pedologically, the origins of most of which we little understand.

At inland locations, vegetation is no more helpful an indicator of long-term human occupation than are the soils. This is not because soils and vegetation do not reflect human activity. The problem is that ancient human activity has been so intensive throughout Belize that the prospect of disentangling anthropogenic from natural processes by examining soils and/or vegetational communities is daunting. The predicament of isolating natural conditions at inland sites—or of isolating what is sometimes called "undisturbed forest"—is what has prompted me to consider a program of environmental investigation that focuses on coastal environments. This does not mean that anthropogenic processes are distinctive to coastal environments or are more common on the cayes than inland, but only that the products of anthropogenic processes are more easily distinguished in an environment in which the parent material is coral sand, where naturally formed organic layers are relatively thin and where salt-water inundation restricts the diversity and character of natural vegetation.

That anthropogenic factors were significant in the formation of Belize inland forests is not universally recognized owing to the complexities of the forest environment. Even today at the site of Lamanai, despite the fact that the forest reserve is underlain by successive layers of masonry buildings that were part of an urban landscape, naturalists have been known to identify the forest as undisturbed or pristine simply because there are no settlements on the reserve, no agriculture or land clearing is practiced, and the soils, at least at first glance, do not appear different from those naturally formed on limestone.

In the Amazon, the distinctiveness of ADE, its contrast in color and characteristics with the soils around it, facilitates recognition of zones of ancient activity (e.g., Glaser, Guggenberger, and Zech 2004). But where such contrasts are elusive, as is common on the Belize mainland, is their absence synonymous with absence of ancient activity? The source of the contrast—the white coral sands of the cayes against the deep black soils associated with human activity, or terra preta together with terra mulata against the surrounding soils under forest in the Amazon—reflects two things that the Belize cayes and the Amazon have in common: the fact that soil parent materials produce distinctively colored soils in the absence of organics, and the fact that human habitation is not systematically indicated by drastic vertical alterations to the landscape or by masonry architecture. Therefore, the existence of a soil-color contrast does not contribute to the uniqueness of the phenomenon of human influence; it simply contributes to the *recognition* of the phenomenon of human influence.

HOMO DEVASTANS VERSUS THE ECOLOGICALLY NOBLE SAVAGE?

If ADE is potentially unique, then this recognition should form part of the research question, and comparisons to test a hypothesis of uniqueness ought to expand to areas outside the Amazon. If ADE is not considered unique (Woods 1995:163), the question remains why dark-earth research has been so prominent in the Amazon but not, for example, in the Maya region. Is it simply that the phenomenon of dark earth is more easily recognizable as a result of the attributes of the Amazonian environment and soils? Or do the history and character of research in both these humid tropical regions have something to do with the way in which dark earths have been anchored to the Amazon experience, whereas collapse and soil erosion have been tied to the Maya?

Although ADE was reported in the literature as early as 1865 (Myers et al. 2003:22), much of its description as a distinctive phenomenon or set of phenomena (see Myers et al. 2003:22–24) took place in the period in which Amazonian societies were viewed as having low population density and low social complexity (Meggers 1954, 1971) and hence without the power ascribed to urban populations to effect large-scale deforestation or soil degradation. Research on ADE has since provided evidence that shows the inaccuracy of claims of low population and low social complexity (Erickson 2003:457), but discussions of how landscapes have been shaped reflect the Amazonian contribution to biodiversity (Balée 1994; Clement, McCann, and Smith 2003; Hecht 2003) and not the potentially deleterious impact that large populations may have had. This is *not* to argue for degradation in the Amazon; it is to highlight what seem to me to be implicit biases in our approaches. It is almost as if Balée's (1998) two doctrines concerning human nature were at work: the ecologically noble savage in the Amazon, but *Homo devastans* in the Maya lowlands.

I am by no means denying the Amazonian contribution to biodiversity, but if Amazonian populations were high and contributed to soil cultivability, then the Maya may have contributed likewise. Excavations at Lamanai coupled with limnological studies do not substantiate collapse or soil erosion and forest degradation (Graham 2004; Pendergast 1986; Breen 2002). This is not to say that urban densities in the Neotropics did not produce eroded soils, but only that our projections based on erosion are simplistic. Not least important, our models for settlement growth and its impact derive from the courses toward urban life followed in cities in the Middle East, northern China, and northern India, where humans and their grazing animals turned forests into grasslands, and where erosion and weathering ultimately produced the treeless landscapes familiar to us today in the Tigris-Euphrates region, in Pakistan, and along the silt-laden waterways of north China (Graham 1999, 2002). These urban histories have structured our expectations, so we may not be aware of the nature of the complexities in managing humid neotropical environments in which plants, trees, and their products were not simply adjuncts to meat and milk from cattle, sheep, or goats. Combining knowledge of the complexities of human-environmental impact in Amazonia with concomitant studies of cities, settlement, and material remains in the Maya region might be a highly productive route to a better understanding of resource exploitation, landscape management, and contributions to soil-formation processes under conditions of high population density.

WHAT'S CULTURE GOT TO DO WITH IT?

I have drawn attention to the fact that approaches to ADE reflect disciplinary imperatives that are rooted in the culture/nature dichotomy, but I do not mean to imply that this dichotomy is inadequate in any absolute sense. It is simply useful and appropriate in some circumstances and not in others, and there is room in our inquiries for greater cognizance of our terminology—including ADE—as means rather than ends.

Nature has a wide range of meanings (Coates 1998:3); in ADE research, however, it is understood as a physical place or environment with a dynamic that can subsume humans, but can also be extrinsic to humans and hence distinct from culture. Even in this sense, *nature* is used in different ways by different people, but it is almost always a referent for something green that grows or for something alive or abiotic (Gray 2004) that is not under the control—or at least under the complete control—of humans. Battling Mother Nature can mean anything from mowing one's lawn to hunting deer to climbing Everest, but whatever is meant by *nature* can usually be clearly derived from the context of our use of the term because we all have a sense that nature's existence is not dependent on human behavior, although it is certainly affected by it.

Culture, in contrast, is trickier (Borofsky et al. 2001). Culture has an element of time dependency that is quite different from that entailed in the way we use the term *nature.* If we say that something is part of culture, we do so because we already accept that it is something practiced over time by more than one individual. A practice that is cultural is learned and shared. Culture, unlike nature, cannot be identified on the basis of a single observation that does not also include a sense of time. Something is cultural only because there is a knowledge base accumulated diachronically that tells us that the trait or the practice is part of a pattern and not unique. Therefore, to contrast *culture* and *nature* is to oppose unequal terms, which undermines the utility of the dichotomy.

One alternative is to abandon the concept of culture and to reduce human activity to identifiable physical and chemical relationships with the environment (Graham 1998), as most ADE investigations have done implicitly in any case. Any and all human activity, if it effected change in soil-formation processes, can be reduced to or quantified as chemical input or physical residues with interactive potential. Maya plaster can be considered a product of metamorphism; decay of the bodies of the dead can be seen as organic input. Because the cause of metamorphism (human manufacture) and the patterning behind the burials are irrelevant to the longer-term outcome of characterizing and measuring any changes in soil-formation processes, culture is an irrelevant concept *at this level of analysis.*

The concept of culture can become relevant at another level of analysis, however, when we ask detailed questions about human groups and their distinctive patterns of behavior, about time depth, the origins of modern landscapes, and whether the earth in some regions has been positively transformed (increased fertility, greater biodiversity) by past human action. If so, we will then want to tie what we know about physiochemical soil-formation processes to the specifics of human action. Then and only then has "culture" got to do with anything because we will need to know what people did and whether they did it consistently in order to repeat what turned out to be useful and reject what did not.

A NEOTROPICAL FRAMEWORK

The critical next stage is expansion of dark-earth studies to include other zones of the Neotropics. As has been demonstrated in the recent publication by Lehmann, Kern, and colleagues (2003) and Glaser and Woods (2004), this expansion should be coupled with greater awareness of the importance of refining archaeological interpretations of human activities, expansion of scientific studies to include—in addition to soil studies (chemistry, micromorphology, macro- and microbotanical remains)—surveys of vegetational communities; studies of nutrient cycling, waste management, and land use; and, not least, greater awareness of the importance of trees in pre-Columbian lifeways. Only with this sort of fine-tuning can dark-earth research affect assessments of human-environmental interaction and contribute to global knowledge of sustainable land use.

NOTE

1. Soil, the genesis of which is influenced by human activity (N. Smith 1980).

REFERENCES

Anderson, A. B., and D. A. Posey. 1989. Management of a tropical scrub savana by the Gorotire Kayapó of Brazil. In D. A. Posey and W. Balée, eds., *Resource Management in Amazonia: Indigenous and Folk Strategies,* 159–173. Advances in Economic Botany no. 7. Bronx: New York Botanical Garden.

Andrade, A. 1986. *Investigación arqueológica de los antrosoles de Araracuara.* Bogotá: Fundación de Investigaciones Arqueológicas Nacionales, Banco de la República.

Arrhenius, O. 1955. The Iron Age settlements of Gotland and the nature of the soil. In M. Stenberger, ed., in collaboration with O. Klindt-Jensen, *Vallhagar: A Migration Period Settlement on Gotland, Sweden,* 2 vols., 2:1053–1064. Copenhagen: Ejnar Munksgaards Forlag.

——. 1963. Investigation of soil from old Indian sites. *Ethnos* 28:122–136.

Balée, W. 1989. Cultura na vegetação da Amazônia Brasileira. In W. Alves Neves, ed., *Biologia e ecologia humana no Amazônia: Avaliação e perspectivas,* 95–109. Belém, Brazil: Museu Paraense Emílio Goeldi.

——. 1992. People of the fallow: A historical ecology of foraging in lowland South America. In K. H. Redford and C. Padoch, eds., *Conservation of Neotropical Forests,* 35–57. New York: Columbia University Press.

——. 1994. *Footprints of the Forest: Ka'apor Ethnobotany—the Historical Ecology of Plant Utilization by an Amazonian People.* New York: Columbia University Press.

——. 1998. Historical ecology: Premises and postulates. In W. Balée, ed., *Advances in Historical Ecology,* 13–29. New York: Columbia University Press.

Balée, W., and A. Gély. 1989. Managed forest succession in Amazonia: The Ka'apor case. In D. A. Posey and W. Balée, eds., *Resource Management in Amazonia: Indigenous and Folk Strategies,* 129–158. Advances in Economic Botany no. 7. Bronx: New York Botanical Garden.

Ball, J. W., and R. G. Kelsay. 1992. Prehistoric intrasettlement land use and residual soil phosphate levels in the upper Belize Valley, Central America. In T. W. Killion, ed., *Gardens of Prehistory,* 234–262. Tuscaloosa: University of Alabama.

Borofsky, R., F. Barth, R. A. Shweder, L. Rodseth, and N. M. Stolzenberg. 2001. WHEN: A conversation about culture. *American Anthropologist* 103 (2): 432–446.

Breen, A. M. 2002. Holocene environmental change: A Palaeolimnological study in Belize. Ph.D. diss., University of Edinburgh.

Camargo, F. n.d. Estudo de alguns perfis de solo coletados em diversas regiões de Hiléia. Mimeo (Embrapa Library, Belém, Brazil).

Clement, C. R., J. M. McCann, and N. J. H. Smith. 2003. Agrobiodiversity in Amazônia and its relationship with dark earths. In J. Lehmann, D. C. Kern, B. Glaser, and W. I. Woods,

Amazonian Dark Earths: Origin, Properties, Management, 159–178. Dordrecht, Netherlands: Kluwer Academic.

Coates, Peter. 1998. *Nature.* Berkeley: University of California Press.

Cunha Franco, E. 1962. As terras pretas do Panalto de Santarém. *Revista da Sociedade dos Agronomos e Veterinários do Pará* 8:17–21.

Deagan, K., and J. M. Cruxent. 2002. *Archaeology at La Isabela: America's First European Town.* New Haven, Conn.: Yale University Press.

Erickson, C. L. 1995. Archaeological methods for the study of ancient landscapes of the Llanos de Mojos in the Bolivian Amazon. In P. W. Stahl, ed., *Archaeology in the Lowland American Tropics,* 66–95. Cambridge, U.K.: Cambridge University Press.

——. 2003. Historical ecology and future explorations. In J. Lehmann, D. C. Kern, B. Glaser, and W. I. Woods, eds., *Amazonian Dark Earths: Origin, Properties, Management,* 455–500. Dordrecht, Netherlands: Kluwer Academic.

Falesi, I. 1974. Soils of the Brazilian Amazon. In C. Wagley, ed., *Man in the Amazon,* 201–229. Gainesville: University of Florida Press.

German, L. 2003. Ethnoscientific understandings of Amazonian Dark Earths. In J. Lehmann, D. C. Kern, B. Glaser, and W. I.Woods, eds., *Amazonian Dark Earths: Origin, Properties, Management,* 179–201. Dordrecht, Netherlands: Kluwer Academic.

Glaser, B. 1999. Eigenschaften und Stabilität des Humuskörpers der Indianerschwarzerden Amazoniens. Ph.D. diss., University of Bayreuth, Beyreuth, Germany (Bayreuther Bodenkundliche Berichte 68).

Glaser, B., G. Guggenberger, L. Haumaier, and W. Zech. 2000. Burning residues as conditioner to sustainably improve fertility in highly weathered soils of the Brazilian Amazon region. Paper presented at the conference "Sustainable Management of Soil Organic Matter," British Society of Soil Science, Edinburgh, September 15–17.

Glaser, B., G. Guggenberger, and W. Zech. 2004. Identifying the Pre-Columbian Anthropogenic Input on Product Soil Properties of Amazonian Dark Earths (Terra Preta). In B. Glaser and W. I. Woods, eds., *Amazonian Dark Earths: Explorations in Space and Time,* 145–158. Berlin: Springer-Verlag.

Glaser, B. and W. I. Woods, eds. 2004. *Amazonian Dark Earths: Explorations in Space and Time.* Berlin: Springer-Verlag.

Glaser, B., G. W. Zech, and W. I. Woods. 2004. History, Current Knowledge and Future Perspectives of Geoecological Research Concerning the Origin of Amazonian Anthropogenic Dark Earths (Terra Preta). In B. Glaser and W. I. Woods, eds., *Amazonian Dark Earths: Explorations in Space and Time,* 9–17. Berlin: Springer-Verlag.

Glossary of Soil Science Terms. 1987. Madison, Wisc.: Soil Science Society of America.

Graham, E. 1989. Brief synthesis of coastal site data from Colson Point, Placencia, and Marco Gonzalez, Belize. In H. McKillop and P. F. Healy, eds., *Coastal Maya Trade,* 135–154. Occasional Papers in Anthropology no. 8. Peterborough, Ontario: Trent University.

——. 1994. *The Highlands of the Lowlands: Environment and Archaeology in the Stann Creek District, Belize, Central America.* Monographs in World Archaeology no. 19. Madison, Wisc., and Toronto: Prehistory Press and the Royal Ontario Museum.

——. 1998. Metaphor and metamorphism: Some thoughts on environmental meta-history. In W. Balée, ed., *Advances in Historical Ecology,* 119–137. New York: Columbia University Press.

——. 1999. Stone cities, green cities. In Elisabeth A. Bacus and Lisa J. Lucero, eds., *Complex Polities in the Ancient Tropical World,* 185–194. Archaeological Papers of the American Anthropological Association no. 9. Arlington, Va.: American Anthropological Association.

————. 2002. Maya cities and the character of a tropical urbanism. In P. Sinclair, ed., *The Development of Urbanism from a Global Perspective*. Uppsala: Institut für arkeologioch antik historia, Uppsala Universitet. Available at: www.arkeologi.uu.se/afr/projects/book/graham.pdf.

————. 2004. Lamanai reloaded: Alive and well in the Early Postclassic. In J. Awe, J. Morris, and S. Jones, eds., *Archaeological Investigations in the Eastern Maya Lowlands*, 223–241. Research Reports in Belizean Archaeology, vol. 1. Belmopan, Belize: Institute of Archaeology, National Institute of Culture and History.

Graham, E., and D. M. Pendergast. 1989. Excavations at the Marco Gonzalez Site, Ambergris Caye, Belize. *Journal of Field Archaeology* 16: 1–16.

Graham, E., D. M. Pendergast, J. Calvera R., and J. Jardines M. 2000. Excavations at Los Buchillones, Cuba. *Antiquity* 74:263–264.

Gray, M. 2004. *Geodiversity: Valuing and Conserving Abiotic Nature*. Chichester, U.K.: John Wiley and Sons.

Greenland, D. J. 1994. Soil Science and Sustainable Land Management. In J. K. Syers and D. L. Rimmer, eds., *Soil Science and Sustainable Land Management in the Tropics*, 1–15. Wallingford, U.K.: CAB International and British Society of Soil Science.

Hammond, N. D. C. 1982. *Ancient Maya Civilization*. New Brunswick, N.J.: Rutgers University Press.

Hecht, S. B. 2003. Indigenous soul management and the creation of Amazonian Dark Earths: Implications of Kayapó practices. In J. Lehmann, D. C. Kern, B. Glaser, and W. I. Woods, eds., *Amazonian Dark Earths: Origin, Properties, Management*, 355–372. Dordrecht, Netherlands: Kluwer Academic.

Hiraoka, M., S. Yamamoto, E. Matsumoto, S. Nakamura, I. C. Falesi, and A. R. C. Baena. 2003. Contemporary use and management of Amazonian Dark Earths. In J. Lehmann, D. C. Kern, B. Glaser, and W. I. Woods, eds., *Amazonian Dark Earths: Origin, Properties, Management*, 387–406. Dordrecht, Netherlands: Kluwer Academic.

Ingold, T. 2001. From complementarity to obviation: On dissolving the boundaries between social and biological anthropology, archaeology, and psychology. In S. Oyama, P. E. Griffiths, and R. D. Gray, eds., *Cycles of Contingency: Developmental Systems and Evolution*, 255–279. Cambridge, Mass.: MIT Press.

Johnston, K. J. 2002. Protrusion, bioturbation, and settlement detection during surface survey: The lowland Maya case. *Journal of Archaeological Method and Theory* 9 (1): 1–67.

————. 2004. The "invisible" Maya: Minimally mounded residential settlement at Itzan, Peten, Guatemala. *Latin American Antiquity* 15 (2): 1–29.

Kämf, N., W. I. Woods, W. Sombroek, D. C. Kern, and T. J. F. Cunha. 2003. Classification of Amazonian Dark Earths and other ancient anthropic soils. In J. Lehmann, D. C. Kern, B. Glaser, and W. I. Woods, eds., *Amazonian Dark Earths: Origin, Properties, Management*, 77–102. Dordrecht, Netherlands: Kluwer Academic.

Kellman, M. 1979. Soil enrichment by neotropical savanna trees. *Journal of Ecology* 67: 565–577.

Kern, D. C. 1988. *Caracterização pedológica de solos com Terra Preta Arqueológica na Região de Oriximiná, Pará*. Pôrto Alegre, Brazil: UFRGS, Faculdade de Agronomia. Tese (Mestrado).

Kern, D. C., G. D'Aquino, T. Ewerton Rodrigues, F. J. Lima Frazão, W. Sombroek, T. P. Myers, and E. Góes Neves. 2003. Distribution of Amazonian Dark Earths in the Brazilian Amazon. In J. Lehmann, D. C. Kern, B. Glaser, and W. I. Woods, eds., *Amazonian Dark Earths: Origin, Properties, Management*, 51–75. Dordrecht, Netherlands: Kluwer Academic.

Kern, D. C., and N. Kämpf. 1989. Antigos assentamentos indígenas na formação de solos com Terra Preta Arqueológica na Região de Oriximiná, Pará. *Revista Brasiliera de Ciência do Solo* 13:219–225.

Kern, D. C., M. Lima da Costa, and F. J. Lima Frazão. 2004. Evolution of the Scientific Knowledge Regarding Archaeological Black Earths of Amazonia. In B. Glaser and W. I. Woods, eds., *Amazonian Dark Earths: Explorations in Space and Time*, 19–28. Berlin: Springer-Verlag.

Kidder, A. V., and G. F. Ekholm. 1951. *Some Archaeological Specimens from Pomona, British Honduras*. Notes on Middle American Archaeology and Ethnology vol. 4, no. 102. Washington, D.C.: Carnegie Institution.

King, R. B., I. C. Baillie, T. M. B. Abell, J. R. Dunsmore, D. A. Gray, J. H. Pratt, H. R. Versey, A. C. S. Wright, and S. A. Zisman. 1992. *Land Resource Assessment of Northern Belize*. Natural Resources Institute Bulletin no. 43, vol. 1. Kent, U.K.: Overseas Development Natural Resources Institute.

Kristiansen, S. M., and W. Amelung. 2001. Abandoned anthills of *Formica polyctena* and soil heterogeneity in a temperate deciduous forest: Morphology and organic matter composition. *European Journal of Soil Science* 52:355–363.

Lambert, J. D. H., and J. T. Arnason. 1982. Ramon and Maya ruins: An ecological, not an economic, relation. *Science* 216:298–299.

Lehmann, J., D. C. Kern, B. Glaser, and W. I. Woods, eds. 2003. *Amazonian Dark Earths: Origin, Properties, Management*. Dordrecht, Netherlands: Kluwer Academic.

Lehmann, J., J. Pereira da Silva Jr., C. Steiner, T. Nehls, W. Zech, and B. Glaser. 2003. Nutrient availability and leaching in an archaeological anthrosol and a ferralsol of the central Amazon basin: Fertilizer, manure, and charcoal amendments. *Plant and Soil* 249:343–357.

Marcus, J. 1982. The plant world of the sixteenth- and seventeenth-century lowland Maya. In K. V. Flannery, ed., *Maya Subsistence*, 239–273. New York: Academic Press.

Mazzullo, S. J., C. S. Teal, and E. Graham. 1994. Mineralogic and crystallographic evidence of lime processing, Santa Cruz Maya site (Classic to Postclassic), Ambergris Caye, Belize. *Journal of Archaeological Science* 21:785–795.

McCann, J. M. 1994. *Terra preta do índio* in lower Amazonia: Fertile ground for *caboclo* agriculture and theoretical debates in cultural ecology. Paper presented at the joint meeting of the Sociedad Mexicana de Geografía y Estadistica XIV and the Conference of Latin Americanist Geographers XX, 26–30 September, Ciudad Juárez, Mexico.

McCann, J. M., W. I. Woods, and D. W. Meyer. 2001. Organic matter and anthrosols in Amazonia: Interpreting the Amerindian legacy. In R. M. Rees, B. C. Ball, C. D. Campbell, and C. A. Watson, eds., *Sustainable Management of Soil Organic Matter*, 180–189. Wallingford, U.K.: CAB International.

Meggers, B. 1954. Environmental limitation and the development of culture. *American Anthropologist* 56:801–824.

———. 1971. *Man and Culture in a Counterfeit Paradise*. Chicago: Aldine-Atherton.

Melville, E. G. K. 1994. *A Plague of Sheep*. Cambridge, U.K.: Cambridge University Press.

Mora C., S. 2003. Archaeobotanical methods for the study of Amazonain Dark Earths. In J. Lehmann, D. C. Kern, B. Glaser, and W. I. Woods, eds., *Amazonian Dark Earths: Origin, Properties, Management*, 205–226. Dordrecht, Netherlands: Kluwer Academic.

Mora C., S. L. Fernanda H., I. Cavelier F., and C. Rodriguez. 1991. *Cultivars, Anthropic Soils, and Stability: A Preliminary Report of Archaeological Research in Araracuara, Colombian Amazonia*. University of Pittsburgh Latin American Archaeology Reports no. 2. Bogotá and Pittsburgh: Programa Topenbos and Department of Anthropology, University of Pittsburgh.

Myers, T. P., W. M. Denevan, A. WinklerPrins, and A. Porro. 2003. Historical perspectives on Amazonian Dark Earths. In J. Lehmann, D. C. Kern, B. Glaser, and W. I. Woods, eds., *Amazonian Dark Earths: Origin, Properties, Management,* 15–28. Dordrecht, Netherlands: Kluwer Academic.

Neves, E. G., J. B. Petersen, R. N. Bartone, C. A. da Silva. 2003. Historical and socio-cultural origins of Amazonian Dark Earths. In J. Lehmann, D. C. Kern, B. Glaser, and W. I. Woods, eds., *Amazonian Dark Earths: Origin, Properties, Management,* 29–50. Dordrecht, Netherlands: Kluwer Academic.

Oldeman, L. R., R. T. A. Hakkeling, and W. Sombroek. 1990. *World Map on the Status of Human-Induced Soil Degradation.* Wageningen, Netherlands: International Soil and Reference Information Centre.

Oliver, José R. 1992. Donald Lathrap: Approaches and contributions in New World archaeology. *Journal of the Steward Anthropological Society* 20:283–345.

Oliver, José R. 2000. Gold symbolism among Caribbean chiefdoms: Of feathers, *çibas,* and *guanín* power among Taino elites. In C. McEwan, ed., *Precolumbian Gold in South America: Technology, Style, and Iconography,* 196–219. London and Chicago: British Museum Press and Fitzroy Dearborn.

Pearsall, D. M. 1995. Domestication and Agriculture in the New World Tropics. In T. D. Price and A. B. Gebauer, eds., *Last Hunters, First Farmers,* 157–192. Santa Fe, N.M.: School of American Research Press.

Pearson, J. L. 2001. *Shamanism and the Ancient Mind: A Cognitive Approach to Archaeology.* Walnut Creek, Calif.: Altamira Press.

Pendergast, D. M. 1981. Lamanai, Belize: Summary of excavation results 1974–1980. *Journal of Field Archaeology* 8 (1): 29–53.

———. 1986. Stability through change: Lamanai, Belize, from the ninth to the seventeenth century. In J. A. Sabloff and E. Wyllys Andrews V, eds., *Late Lowland Maya Civilization: Classic to Postclassic,* 223–249. Albuquerque: University of New Mexico Press, School of American Research.

Pendergast, D. M., E. Graham, J. Calvera R., and J. Jardines M. 2001. Houses in the sea: Excavation and preservation at Los Buchillones, Cuba. In B. A. Purdy, ed., *Enduring Records: The Environmental and Cultural Heritage of Wetlands,* 71–82. Oxford: Oxbow.

———. 2002. The houses in which they dwelt: The excavation and dating of Taino wooden structures at Los Buchillones, Cuba. *Journal of Wetland Archaeology* 2:61–75.

Peters, Charles M. 1983. Observations on Maya subsistence and the ecology of a tropical tree. *American Antiquity* 48 (3): 610–615.

Pohl, M. D. 1990. The Rio Hondo Project in northern Belize. In M. D. Pohl, ed., *Ancient Maya Wetland Agriculture,* 1–19. Boulder, Colo.: Westview Press.

Preu, C. 1991. Human impact on the morphodynamics of coasts—a case study of the SW coast of Sri Lanka. In W. Erdelen, N. Ishwaran, and P. Müller, eds., *Tropical Ecosystems: System Characteristics, Utilization Patterns, and Conservation Issues,* 121–138. Proceedings of the International and Interdisciplinary Symposium, Saarbrücken, Germany, 14–18 June 1989. Weikersheim, Germany: Verlag Josef Margraf.

Price, H. W. 1899. Excavations on Sittee River, British Honduras. *Proceedings of the Society of Antiquaries* 17:339–344.

Puleston, D. E. 1968. *Brosimum alicastrum* as a subsistence alternative for the Classic Maya of the central southern lowlands. Master's thesis, University of Pennsylvania.

Roosevelt, A. C. 1989. Resource management in Amazonia before the Conquest: Beyond ethnographic projection. In D. A. Posey and W. Balée, eds., *Resource Management in Amazonia: Indigenous and Folk Strategies,* 30–62. Advances in Economic Botany no. 7. Bronx: New York Botanical Garden.

——. 1991. *Moundbuilders of the Amazon: Geophysical Archaeology on Marajo Island, Brazil.* San Diego: Academic Press.

Ruivo, M. L. P., M. A. Arroyo-Kalin, C. E. R. Schaefer, H. T. Costi, S. H. S. Arcanjo, H. N. Lima, M. M. Pulleman, and D. Creutzberg. 2003. The use of micromorphology for the study of the formation and properties of Amazonian Dark Earths. In J. Lehmann, D. C. Kern, B. Glaser, and W. I. Woods, eds., *Amazonian Dark Earths: Origin, Properties, Management,* 243–254. Dordrecht, Netherlands: Kluwer Academic.

Sanders, W. T. 1977. Environmental heterogeneity and the evolution of lowland Maya civilization. In R. E. W. Adams, ed., *The Origins of Maya Civilization,* 287–297. Albuquerque: University of New Mexico Press.

Schiffer, M. B. 1972. Archaeological context and systemic context. *American Antiquity* 37:156–165.

——. 1987. *Formation Processes of the Archaeological Record.* Albuquerque: University of New Mexico Press.

Silva, F. A. 2003. Cultural behaviors of indigenous populations and the formation of the archaeological record in Amazonian Dark Earth: The Asurini Do Xingú case study. In J. Lehmann, D. C. Kern, B. Glaser, and W. I. Woods, eds., *Amazonian Dark Earths: Origins, Properties, Management,* 373–385. Dordrecht, Netherlands: Kluwer Academic.

Smith, H. H. 1879. *Brazil: The Amazons and the Coast.* New York: Charles Scribner's Sons.

Smith, N. 1980. Anthrosols and human carrying capacity in Amazonia. *Annals of the Association of American Geographers* 20 (4): 553–566.

Sombroek, W. G. 1966. *Amazonian Soils: A Reconnaissance of the Soils of the Brazilian Amazon Region.* Wageningen, Netherlands: Centre for Agricultural Publication and Documentation.

Sombroek, W. G., D. C. Kern, T. Rodrigues, M. da Silva Cravo, T. Jarbas Cunha, W. Woods, and B. Glaser. 2002. Terra preta and terra mulata: Pre-Columbian Amazon kitchen middens and agricultural fields, their sustainability and their replication. In R. Dudal, ed., *Symposium 18, Anthropogenic Factors of Soil Formation, Seventeenth World Congress of Soils Science.* Bangkok: Transaction (CD-ROM).

Stahl, P. W., ed. 1995. *Archaeology in the Lowland American Tropics.* Cambridge, U.K.: Cambridge University Press.

——. 2002. Paradigms in paradise: Revising standard Amazonian prehistory. *Review of Archaeology* 23 (2): 39–51.

Terry, R. E., P. J. Hardin, S. D. Houston, S. D. Nelson, M. W. Jackson, J. Carr, and J. Parnell. 2000. Quantitative phosphorus measurement: A field test procedure for archaeological site analysis at Piedras Negras, Guatemala. *Geoarchaeology* 15 (2): 151–166.

Turner, B. L., II, W. C. Clark, R. W. Kates, J. F. Richards, J. T. Mathews, and W. B. Meyer, eds. 1990. *The Earth as Transformed by Human Action: Global and regional Changes in the Biosphere over the Past 300 Years.* Cambridge, U.K.: Cambridge University Press.

Woods, W. I. 1977. The quantitative analysis of soil phosphate. *American Antiquity* 42 (2): 248–252.

——. 1984. Soil chemical investigations in Illinois archaeology: Two example studies. In J. B. Lambert, ed., *Archaeological Chemistry III,* 67–77. Advances in Chemistry Series no. 205. Washington, D.C.: American Chemical Society.

——. 1995. Comments on the black earths of Amazonia. *Papers and Proceedings of Applied Geography Conferences* 18:159–165.

——. 2003. Development of anthrosol research. In J. Lehmann, D. C. Kern, B. Glaser, and W. I. Woods, eds., *Amazonian Dark Earths: Origin, Properties, Management,* 3–14. Dordrecht, Netherlands: Kluwer Academic.

Woods, W. I., and J. M. McCann. 1999. The anthropogenic origin and persistence of Amazonian Dark Earths. *Yearbook Conference of Latin Americanist Geographers* 25:7–14.

Woods, W. I., J. M. McCann, and D. W. Meyer. 2000. Amazonian Dark Earth analysis: State of knowledge and directions for future research. *Papers and Proceedings of the Applied Geography Conferences* 23:114–121.

Woods, W. I., C. L. Wells, D. W. Meyer, and H. W. Watters Jr. 1996. Comments on landscape and soils in the upper Belize Valley. Paper presented at the Latin Americanist Geographers Congress, 4 January, Tegucigalpa, Honduras.

Wright, A. C. S., D. H. Romney, R. H. Arbuckle, and V. E. Vial. 1959. *Land in British Honduras.* Report of the British Honduras Land Use Survey Team. London: Her Majesty's Stationery Office.

Zech, W., L. Haumaier, and R. Hempfling. 1990. Ecological aspects of soil organic matter in tropical land use. In P. MacCarthy, C. E. Clapp, R. L. Malcolm, and P. R. Bloom, eds., *Humic Substances in Soil and Crop Sciences: Selected Readings,* 187–202. Madison, Wisc.: American Society of Agronomy and Soil Science Society of America.

Zech, W., E. Pabst, and G. Bechtold. 1979. Analytische Kennzeichnung der Terra Preta do Indio. *Mitteilungen der Deutschen Bodenkundlichen Gesellschaft* 20:709–716.

Zeidler, J. A. 1995. Archaeological survey and site discovery in the forested Neotropics. In P. W. Stahl, ed., *Archaeology in the Lowland American Tropics,* 7–41. Cambridge, U. K: Cambridge University Press.

3

DOMESTICATED FOOD AND SOCIETY IN EARLY COASTAL PERU

CHRISTINE A. HASTORF

THE RECENT LITERATURE, including the chapters in this volume, debate about the extent of human impact on post-Pleistocene Amazon basin ecology. Discussion has included the landscape as a form of built environment and to what extent humans have altered this vast South American ecology over the past 10,000 years. Over this time span, the environment shifted to the modern condition. In the past, the annual cycle was more seasonal and moist in the low-lying tropical forest areas than today. Intimately linked to these environmental changes are the changes that occurred within the human groups themselves. These interactive influences lead us to a series of related questions about human-plant relationships. Historical ecology highlights these long and intimate interactions humans have had with their landscape. How have these human-plant interactions shaped not only the tropical forest but also the selection and ultimate domestication of certain plants within that landscape? What can the history of the domestic plants and their movements illuminate about this domestication process as it is intertwined with the domestication of humans? This dynamic bridge between environment and culture is central to the historical ecology enterprise. The first question is one of the major thrusts of this volume. The second question is the focus of this chapter.

My thesis is anthrocentric in that it places humans as active agents in plant domestication, movement, and adoption. I assume that humans have always had an impact on their environment, just as the environment has continuously crafted humans (Crumley 1994; Lentz 2000; Zimmerer 1994). Thus, any historical trends suggested in the presence and frequencies of specific plants that have become particularly intimate with humans reflect the history of human interest *as well as* human impact on the plant taxa. There is a constant and close relationship between people and their food items. People search for food, yet are also food for other forest dwellers (Viveiros de Castro 1992). Such a dynamic should have alerted archaeologists years ago to the active role humans have in plant domestication and spread, yet there has been some reticence when it comes to the discussion of this side of plant food in the archaeological literature (Pollan 2001). One line of inquiry concerns the choices people have made about their food procurement and what that can tell us about the dynamics of these interactions. In some settings within South America, human-plant interactions led to plant domestication; in others it did not. In this chapter, I focus not on the ecological and biological causes of plant domestication in South America, which has a fruitful literature, but more on the cultural implications for the adoptions of these plants in areas outside of their place of domestication. In tracking plant adoption, we must assume that the plants have been domesticated and also traded or carried across the landscape. These activities are therefore implicit within this discussion. Here I focus on the taking up of "foreign" food plants as a cultural act.

By doing so, I am emphasizing the active roll humans had in the plant movements across the Americas. The literature on plant adoption and migration, often framed in terms of crop diffusion (Sauer 1952), rarely locates the human participants as the active agents. Ellen Messer's (1997) work on the potato introduction in Europe is a notable exception. The cultural influences of such plant movements form the thesis I explore here.

Cultural choices related to domestication, movement, and adoption made by individuals over the years are reflected in plant distributions over space and time. I am proposing that across South America, the reasons for plant domestication, movement, and adoption of cultigens were as much cultural as economic or ecological. Over the past 7,000 years, specific plants spread across the continent at different speeds and with varying levels of popularity. What might these plants have meant to the inhabitants that encouraged them as they devoted more time to them? How were the plants brought into the cultural world? Historical ecology allows us to bring the humans as well as the plants central stage and to make their actions visible within the greater landscape. It provides a way to rework the embedded yet artificial division between nature and culture, due to our Enlightenment intellectual history, that has so hindered a more interactive way of thinking about this dynamic relationship (Ingold 1993; Thomas 2004). The active cultural choices about plant interaction reflect the knowledge, perception, and ecology in this dialectic. How did people-plant interrelations move beyond plant physiology and life cycles? Here I discuss four cultural aspects of plant use that participate in plant domestication, movement, and adoption processes. I introduce each influence and then apply it to the early plant data from Peru. These four aspects are: the creation of identity, development of a horticultural mindset, women's roll in creating descent groups, and gifting.

One of the clearest situations in which to track plant adoption is the dry west coast of Peru, where virtually all domesticates are nonlocal plants, introduced from elsewhere. By highlighting the human choices involved in plant movement and adoption, we can better understand the dialectical relations between human acts and acts of nature, which are made manifest on the landscape (Crumley 1994). The plants can grow outside their natural habitat when humans tend them. However, in order to move great distances quickly, they usually need to be carried and replanted.

People were living throughout the South American continent by 8000 BC and possibly earlier (Roosevelt et al. 1996). The inhabitants were initially full-time foragers, albeit sparsely scattered. Over time, they developed relationships with the local fauna and flora. As the environment and climate changed, new resource extraction possibilities became available to them. These resources directed people to certain plants and in some settings led to the plants' incorporation into human settlements and house gardens, and, for some plants, to the plants' domestication (Lathrap 1977; Piperno and Pearsall 1998; Sauer 1952). Why this happened at the juncture between the Pleistocene and the Holocene is cogently discussed by Piperno and Pearsall (1998). Overall, the moist tropics held an abundance of food sources that allowed people to select from a broad range of taxa, environments, and foodways.

The study of plant domestication in the moist tropics has been difficult due to the problems of preservation and visibility. Piperno and Pearsall are dedicating their productive careers to compiling and synthesizing data from South America. Their book *The Origins of Agriculture in the Lowland Neotropics* (1998) presents a detailed survey of initial American plant domestication in the neotropical lowland moist tropics, beginning with tubers, then dealing with fruit trees and seeds. They have identified 10 crop-domestication centers based on archaeological evidence, environmental conditions, and wild progenitors (Piperno and Pearsall 1998:164). By 1000 BC, domesticated plants were widespread across the South American continent, albeit in patchy distributions far beyond the original loci of origin. The authors in this volume are adding to this picture of environmental transformation into cultural landscapes through agroforestry, gardening, earthworks, and settlements that ultimately created dark-earth and other cultural constructions (Denevan 2002).

The recent data presented about the South American built environment suggests there was simultaneous early human impact in both the Andean highlands, best illustrated by Erickson's (2000, 2001) work in the Titicaca basin on raised fields, terraces, and sunken gardens, and along the arid western coast of Peru with its intensive irrigations systems (Moseley 1975, 1983; Shady, Haas, and Creamer 2001; Wilson 1988). The agricultural field evidence demonstrates that intensive agricultural systems were present across the continent by AD 1000, long before European contact (Denevan 2002). These built environments produced food and the propagation of domestic plants, but they also reflected the inhabitants' perception of their own world, that of intimate and intensive interaction, whether on riverbanks or on hillsides.

The moist lowland tropics, the dry highlands, and the arid coast represent three major environmental zones within the environmental extremes that make up South American geography. In Western scientific study, these three zones are treated as completely separate entities. The difference in their histories, their overlapping relations, and what role each played in the domestication, movement, and adoption of plants is an illuminating story in historical ecology. Here I investigate one aspect of the relationship between the moist lowland tropics and the arid coast, in particular the central coast of Peru, in terms of the domestication, movement, and adoption of plants, which should offer some new insights into the cultural developments of these two zones. The story of Pacific coastal plant adoption illuminates this investigation. At first glance, these two zones are very different, as reflected in the history of investigations of them. Scholars are perfectly willing to accept that intensive cropping and irrigation systems existed throughout the arid coast, yet for the moist lowland tropics, an equivalent acceptance of extensive, let alone intensive, agriculture has been difficult (Meggers 1971; Moseley 1999).

Archaeological evidence demonstrates that there has been long human occupation and plant use in the moist tropical lowlands of the northern Amazon basin as far back as 5,000 years ago, but that domestic plants are relatively recent on the arid coast, becoming common only around 1000 BC (Oyuela-Caycedo 1996; Piperno and Pearsall 1998). Most archaeologists focus on the large-scale complex societies along the western coast , acknowledging that these large polities have been there for millennia. Yet the Amazon basin, the location where most crops were domesticated, has usually been considered a place of scant population and of small-scale, even isolated, societies with little large-scale political complexity. This disjuncture is highlighted especially because the western coast has no native domesticates. These archaeological histories illustrate the dissonance between the history of investigation and the perception of the two areas.

These traditional views describe the two regions as disconnected universes, with little or no interaction. Donald Lathrap (1973) and colleagues have shown that this is not the case. The ecotone of the two regions, or the eastern slopes of the Andes, is one of the least investigated areas of South America (Kojan 2002). If we accept the notion that the Amazonian environment does not a priori limit cultural development (Balée 1989; Denevan 2002; Rival 1998a:233), we can begin to accept that long-lived and even large complex societies must have flourished there (Erickson, chapter 8, and Heckenberger, chapter 10, this volume). In addition, both regions share the same plant foods. How, when, and why did these moist tropical plants of the Amazon basin find their way into the lives, gardens, and stomachs of the inhabitants of the dry western coast and eventually become the foundation of large polities?

HUMAN-PLANT RELATIONSHIPS

In my investigation of plant movements from the moist tropical forests and their eventual adoption into the dry coastal environment, the social and cultural concepts *enculturation* and *two-way domestication* are important. First, we need to accept that people are curious and interactive with their environment. As a core idea in historical ecology, we assume that people influence and enculturate (naturalize) the plants and animals with whom they live and upon whom they rely. There are many ethnographic examples of this anthropomorphizing relationship (Balée 1994; Descola 1996; C. Hugh-Jones 1996; Viveiros de Castro 1992). These assumptions allow us to examine the influences that plants and animals have had on human existence and social formation, beyond their basic role in subsistence as a food supply. I also explore the notion that plant domestication and the planting of domesticated plants are associated with the concept of the cultivating society or social domestication (Cauvin 2000; Hodder 1991). This approach to domestication focuses on the intimate role of plant-human relationships within the formation of societies across the continent. By exploring which and how specific plants moved long distances across the continent, we can begin to elucidate what role the plants played in the formation of these societies. Most scholars agree that early plant movements did not have a purely calorific or demographic cause. We know that such movements were not steady or directional. Their histories contain clues to the meanings these plants held for people and in turn illustrate something about the past human societies that nurtured them as well as consumed them. Several social and cultural concepts may help us in understanding why specific plants were adopted by specific groups.

Given that humans have manipulated their plant neighbors ever since they arrived in South America, what do the changes in plant use tell us about social changes? Amazonian and Andean people have an intimate relationship with the living things among which they reside. This relationship is complex because some taxa are broadly distributed across the continent, whereas others are locally, but intensively propagated and utilized. It may exist in part because many people believe that plants are sentient beings and capable of feelings and thoughts (Allen 1988; Descola 1992). Therefore, to use and tend a plant that then is consumed or made into a tool implies a mutual relationship, with its own obligations and requirements. Although such an intimate social relationship is easier to envision with animals given that they are more like humans in their movements and life cycles (Bloch 1998; Cormier, chapter 11, this volume; Descola 1992; C. Hugh-Jones 1996; Lévi-Strauss 1969), humans do register plant growth and their living qualities. The recent volumes by Rival (1998b) and Simoons (1998) on tree symbolism and the role of trees in ritual and religious life illustrate how plants participate in social-identity formation. Building on this interactive notion, I assume that the plants people lived among and used in the past had a social significance, at times making them intimately part of the group or at other times ostracizing them. How this interaction between plants and people changed is a story not just of domestication, but also of need, mutual impact, socialization, and meaning formation—like the *mulefa* and their seed pod trees in the third book of Philip Pullman's His Dark Materials trilogy (2000).

All societies have names and therefore meanings for the plants they use. Through this recognition, plants become linked into wider meaning structures that help people navigate through their natural and social worlds (Ortiz 1994). Brown (1985) has noted that foragers classify the plants they use differently from farmers, suggesting a qualitative change with this different relationship with plants. Foragers' interactions with plants are usually less intensive than horticulturalists', in part by definition but also because foragers interact differently in the plants' life cycles. The foraging Kumeyaay of southern California were particularly interactive with and manipulative of their local wild taxa (Shipek 1989). Their interactions with plants was more intensive than their foraging neighbors' interactions because of their proactive role with some plants. Several centuries before migrating to California, they had been farmers in Colorado (Steven Shackley, personal communication, 2000). This history of different plant-human interaction was still present in their daily practice, so that they acted differently from their permanently foraging neighbors. Loretta Cormier (chapter 11, this volume) records a similar history for the Guajá foragers of Brazil based on the fact that these foragers are quite intimate with monkeys. She believes that they used to farm before they were pushed off of their land. These and other ethnographies illustrate how long-lived worldviews impact human-plant relationships. These long-term relationships can be detected in a culture's classification and symbolism of plants (C. Hugh-Jones 1979; Weismantel 1988). If we keep in mind that these types of meanings were probably present in the deep past of human societies, we can begin to place plants and their domestication more actively within each past human society. Investigating this long-term human-plant relationship will help in our understanding of the historical ecological beginnings of domestication. Scholars who engage in such investigations, like other historical ecologists, see *human society* as a narrow term because these societies also included plants and animals in their world.

THREE DISTINCT ZONES

Although the three major zones of central South America have different ecologies and histories, we know the people inhabiting each zone were intertwined with the people of the other zones through the plants they shared. In each zone, there is an early and long record of human impact on the environment that shows both nurturing and destruction. After presenting a brief overview of plant and landscape manipulation in these regions, I illustrate the cultural nature of plant use with the clearest example of timing and scale of domestic plant uptake along the west-central coast of South America.

There has been a large and prolific debate about the extent of anthropogenic landscapes and plant manipulation in the Amazon basin, in which it ranges from pristine wilderness to a cultivated orchard. The recent synthesis of evidence of crop domestication in the Amazon and northern South American lowlands shows us that cultivation in small plots or house gardens next to residences began between 9000 and 8000 BP (Piperno and Pearsall 1998). Based on site placement and other evidence (Roosevelt 1984), it is believed that by 7000 BP (5000 BC) larger-scale food production emerged in fields more removed from residences. Many centers of plant diversity could be the locations where intensive plant use began (Piperno and Pearsall 1998; Clement 1999b). Balée (1989) estimates that approximately 12 percent of the Amazon is anthropogenic, suggesting that although the human footprint on the vegetation is substantial, people have not completely reconfigured the basin's landscape. Others believe that humans have influenced a much larger extent (Denevan 2002; Erickson, chapter 8, this volume).

There is much less debate about the level of human impact along the dry western coast of South America. The plant distribution along the coast is heavily anthropogenic, probably as much as 90 percent altered by humans. Its dryness makes this impact more visible. The seasonal *lomas* (seasonal cloud forests along the foothills of the Andes in from the coast that catch the winter fog in from the ocean),[1] and the *Prosopis* forests held local vegetation. Most nonlocal domestic plants require environments wetter than these two environments are. Domestic plants are grown using irrigation along the rivers or near springs. Even the plants that could have been domesticated close to the coast—cotton, gourd, and the jack bean—were restricted to the riverbanks for survival (Piperno and Pearsall 1998). Despite the desert conditions, there is no evidence of food stress during these early times. The cool waters of the Humbolt Current have provided the inhabitants with rich marine life for the past 6,000 years (Sandwiess et al. 1989).

The western Andean highland mountains contained other indigenous food plants that became staple domesticates for the people who lived in and passed through these mountains. The highland inhabitants expanded the useful food-producing microenvironments, especially in the valleys and hillsides (Hastorf 1993). The highland crops of potato (*Solanum tuberosum* and *andigenum*), quinoa (*Chenopodium quinoa*), and lupine (*Lupinus mutabilis*) spread both north and south from northern Bolivia. Some plants, such as the *Phaseolus* beans, moved up from the eastern slope of the Andean mountains into the intermontane valleys and plateaus and eventually to the arid coast (Gepts 1990); and maize entered from the north along the coast as well as along the eastern and western slopes of the Andes (Piperno and Pearsall 1998). Today the highlands are a built landscape covered with fields, canals, sunken gardens (*qochas*), raised fields, and terraces (Denevan 2002). There is not a vista that one looks out across where one does not encounter the outlines of field or canals. The ubiquitous herds of camelids and other domestic animals reflect the intensive anthropogenic manipulation of the higher grassland pastures . The higher environments can be roughly estimated at close to 50 percent anthropogenic, using Balée's (1989) scale of influence.

THE WEST-CENTRAL COAST

More than 30 valleys flow in a parallel formation down from the western mountain slopes of the west-central coast of South America to the ocean. The distances between these parallel valleys vary but tend to be about 40 kilometers apart. The northern rivers have permanent and more abundant water during the highland rainy season than the southern rivers. Despite differing amounts of seasonal water flow, the river valleys are basically redundant with coastal access at their mouths, regular water availability, and periodic cloud forest (lomas) resources nearby on the foothills. The valley productivity is excellent, with vast irrigation systems that yield virtually anything people want to produce, including rice and sugar cane, traditionally wet-climate crops.

By 4000 BC, the western coastline had stabilized to essentially the modern condition. In this new Holocene environment, the seasonality became more pronounced. Conditions included little to no regular coastal rainfall outside of the winter fog belt, more water in the rivers during the highland summer rains, and a more prolonged El Niño pattern of regular if stochastic cycles of rain and flooding than what is experienced today (Enfield 1992; Martin et al 1993). By this time, the domestic crops were "on the move," beginning to be found outside of their loci of domestication (Sauer 1952).

In every valley, agricultural evidence exists in canals, dams, and fields built and rebuilt over several thousand years. We now know that this riverine, terrestrial focus occurred later. The first adaptations that we can discern before and for a millennium after the coastline stabilized (between 4000 and 1500 BC) demonstrate that the inhabitants focused on the rich marine life with later additions of the seasonal lomas biota (Benfer 1982; Quilter 1989; Sandweiss et al 1989). The earliest levels of the Ring Site (Sandweiss et al. 1989) date to the Early Archaic period, where subsistence appears to be largely based on marine resources (10,000–8000 BC). Later, there was increasing subsistence reliance on the resources available at the river mouths, where both coastal and riparian resources could be exploited. This coastal focus continued well into the Holocene and in fact continues in part today with fisherfolk who exchange their catch for agricultural crops produced by inland farmers. There is evidence for increasing use of the rivers and agriculture along the coastal valleys between 6000 and 1000 BC, with settlements moving inland to focus on the river water (Benfer 1982; Quilter 1989; Quilter et al. 1991). Huge settlements and vast irrigation systems have existed up and down these dry coastal valleys, with the earliest dating to 4000 BC (Engel 1970; Shady, Haas, and Creamer 2001).

There are no major topographic barriers between these valleys. Movement north and south along the coast was probably easy, although freshwater would have been carried. Although we do not know how far back ocean-going boats operated, rafts were an important form of coastal communication. In tracking the beginning of agriculture on the Peruvian coast, it is striking to note that almost all of the domesticated plants were nonlocal. In addition, we find that hundreds of years passed before certain coastal valley residents adopted the crops that their neighbors were growing and consuming in valleys to the north and south (Hastorf 1999). This early plant-use evidence reflects selective uptake strategies by the inhabitants in different valleys, with crops entering the region in a nonuniform, nondirectional, and nonclustered pattern.

How and why might this have happened?

PATTERNS OF EARLY PLANT USE ON THE COAST

Based on archaeological work, all three South American zones are known to have received long-term impact by the pre-Colombian population although we know more about human impact in the highlands and on the coast than in the Amazonian forests. It is in the Amazon basin where we are learning about an important set of human-plant interactions from the scholarship over the past twenty years, as this present volume illustrates. If we assume that this interaction has been occurring for thousands of years between all three regions. Especially given that most of the plants on the dry western coast have come from the highlands, moist tropics, and eastern slopes (the north, southeast, and east), it is difficult not to accept that the Amazon developed horticulture first (Clement 1999b; Lathrap 1973, 1977; Piperno and Pearsall 1998; Roosevelt 1984). Historical, ethnographic, and archaeological data suggest that these three regions were regularly interacting and sharing culture, artifacts, and plants. What is laid bare in the arid western valleys is that coastal agriculture was built on the introduction and adoption of nonlocal exotic plants rather than through a long-term interaction with local plants that characterizes the other two regions. These plants had to have been brought from the Amazon or from other centers of domestication and adopted to the local growing conditions. *Therefore, the coast is the truly the ultimate anthropogenic environment.*

Although the Peruvian coast is not unique in having a long sequence of plant introductions, the excellent preservation and lack of locally domesticated food plants make it an important and unique case study for plant movement and adoption. For the earlier inhabitants, the nonlocal domesticated plants competed with marine animal resources as foodstuffs. Thus, coastal plant movement and adoption is a particularly interesting study of both cultural and environmental formation because we can track how people actively brought in and nurtured plants that participated in forming some of the longest-lived and largest-scale societies ever known in the Americas. We know that most plant domesticates along the western coast came from the greater Amazon basin, such as achira, beans, chili peppers, manioc, and guava, to name a few. Others, such as potatoes, came from the highlands. People carried, traded, and received gifts of these plants across the highlands and along the coast from the plants' places of domestication. This process, occurring over thousands of years, was due to exchange more than to migration or to what is traditionally called "diffusion" in the literature. I prefer not to use the term *diffusion* because it has a passive connotation, and I am confident that people knew they were carrying and or trading crop plants when they traveled and traded them.

Migration is the large and small movements of peoples carrying crops with them. Archaeologically, this migration should show crop and/or human replacement or the addition of new human populations bringing their plants. This pattern is not evident along coastal Peru. Traditional discussions of diffusion ignore process and human agency, although more recent discussions are sensitive to these issues. Thinking about the types of diffusion that made the movement of some crops possible leads us to ask about the range of social and economic exchanges that must have occurred in the past. Although the crops were nonlocal, the technology and knowledge to farm them (spring flooding, irrigation, and sunken gardens) must have been a local invention or traveled with the crops from the Andean highlands.

To begin to understand this botanical and human history, we need to focus on the less-visible part of the process and think about why people might have nurtured specific plants and why they transplanted them hundreds of kilometers across the continent. I believe this process reflects early social identity and group formation as an intimate part of the domestication process.

WHY DID PLANTS MOVE?

Our question becomes: Why did many of more than 100 different plants expand out of their diverse home ranges and into new territory, given that they, as fragile domesticates, required human help to move and to reproduce (Clement 1999a)? What led people to carry, trade, gift, or steal these plants? Part of the answer lies in why these plants received more attention than other plants in the first place. Were they simply fast-growing annuals that like disturbed conditions, as Anderson (1952), Harlan (1975), and Ingold (1996) have suggested? This is probably true for some plants, but not for all, and certainly not for trees. Heiser (1965) proposed that these disturbed habitat–loving plants would essentially spring up around the campsites along a seasonal round, making themselves more available with a return visit. This model is also a "phyto-centric" type. Models for domestication are vast and cover the full range from unintentional ecological to intentional to proactive and cognizant, as we know so well from Monsanto's genetically modified crop development of today. Space does not allow for a discussion of these models in this chapter, however, so I focus on the ideas that expressly relate to the concepts I am developing for the arid coast of Peru.

Farrington and Urry (1985) revitalize Sauer's (1952) anthropocentric model, which indicates that the flavorful, spicy characteristics of certain plants initially grabbed people's attention so that they brought these plants under cultivation. Iltis (2000) and Smalley and Blake (2003) have recently proposed a version of this model even for maize. Following along with the notion of plant notoriety based on their desirable flavors and textures, we can propose that the early cultivated plants were adopted not only because they enticed people, but because of special meanings and memories of other people, places, and times. Special qualities attributed to certain plants can be based on the memories associated with their participation in family life as well as on their tastiness and their physical characteristics. Domestication, movement, and adoption of specific plants could have been a part of the history of people's curiosity about the world around them.

People must give significance to, identify, label, and name things before they can interact with them. Although they need to eat every day, they are always selective (Rozin 1982). No group in the world eats purely based on economic efficiency; there are always historical influences, traditions, and tastes that direct food choice (Counihan and Van Esterik 1997; Fischler 1988; MacBeth 1997). The plants that became more significant in a particular human society were reactive and tasty and had special attributional connotations. Hunger drove people to search for food, but which items they chose was determined by what they thought about the plants and animals they encountered. How did this curiosity channel people to move beyond the plant physiology and life cycles and to adopt some plants into their diet as well as into their society? The four cultural aspects of plant use I mentioned earlier—identity creation, development of a horticultural mindset, women's roll in creating descent groups, and gifting—were part of the plant expansion process and can be applied to the early plant data from Peru

CREATING IDENTITY

Social identity is associated with foods and food preferences. As people and families begin to identify with a place, they begin to link the food they eat there with the place (Appadurai 1981; Douglas 1973, 1984). Certain food combinations have signified meanings, and these meanings are associated with events, people, and places where they are consumed (Falk 1991:773). Foods, along with family histories and cultural knowledge, are social markers of group affiliation, signifying culture even when other material signs are not visible, as in the archaeological record. Foods are used in the creation of political difference as well (Dietler 1996; Hastorf and Johannessen 1993; Wiessner and Schiefenhövel 1996). In addition, specific dishes or foodstuffs are often mentioned when signifying differences between one group and another through feasts, fasts, taboos, dishes, agricultural styles, and taste preferences (Douglas 1973; MacBeth 1997).

I have proposed elsewhere how plants that accompanied people into new places and new communities in the past were meaningfully part of the group's cosmology and self-identity (Hastorf 1998). Laura Rival presents this same position for the contemporary Ecuadorian Amazon with the nurturing of the peach palm by the Huaorani (1998a:240; Rival and Whitehead 2001). She outlines how groves of this long-lived palm are nurtured not only for the edible fruit and the useful boles, but also for a sense of kin continuity, a sense of security and socialization. These palms are gifts from the deceased ancestors, linking the past to the future while also providing a meaningful symbolic link between people and plant. During fruiting time, scattered groups of Huaorani come together for months among the group's groves to harvest and consume the fruits while nurturing the plants, as if the plants were the ancestors themselves and the Huaorani were eating the ancestor's fruit. Such nurturing in the past might have led to transitional domestication, as described about monkeys in Cormier's example (chapter 11, this volume). Amazonian crops such as peach palms do not prosper without such attendance (Clement, chapter 6, this volume).

Descendant plants acquire value because they embody knowledge and memory about the past (Connerton 1989). As a mnemonic, such plants, associated with the ancestors, can symbolize personal and family experience. Remembering genealogical histories of specific plants make them meaningful to the group and thus part of it (Weiner 1992:38). Therefore, when transported and propagated, these plants can evoke links to the past plant and therefore to the past human community. The living plant is imbued with history, making food from the propagated plants the embodiment of the past in the present.

Thus, descendant plants help maintain community over generations through the activities associated with them and the meanings that these actions generate, such as the peach palm for the Huaorani or the chili pepper for the Barasana (S. Hugh-Jones 2001). Farther north in the Colombian tropics, the Barasana foragers have adopted the chili pepper as this descendant plant, where family histories are associated with every variety (Hastorf 2003). Such plants become part of a group's cosmology and self-identity, so that group names are sometimes derived from their signifying cuisine (Descola 1992).

Creating identity with plants and especially with trees simultaneously creates territoriality through their "rootedness" and immobility in space while they are alive. During plant domestication, movement, and adoption, people began to focus on selected plants and animals. Foragers began repeatedly to associate themselves with certain plants. Groups differentiated themselves by their perceptions of their local environments and what they used in them. Boundaries were created based on food use, cultural knowledge, and plant distribution across the landscape. In the case of the Huaorani, people began to live seasonally among the peach palms, bringing together smaller kin groups into a larger self-identifying entity during the time of the peach palm harvest. Increased horticultural activity, as seen with the Huarorani's manioc parties, parallels the nurturing of more intensive social relationships. Based on this perspective, we might propose that the nonlocal plants nurtured by groups early on along the arid western coast had developed some special meanings and identities because of their previous associations with places, events, and descent groups.

DEVELOPING A HORTICULTURAL MINDSET

Both Balée (1989) and Ingold (1993) have pointed out that people have always "played their part" in the ongoing transformation of the natural world. People have lived *with* their fellow plants and animals, not just among them. The world has always been a garden in the sense that humans (and other animals) have interacted with all parts of the ecosystem, albeit at different interactive intensities (as with the Kumeyaays' and their neighbors' relationships with plants). Being omnivores with a strong sense of curiosity, people have shown a propensity to consume new things (Milton 1987). Yet, as Rozin (1976) surmises, this exploration causes anxiety due to a fear of ingesting new or foreign foods, not knowing if the item to be tested is poisonous or even tasty. Based on this dissonance, Rozin (1982) proposes the *omnivore's paradox* to account for our unique food habits. Through experimentation, humans have learned what is edible and what is not. This exploration and improvisation has developed the construction of nature both physically and mentally (Latour 1999; LeFebvre 1991; Sperber 1980). Wagner suggests that social groups "create their universe ... by constantly trying to change, readjust, and impinge upon it, in an effort to knock the conventional off-balance, and so make themselves powerful and unique in relation to it" (1975:88–89). Thus in some ways, our evolution and our curiosity have opened up a series of adaptive paths that include anthropomorphizing the world around us.

Balée (1994) and colleagues have recorded this systematic tinkering in the Amazon by farmers and foragers. Economic species distributions in forests today are largely due to human intentionality and manipulation over millennia. Plant distributions are the result of many years of people carrying clippings, root stocks, and seeds as they hunt, visit relatives, go on trading or war trips, and journey to new territories, or simply placing plants where they are more useful when people move to a new place (Posey 2002; Posey and Balée 1989; Rival 1998a). Recorded on the landscape are many examples of foragers nurturing and actively manipulating encountered ecological zones to aid access to plants and other resources, creating different levels of intensity and types of plant interaction (Balée 1994). Some interactions led to the creation of new microenvironments and new combinations of taxa; others resulted in shifts in taxa frequencies. These different consequences of interaction reflect the type and approach people have had in these environments. Historical ecology expressly studies, records, and interprets such histories.

Florence Shipek (1989) has noted the Kumeyaay's impact on the diverse plant communities throughout their territory in southern California. Although the Kumeyaay are classified as foragers, we have learned from archaeology that they were farmers in their recent past, and this history is reflected in how they currently interact with their local biota (Steven Shackley, personal communication, 2000). In their classificatory system and plant use, we can see a horticultural approach in lieu of a foraging one (Brown 1985). As discussed in the earlier section on human-plant relationships, foragers manipulate the plants quite actively throughout their territory, increasing biodiversity and transplanting taxa to desired locations, as recorded for the Amazonian foraging groups (Balée 1994; Rival 1998b). These different ways of living and viewing the world have associated physical impacts on the ecology. It is probable that a horticultural mindset, with its more overt interaction with plants must develop before a group will take on farming in a sustained manner. Such a mindset and its accompanying behaviors will be reflected in a group's plant terminology as well as in its interactions with plants. We should be able to identify some reflection of this mindset in the archaeological record with systematic data and careful scrutiny. I am proposing that this horticultural mindset must develop before plants are actively transplanted and maintained.

WOMEN'S ROLE IN CREATING DESCENT GROUPS

Ingold defines the essence of domestication as the constant human involvement with fast-growing plants (1996:21). In the same way, it is the constant involvement with children that brings people into the cultural world of kin relations and into larger societal interaction. This interactive and recurring social dynamic, as much as the need for food, encourages women and men to monitor, pick, weed, water, watch, trim, harvest, tend, thin, and propagate plants. These actions, performed by people on plants, are culturally constituted as well as culturally valued. As Ingold says, "There is far more to gardening than the mere production of food.... [F]or practitioners, growing crops and raising animals are not just ways of producing food; they are forms of social life" (1996:24).

In most times and places in the world, it is almost certainly women who, if not gathering food ingredients, process and organize food consumption and distribution to their families. Women produce food and children and through these actions create families. In turn, they nurture their families as well as their social world with every meal they prepare. They eventually bring the plants that are especially savored and special into their families and communities to be cared for like their children. Women have several sets of offspring then: the plants and animals in their gardens and the children in their shelters. In Papua New Guinea, for example, women nurse the pigs, raising them like their own children until they are taken away for alliance-building feasts and gifts (Rappaport 1976).

It is not surprising that in many horticultural societies in the Amazon most crops and kitchen gardens are women's domain; women are the "carers" (Descola 1996; C. Hugh-Jones 1979; Kahn 1986). Today in many cultures, as in the past, decisions about whom, what, and when to nurture are at the center of a woman's daily life. In the past, nurturing was ongoing, and that eventually led to planting and tending in certain environmental and social settings. In this way, women probably selected plants to be propagated in new places, especially those brought back into the house gardens to be watched over and grown. Thus, they were intimately and continuously involved in the activities that led to plant cultivation (Watson and Kennedy 1991).[2]

Women are the primary experimenters and collectors of plants for the Barasana of northwest Amazonia in Colombia, a horticultural group (S. Hugh-Jones 2001). Stephen Hugh-Jones notes that women are constantly bringing back cuttings during foraging trips as well as exchanging plants with friends and kin, thus adding to the local kitchen gardens that surround their houses. Their keen and sensitive observation informs them where plant taxa are growing as well as these plants' life cycles and their subtle differences through the seasons. Many plants with which they interact are not domesticated, but they tend them for food, medicine, contraceptives, drugs, and containers (gourds), and out of sheer interest. Women also interact with their neighbors and kin by giving plants as gifts. A woman's garden therefore evolves through time, adding genera and varieties from family and friends, from near and far. Each new plant comes with a personal story and a social relationship.

Within their gardens, Barasana women tend specific varieties that they inherit and pass down along family lines. These plants embody the family over generations. Gourds, with their long vines of fruit, represent a lineage with its daughters and sons. The Barasana are patrilineal, meaning that women must move house when they marry. Hugh-Jones notes, however, that chili pepper varieties in particular are inherited along female lines and cuttings are taken when women move to their marriage home. Today, when women cook and feed dishes with chili peppers to their families, every consumer is brought corporeally into association with his or her family. Chili peppers therefore represent individual female lineages and are a symbol of this generational relationship. Neighbors recognize each variety as a specific family's plant, with its own history and kin links. To walk through a Barasana woman's garden is to view her daily life, her ancestral kin group, and a history of her social relations. Peppers are also the binding agent of male (fish) and female (manioc) Barasana meals, forming the glue that completes the meal (C. Hugh-Jones 1995:231). With sexual connotations in their shape, piquant taste, and binding qualities, they have multiple levels of meaning and significance for the group. Is it surprising, then, that one of the first long-distance cultivated plants in the sites along the Peruvian coast was the chili pepper?

In small-scale, exogamous societies of the past, like today, it is likely that specific plants accompanied women as they changed communities or kin groups with marriage. If certain plants reminded them of their own family, kin group, childhood ritual events, and previous residences, women may have wanted to have these plants with them to help remember, symbolically as well as physically, their family ties in their days, their meals, and their bodies. When self-definition often comes from one's relationship with one's family, reassurance is gained through eating food derived from that family. This form of plant movement across the local landscape by women provides another dimension to how and why the early domesticates might have arrived on the coast of Peru.

THE SOCIAL GIFT

Although humans carry plants across the landscape for many reasons, one main reason is to share the plant with others. Intrafamily and interfamily gifts create and maintain society (Mauss 1990). Although it is impossible to substantiate this claim, food probably has been the most common gift item throughout human (pre)history. Because food is biologically necessary, these exchanges have been essential for family maintenance as well as for social bonding. I assume, therefore, that food exchange was at the core of social life in the past. The bond gained from eating together is a building block in all social life, often an emblem of acceptance and trust. As Mary Douglas ([1972] 1997) notes, there are increasing levels of social intimacy in different forms of food sharing. For example, in Western society, eating dinner together is more intimate than having drinks, which in turn is more intimate than drinking coffee together in a café. We know that a range of food sharing also exits across South America (DeBoer 2001).

Two types of gift giving are linked to plant foodstuffs. The first type is the giving of the plants themselves—seeds, flowers, fruit, or cuttings. If the plant survives, it is an offering that keeps on giving. This form of organic gift is usually at the individual level. This live-gift giving will result in the movement of plants across the landscape. The second type of gift giving, food exchange, is less long lived. Food sharing can occur at many scales, from individual meals to large group feasts. Feasting rarely results in plant movements.

Archaeologists have focused most often on the presence of nonlocal goods as indicators of exchange and alliance building because such goods are visible in the archaeological record (c.f. Lathrap 1973). It is reasonable to suggest that in the past most gifts were of things that people had fairly easy access to and with which they were familiar. Such items are often the most invisible in the archaeological record, such as stories, dances, myths, songs, or things made of animals, plants, and especially plants and food (Hastorf 2001). Balée notes that the Ka'apor request from visitors viable seeds of plants that they like (1994:153). I myself have been known to do this. Although these types of exchanges are usually invisible archaeologically, the movement and adoption of plants tells us that these events occurred. The plants on the western coast confirm that the people of the Amazon basin have a long history of interaction with their western neighbors.

The second form of gifting is well illustrated in Rival's example when she notes that "foraging" Amazonian people, the Huaorani of Ecuador, cultivate plants such as manioc to feed their visitors and their enemies at feasts, although they themselves do not consume it as part of their regular diet (Rival 1998a:243; Rival and Whitehead 2001). Cultivation for the Huaorani is only to provision special meals at social events. Like Braidwood and colleagues (1953), Bender (1978), and Hayden (1990) in their domestication models, Rival also proposes that as groups maintain wider social networks, they become more dependent on horticulture and trade of foodstuffs. These foods are needed to provision feasts. Feasting has been proposed as the ultimate social glue in maintaining neighborliness as well as a key element in the increase of social power (Dietler and Hayden 2001). Starchy staple production increases with larger permanent settlements, but also with increased political positioning (Young 1971). Michael Young (1971) cogently demonstrates how staple produce on the Goodenough

Islands was used not only to position male household heads within the group, based on the size of their rotting yams after harvest, but also to escalate food presentation to visiting neighbors at feasts. Such political food giving is an important dimension of plant movement that could have occurred regularly in the past and that probably participated in the political development across South America. If such gastropolitical trajectories existed in the past, then increased staple farming might suggest more social interaction, more intra- and intergroup politics, as well as more mouths to feed.

These two types of food gifts, that of plant exchanges and that of food, are registered in the archaeological record in the plants. The first type is probably ubiquitous—exchanges between family and friends, gifts as offerings in hospitality and recognition of a relationship. Food giving can reflect more escalatory levels of feasting that holds an element of political tension in the food performance (Appadurai 1981).

DOMESTICATED PLANT EVIDENCE ON THE PERUVIAN COAST

If the earliest plants are not exotic, morphologically altered, or densely deposited in discrete contexts, archaeologists have difficulty identifying early agricultural practices directly in the archaeological record. Places such as the arid Peruvian coast, where virtually all cultivated plants are nonlocal, make visible this often ephemeral phenomenon. What was it about these plants and their associated meanings that prompted the coastal inhabitants to begin cultivation, given that they had rich marine resources and cloud-forest vegetation on the coast? These were not hungry people. The El Niño oscillation also cannot provide the explanation because farming would have been disrupted by this shifting weather pattern. I argue here that the exotic plants were brought to and nurtured on the coast due to other causes and circumstances. The four cultural aspects, outlined in the previous section, would have participated in the movement to and adoption of nonlocal crops on the coast.

The earliest coastal cultivated plants were not carbohydrate-rich staple crops, but rather industrial and flavorful plants—beans, gourds, squash, and guava fruit trees eaten as greens, seeds, and fruit. Starchy plants regularly entered the middens several millennia later. Full-scale agriculture on the coast with a regular array of 15 to 20 cultivated plants occurred only by the end of the Initial period, after 1500 BC. By that time, irrigation systems were associated with monumental civic architecture (Moseley 1999; Shady 1997; Shady, Haas, and Creamer 2001). All of the crop plants that eventually entered the region can be grown and cultivated throughout the coastal valleys, supporting the thesis that this entry was more cultural and political than ecologically driven.

Given that the river floodplains, the springs, and the seasonal cloud forests are the only places for terrestrial plant and animal growth on the arid western coast, people clearly began manipulating these locations early on to develop their carrying capacities. What was involved in this process of crafting the coastal landscape with these foreign plants? For millennia, small groups moved up and down the valleys from the highlands to the coast hunting, fishing, and foraging, following the seasonal availability of resources and water. Settlements occurred in prime locations along the seashore and by the cloud forests. The early inhabitants exploited the marine, cloud forests, and riparian zones (Benfer 1982; Quilter 1989). The archaeology of such early sites as La Paloma around 4000 BC tells us that people were residing along the coast and slightly inland for millennia with few imported plants. With approximately 30 sites that have some botanical remains on or near the coast and that span these 5,000 years, the data are in no way complete or systematic. Over the years, many different collection strategies have been implemented at the various excavations, from judgmental, grab sampling to screening of some contexts and systematic sampling (Popper and Hastorf 1988), making it impossible to compare the plant evidence quantitatively. Therefore, the plant taxa presences are noted but not quantified. The archaeological taxa patterns, however, provide a general history of plant entrance into the region.

To view these cultural developments of early plant entry along the coast of Peru, I plotted twenty-eight crop taxa by site and phase. I placed them in five temporal phases spanning from the earliest crop evidence to large ceremonial centers with their semiurban ways of life, between 8000 and 1400 BC (table 3.1). The archaeological plant data are synthesized from the published literature, primarily Pearsall 1992 and 2003, Piperno and Pearsall 1998, and Towle 1961. I have dated the plants based on the phase in which the site's occupation predominantly spanned. Table 3.2 lists the taxa names, plant types, and zones where the plants have been encountered. These plants include some local plants, *Begonia* on the coast and potato (*Solanum* spp. L.), oca (*Oxalis* sp.), Ulluco (*Ullucus* sp.), and quinoa (*Chenopodium* spp.) from the highlands. The remaining plants were introduced from the moist tropics to the east and north. These plants include herbaceous, perennial, and tree species, important characteristics when tracking the people-plant relationship over time and space.

The plant forms I focused on include *root crops* (achira [*Canna edulis* Kerr.], manioc [or yuca, *Manihot esculenta* Crantz], and sweet potato [*Ipomoea batatas* Lam.]); *food annuals* (chili peppers [*Capsicum frutescens* L., *baccatum* L., *pubscens* Ruiz and Pavon], squash [*Cucurbita, ficifolia* Bouché, *maxima* Duch., *moschata* Poir.], lima bean [*Phaseolus lunatus* L.], common bean [*Phaseolus vulgaris* L.], maize [*Zea mays* L.]); *craft annuals* (bottle gourd [*Lageneria siceraria* (Molina) Standl.], cotton [*Gossypium barbadense* L.]); and *perennial fruit trees*

TABLE 3.1 Temporal Phases

Archaic-Early Preceramic (Phase III)	(8,000–6,000 BC)
Preceramic (Phase IV)	(6,000–4,200 BC)
Middle Preceramic (Phase V)	(4,200–2,500 BC)
Late Preceramic (Phase VI)	(2,5009–2,100 BC)
Initial Period	(2,100–1,400 BC) (Cotton Preceramic)

Source: These phases are updated but draw upon the Lanning (1967:25) and Rowe and Menzel (1967:ii) phasing.

(guava [*Psidium guajava* L.], avocado [*Persea americana* L.], pacae [*Inga feuillei*], *Bunchosia armeniaca,* with *Erythroxylum coca* also included).

Piperno and Pearsall offer a detailed synthesis of the domestication locations of these species (1998:164). I also draw from other works on the plant origins and distributions (Bergh 1976; Clement 1999a, 1999b; Pearsall 1992; Pickersgill 1969; Pickersgill and Heiser 1977). These domesticates came from at least four areas: western Ecuador, lowland northern South America, the southeastern Bolivia–Peru–Gran Chaco region, and the southern Amazon (see Clements 1999a, 1999b for maps). Four plants were domesticated in the moist tropical lowlands of western Ecuador: the jack bean (*Canavalia*), lima bean, (*P. lunatus*), crockneck squash (*C. moschata*), and cotton. One group comes from the moist lowlands of northern South America, including achira, *Bunchosia,* and guava. A larger number of food plants moved out of the southeastern Bolivia–Peru–Gran Chaco region, including the common bean (*P. vulgaris*), squash (*C. maxima*), chili pepper (*Capsicum pubscens*), peanut (*Arachis hypogaea*), and lucuma (*Pouteria lucuma*). Manioc is the only plant from the southern Amazon basin in this early plant group, with *Erythroxylum coca* from midelevation southeastern Colombia. The homeland for *Zea mays* is western Mexico. All of these locations are much wetter than the arid western coast of Peru. Their journeys to the coast are not along one path. These plants could have followed communication routes from the north, the south, and the east.

8000–6000 BC

The first evidence for nonlocal plants in the western region of South America is in the Early Archaic–Preceramic III Phase (8000–6000 BC), which is the earliest post-Pleistocene period (figure 3.1). Most of the coastal sites pertaining to this phase are now under water due to rising sea level around 3000 BC. Only one western coast valley site (La Paloma in the Chilca Valley), along with two highland sites (Tres Ventanas and Guitarrero Cave), have botanical remains that date to this phase. Lima bean, potato, manioc, sweet potato, and wild *Pachyrrhizus* are present. The lima bean comes from the northern Andes, but the other plants are from the east,

TABLE 3.2 Early Andean Plants in the Western Region

SCIENTIFIC NAME	COMMON NAME	PLANT TYPE	EARLY-USE ZONE
Annona cherimolia Mill.	cherimoya	fruit tree	coast
Arachis hypogaea L.	peanut	pulse	coast
Begonia geraniifolia L.	begonia	root	coast
Bunchosia armeniaca, spp.	ciruelo del fraile	fruit tree	coast
Canavalia plagiosperma Piper	jack bean	pulse	coast, low mountains
Canna edulis Kerr.	achira	root	coast
Capsicum chinense Jacq., *frutescens* L., *baccatum* L., *pubscens* Ruiz & Pavon	chili pepper, aji	spice	coast, low mountains
Chenopodium quinoa Willd.spp.	quinoa	pseudocereal	mountains
Cucúrbita ficifolia Bouché, *maxima* Duch., *moschata* Poir.	squash	vegetable	coast, low mountains
Cyclanthera cf. *pedata*	caigua	vegetable	coast
Erythroxylum spp. Lam., Plowman, Morris.	coca	drug	coast, east
Gossypium barbadense L.	cotton	seed, fiber	coast
Inga feuillei Willd.	pacae	fruit tree	coast
Ipomoea batatas (L.) Lam.	sweet potato	root	coast
Lagenaria siceraria (Molina) Standl. gourd	container		coast
Lepidium spp (cf *meyenii*) Walp.	maca	root	mountains
Pouteria lucuma	lucuma	fruit tree	coast
Manihot esculenta Crantz.	manioc	root	coast
Oxalis sp. (cf. *tuberosa*) Mol.	oca	root, veg.	mountains
Pachyrrhizus sp. (cf. Erosus [L.])	yam bean	root	coast
Passiflora spp.	tumbo	fruit vine	coast
Persea americana L.	avocado	fruit tree	coast
Phaseolus lunatus L.	lima bean	pulse	coast, low mountains
Phaseolus vulgaris L.	common bean	pulse	coast, mountains
Psidium guajava L.	guava	fruit tree	coast
Sapindus spp.	soap		coast, mountains
Solanum tuberosum L.	potato	tuber	mountains
Zea mays L.	maize	grain	coast, low mountains

Source: From Clement 1999a, 1999b; Lentz 2000; Pearsall 1992; Piperno and Pearsall 1998; Towle 1961.

from the Amazon basin. Tubers are dominant in these western valley site, which is intriguingly similar to the Tres Ventanas site located on the slopes in the upper Chilca River valley. Pearsall has studied the tubers from these excavations and believes that the Tres Ventanas tubers are wild (personal communication, 2000). Highland sites have always emphasized tuber crops. I have included the plants (early chili peppers and fruit accompanying the highland tubers) recovered from the Guitarrero Cave site in the Callejon de Huaylas intermontane valley, although the stratigraphy is mixed, making the dates questionable. These taxa suggest contacts to the east for the peppers and fruit. The plants from these three sites inform us that people here in the past were moving, planting, and consuming tubers, tasty peppers, and beans. More absolute dating must be done on these important plants. Regular movement back and forth is likely still across the mountains between the coast and the eastern slopes.

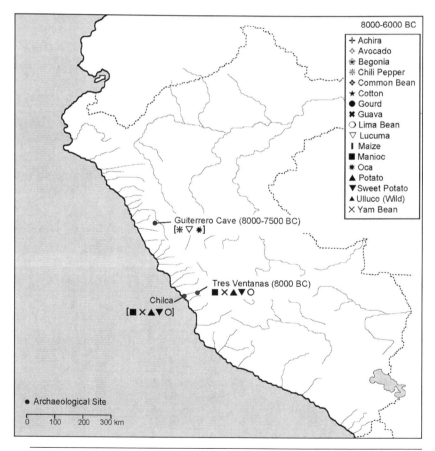

FIGURE 3.1 Twenty-seven domestic plant taxa distributions of the Preceramic III (8000–6000 BC) from the central Andean region, after Lanning 1967; Pearsall 1978, 1992; Piperno and Pearsall 1998; and Towle 1961.

6000–4200 BC

Between roughly 6000 and 4200 BC, the plant evidence demonstrates that lima bean became the first common crop grown along the coast (figure 3.2). The seacoast was stable, but the El Niño climatic swings began. To the north, western Ecuador claims the first domestic crop complex of the region—with maize, squash, and gourd. Although squash and gourd could have been adopted as containers or for other industrial uses, the maize was an import from Mesoamerica that would have provided a sweet and flavorful addition to the local inhabitants' diet (Iltis 2000; Pearsall and Piperno 1990).

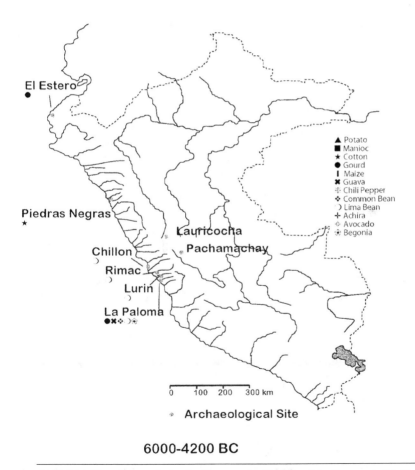

FIGURE 3.2 Twenty-seven domestic plant taxa distributions of the Preceramic IV (6000–4200 BC) from the central Andean region, after Lanning 1967; Pearsall 1978, 1992; Piperno and Pearsall 1998; and Towle 1961.

On the coast, people were eating a combination of indigenous riparian vegetation (including *Prosopis* pods), seasonal cloud-forest plants and animals, and marine animals and plants (Quilter 1989). All coastal sites have lima bean. The most detailed investigation of this time period has been at La Paloma, a site located slightly inland in the Chilca Valley, maximizing several resource zones (Benfer 1982; Quilter 1989). During this phase, the plants gourd, lima bean, and squash, all of northern origin, were adopted by the inhabitants at the La Paloma site. The guava tree also turned up, originally coming from the northern Amazon area. Given that it is extremely unlikely that the guava plant remains found at the site were from a gift exchange, their presence supports the notion that guava trees were planted locally. Guava is an aggressive plant that likes disturbed habitats, suggesting that it would have been the most likely tree to survive the early horticultural conditions on the coast with perhaps little attention (Charles Clement, personal communication, 2003). The fruit with seeds can be preserved as a paste and easily transported. The plant does imply permanence on the landscape, however, for it takes some years for fruit to mature. Although the begonia did not become a common food plant outside of the Chilca Valley, its dense presence at the La Paloma site prompted the investigators to suggest that residents cultivated it.

The few plants identified in coastal sites at this time suggest a modest interest in horticulture, but occurring only with hardy plants. We see local plants becoming common, such as the begonia, but also some plants that were traded in and accepted from the northern Amazon along possible exchange routes through western Ecuador. The residents who were experimenting with keeping plants near their settlements were not living on the coast, but slightly inland. In general, during this time there was little interest in nonlocal crops. It is possible that the coastal dwellers were responsible for the movement of the nonlocal plants through exchange by sea along the coast or perhaps over the mountains from the north.

Of the edible, exotic crops, squash, beans, and guava are savory—two of which are hardy, self-propagating annuals and one is a tree. These three plants are likely candidates for descendent plants. When little else was grown, beans were cultivated. Later on the coast, beans were highly symbolic for the Moche residents on the north coast (Stephen Bourget, personal communication, 2002; Donnan 1976). Beans are iconographically used on the Moche's elaborate pottery, where they are linked to the powers of the moon, rainfall, and fertility. As climbing plants, beans are descendant (kin) plants for various groups in the Amazon today. These plants are hardy and could be left to grow on their own along riverbanks or near other water sources in the same areas in which gourds grow, suggesting that coastal people were still not farmers at 4000 BC.

To muster all potential archaeobotanical data to track the potential forms of early plant movements to the coast, I included two interior valleys with botanical material in this time period—the Zaña Valley and the Ayacucho Valley sites.

Unfortunately, these sites have dating and stratigraphy problems. If the early dating and contexts are correct, however, the northern Zaña Valley has manioc, cotton, squash, quinoa, peanut, and coca (Rossen, Dillehay, and Ugent 1996). Quinoa is from the highlands or could even have been a local, wild variety. The other plants come from three zones, from the north, the northeastern lowlands, and the southeastern eastern slope. These plants either were introduced from nearby Ecuador or over the low pass connecting the eastern moist tropical forests and the western coast. While the dried coca leaves could have been traded from the eastern slopes of the moist tropics, the other plants would have been grown in the Zaña Valley. These data, however, are speculative and are mentioned here only as a possibility, awaiting further dating. The inhabitants of the Ayacucho caves had quinoa, squash, and gourd, if the stratigraphies and dates for these sites are correct. These three plants are hardy, and the squash and gourd could have been brought in as containers. These two highland sites suggest that quinoa was of interest to people at an early time.

4200–2500 BC

In the Middle Preceramic Phase (4200–2500 BC) (figure 3.3), domestic plants were present but rare along the coast, with no more than four taxa found at any site. Only the Chilca Valley sites contained a wider range of taxa. This difference may be due to archaeological collection strategy bias on the central coast or may represent more long-term interest in domesticated plants by the inhabitants of the Chilca Valley. Bean and tree species were introduced from along the coast or over the mountains from the east.

The lima bean still dominates the coastal evidence, indicating that it was truly the first widespread domesticate, rightly receiving its name in English, *lima*, after the central coastal Rimac River valley, where the Spanish capital Lima was placed, from which the bean was exported to the world. Gourd and squash were irregularly present on the central coast. The Yacht Club site on the Ancon Bay near the Chillon Valley yielded guava, chili pepper, and cotton, along with gourd. Begonia and guava were still tended in the Chilca Valley sites, along with new food plants including jack bean, achira, *Pachyrrhizus,* caigua, and the *Inga* tree. This expanded plant list supports the notion that the residents had an increasing interest in horticulture by this time, including cultivation of vegetables, trees, and root crops. Over time, the starchy achira and *Pachyrrhizus* eventually usurped the place of the modest local begonia as the main food crop in the south.

The distribution is sill irregular and patchy along the coast, displaying a pattern of early regional plant-use history. Beans were found at half of the sampled coastal sites up and down the coast; maize is present only in the north; and the nonlocal root crops are most common at the interior sites. The lima bean and

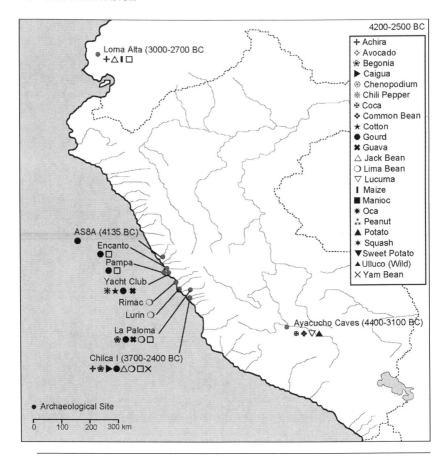

FIGURE 3.3 Twenty-seven domestic plant taxa distributions of the Preceramic V (4200–2500 BC) from the central Andean region, after Lanning 1967; Pearsall 1978, 1992; Piperno and Pearsall 1998; and Towle 1961.

gourd, with their tendrils, dominate the coastal plant list. This suggests that the most common plant cultivation at coastal sites was small-scale horticulture of a plant complex that resided in house gardens. As the people settled on the landscape, slowly adopting gardening to nurture both local and nonlocal plants, they received species from their relatives or as gifts in long-distance trade, or brought them from their previous homes in marriage exchanges. But they did this selectively and did not cultivate all plants that were grown on the coast. We can see the spatial patterning of entry best in this phase, with highland folk most engaged with tuberous production. Did the coastal residents have a horticultural mindset in this phase? Probably not. Only the greater number of taxa at the La Paloma and Chilca sites, which had domesticates for some time, suggest a hint at such a mental shift after centuries of horticulture.

The settlements away from the coast also clarify our picture of early plant movements and adoption. The sites from which we have data show that inhabitants cultivated plants from their part of the continent first. In Ecuador, inhabitants of the inland Loma Alta site cultivated achira, a starchy root food originally from the moist northern Andes. That site also yielded maize, squash, and jack bean, all entering from the north, suggesting a western–eastern interaction. Farther south, the Ayacucho cave sites have interesting plants, most coming from the closest domestication zone, the eastern slopes of Bolivia-Peru. The caves contain the common bean, the local potato, and two trees—lucuma and coca. Lucuma and the common bean originate from the eastern Gran Chaco region far to the southeast. The first evidence of the common bean in the Andean region is found in the Ayacucho caves, suggesting a southeast–northwest exchange path for that important plant. Coca is the odd one out, coming from far to the north. Its history is surely complex, but the leaves could have been traded down the eastern slopes quite early, making it available also from the southeast to the highlands.

2500–2100 BC

It is in the next phase, the Late Preceramic VI (2500–2100 BC), that more nonlocal plants entered coastal foodways, suggesting a shift in daily practices as well as viewpoint about plant-human relations (figure 3.4). Pearsall has plotted the plant taxa richness, which shows an increase around 2500 BC, further supporting this proposition (2003:240). Along the coast, we have evidence of expanded plant adoption and an increasing interest in different horticultural practices required for cultivation of annuals, perennials, tubers, and trees. More nonlocal plants are present at the sampled sites. This is the time when the first civic architecture was being built along the coast, suggesting a greater interest in the social identity and group solidarity accompanying this interest in a different diet as well as in a different lifestyle. The increasing numbers of domestic plants could have been part of these activities, not only through the reformulation of the landscape with more and different gardens, more permanent settlements, and more territories, but also through the new foods and dishes to be consumed at the public ceremonies associated with the civic sites.

The similar distribution of achira and chili pepper at 9 of the 15 coastal sites suggests that they moved and were adopted together, perhaps combined as a filling yet pleasing meal, especially suitable for kin group (commensal) feasts (Dietler and Hayden 2001). The increased presence of tree taxa along the coast suggests larger kitchen gardens, if not orchards. Cotton and gourd, the two craft species in this group of cultivated plants, were now ubiquitous and are found at sites as frequently as food crops such as squash, guava tree, chili pepper, and achira. In addition to their use for bags, rope, and clothing, cotton and gourd could have participated in increased offshore fishing. This trend in increased plant presence

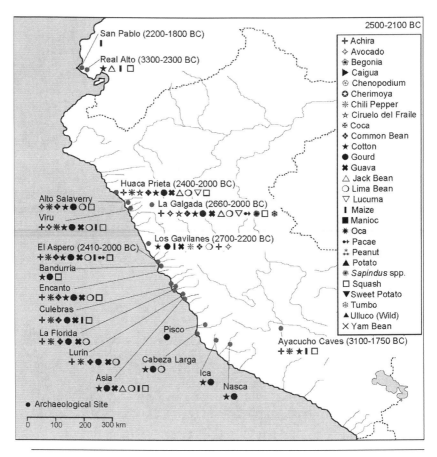

FIGURE 3.4 Twenty-seven domestic plant taxa distributions of the Preceramic VI (2500–2100 BC) from the central Andean region, after Lanning 1967; Pearsall 1978, 1992; Piperno and Pearsall 1998; and Towle 1961.

and indirectly in fishing through the industrial crops supports a major shift in worldview, of life in larger communities, with more focus on local subsistence activities.

Cultural food selection was still operating and visible, as maize entered the coast from western Ecuador in a selective manner, found only at three settlements. It had been cultivated in Ecuador for thousands of years without entering this more arid coast. Although the most northern arid coastal sites did not have maize, it was present in the Viru, El Aspero, and Culebras valleys, all separated by hundreds of kilometers (figure 3.4). It is not found in the neighboring valleys of the sites where it occurred. This distribution illustrates how some communities grew maize, but others did not, suggesting that early on maize was a peripheral condiment. In some ways, maize is the plant that best reflects the selective cultural choice in horticultural activity at this time. Northern coastal residents surely knew about the crop if they traveled to the valleys north or south of their own, yet they chose not to plant or prepare it as a food.

By this phase, the southern common bean and northern lima bean had spread throughout the coast. Guava fruit trees were also ubiquitous. Achira was now the dominant starchy tuber, as manioc (a bitter variety) and sweet potato were dropped from the coastal crop inventory. Cotton, jack bean, and squash would have entered from the north, as all have been found at the Real Alto site in Ecuador. Cotton occurred irregularly within the southern coastal sites, with gaps of three or four valley between each location.[3] After 3,000 years of bottle gourd being cultivated along the coast, the late introduction of cotton in the north does not support the idea that these two crops were part of the same craft strategy to improve fishing.

The plant evidence suggests that each settlement cultivated a unique combination of plant taxa. On the north coast, the Huaca Prieta site has a broad horticultural focus with tree crops, lucuma, guava, and *Bunchosia,* as well as edible annuals, three bean varieties, chili peppers, cotton, and achira (Bird, Hyslop, and Skinner 1985). South of that site, residents of the nearby Moche and Virú valleys shared a different and smaller plant complex, growing gourd, squash, bean, cotton, chili peppers, and avocados. Farther south the sites of Los Gavilanes and Aspero have about ten taxa each, sharing most but not all species, including lucuma, guava, avocado, beans, peppers, and achira. The inhabitants grew several tuberous crops, including sweet potato, manioc, and peanuts. Even farther south, the south-central coastal sites now contain the same plant taxa as the north. Figure 3.4 illustrates how the sites from Bandurria to Asia have a similar nonlocal plant-use pattern of beans, chili peppers, achira, squash, gourd, and cotton. The sites in the Pisco, Ica, and Nasca valleys to the south have only craft plants, a fact that probably reflects in part the biased collection and reporting strategies, but also might suggest their inhabitants' lack of engagement yet with farming.

This Late Preceramic Phase plant pattern illustrates the long-term selective nature of the coastal residents as they initiated plant agriculture. By this time, every excavated coastal settlement had some nonlocal domesticates. However, each community did not grow or trade the same domesticates, nor did any single settlement have all of the domestic plants that were being grown along the coast during this phase. Crop agriculture was still selective and on a small scale. Some plants did co-occur, suggesting some concerns for specific meals.

The first monumental civic centers were built along the coast during this phase, with mounds and plazas (Burger 1992; Shady, Haas, and Creamer 2001). Thus, larger group ceremonies and social cohesion activities could have been encouraged by hosting feasts (c.f. Dietler and Hayden 2001; Pauketat 2000).

Rituals and shared experiences such as feasts encourage community solidarity (Atkinson 1989). Meals prepared for a gathering would be special, exhibiting unique flavors and large quantities, especially of beer, which we know can be made of any starchy plant food, such as maize, quinoa, achira or manioc. The desire for unique meals perhaps encouraged families to grow extra food plants for feasts as well as for gifts. New edible plants entered through the coastal and inland trade networks. Engagement and curiosity about these nonlocal plants was more active at this time. The residents of individual valleys adopted different combinations of plants, however, maintaining distinct plant taxa and perhaps distinguishing their recipes from their neighbors' recipes. These social activities would have contributed to the formation of a horticultural mindset as they slowly became farmers.

The interior Ayacucho caves now have coastal plants with cotton, gourd, and maize, along with the long-cultivated potato from the highlands and common beans from the east, reflecting increased western trade between the coast and the highlands. The La Galgada site (2600–2000 BC), a long-lived residential and ceremonial site near the Santa River valley is on a direct path between the eastern slopes, highlands, and the coast. This warm valley site also has the greatest evidence for cultivation at this time. C. Earle Smith identified 15 domestic plant species (in Grieder et al. 1988:125–151). Orchards may have existed near the community because six tree taxa have been identified from the middens. Four pulses were grown, including the common bean, lima bean, jack bean, and the peanut, all originally from the Amazon basin. Although several new tree crops, along with achira, were found at the La Galgada site, maize is absent, suggesting that it was circulating mainly along the western coast at this time.

The taxa uncovered from this time period on the coast tell us that more crops introduced during this phase came from the east and southeast rather than primarily from the north as in earlier periods. The people of this period were apparently looking east for their social exchanges and identities, as Lathrap (1977) has suggested.

2100–1400 BC

Plant distributions during the Initial Period (2100–1400 BC) (figure 3.5) illustrate continuing selective crop adoption that we observed in the previous period. During this time, more civic monumental centers were constructed, including the sites of El Paraíso, Pampa de los Llamas Moxeke, Las Haldas, and Asia. Throughout the central coast, the most common plants continue to be beans, squash, gourds and cotton, achira, and chili peppers, along with several trees. The rate of nonlocal plant use increased in this period and remained stable over the next 2,000 years (Pearsall 2003:242). Some settlements continued to be conservative in their choice of plants they cultivated, as spatially different plant-use trends continue to be visible in the archaeological data. Avocado remains are found more often in the north-central coastal sites. The central coastal settlements are still not strongly agricultural, although there is evidence for root-based food at the site of Huaynuná, with both potato and sweet potato. Nearby, Pampa de los Llamas Moxeke, the largest site of this time period, has 13 domestic taxa, including three root tubers and four fruit trees. The 11 taxa— including four trees, beans, squash, gourd, cotton, and achira—found at the inland ceremonial site of El Paraiso along the Chillon River still does not evoke a strong emphasis on plant food production (Quilter et al. 1991).

Fruiting trees became important to more communities at this time. Fruit trees, however, did not produce a staple product able to feed the masses. The staple food evidence remained marine resources along much of the Peruvian

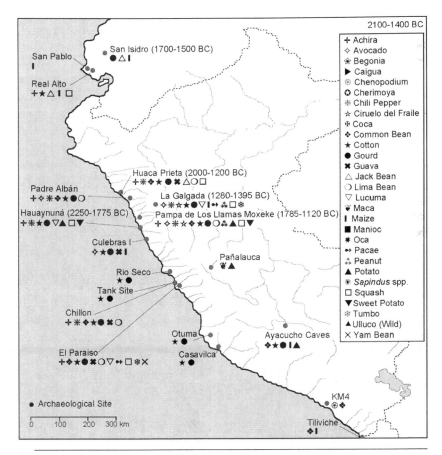

FIGURE 3.5 Twenty-seven domestic plant taxa distributions of the Initial period (2100–1400 BC) from the central Andean region, after Lanning 1967; Pearsall 1978, 1992; Piperno and Pearsall 1998; and Towle 1961.

coast. The new, inland site locations for their large ceremonial centers suggest a selective decision to reorient the worldview of the populace toward plant production along the river valleys. The new nonlocal plant food indicates that special foods were being prepared. These foods could have been produced by intensifying kitchen gardens to gain more food without having to relocate, especially if the required meals occurred only periodically. This new site orientation seems linked to riverine agriculture, suggesting that these ceremonial centers were intertwined with crop production, if only on a small scale initially. Many residents eventually did move inland and reoriented their focus to terrestrial resources, as is evident in the next period.

Figure 3.5 illustrates the plant combinations, perhaps as potential feast ingredients. Achira and chili peppers continue to co-occur throughout the coast. Peppers are known to have special significance throughout the Americas. At the La Galgada site, pepper seeds are found in ritual hearths by 1200 BC (Smith in Grieder et al 1988). In the Late Intermediate period, around AD 1000, the Chirabaya of northern Chile ate chili peppers in their final meals, as found in burial coprolites, whereas none were found in food midden coprolites (Karl Reinhard, personal communication, 2002). This pattern suggests chili pepper's use in ritual or at least as a special food for the ill. Chili peppers have continued to be an important spice, as seen in the Amazon basin example given earlier, and have been rapidly accepted around the globe. Could chilies have been adopted and cultivated by coastal residents as taste treats, spread by women's movements in marriage exchanges, initiating horticulture as they promoted their family foods? The data suggest this trend.

DISCUSSION

In a rich marine environment like the Peruvian coast, farming began not because people were hungry, but for cultural reasons. Tracing the plant evidence along the western coast of Peru through time and space while applying the cultural concepts I proposed in an earlier section allows us to gain a more active, agent-based view of the onset of agriculture outlining the long and intimate interactions humans have had with their landscape. Flavorful, annual beans were one of the first plants to be cultivated and are relatively common at the coastal sites. Beans grow quickly on a stalk and produce many offspring. As I argued earlier, they could easily have been a metaphor for a family history as they were taken to new places and propagated. The crops present before 4000 BC are not the staple foods of hungry people. Nor did crops move or become adopted as packages, except perhaps achira and chili pepper. Notable in these plant distributions is the lack of a single route of entry or origin. Plants entered the coast from the north, east, and southeast, where they were originally domesticated and

cultivated. The *Phaseolus* beans clearly demonstrate the complex routes of introduction: *lunatus* from the north and *vulgaris* from the southeast.

Cultural selection is evident as neighboring settlements adopted crops differentially. Some valleys' settlements have maize, whereas others have chili peppers,
differences that may reflect separate cultural identities, expressed in crop cultivation and consumption. Differences are also suggested by the varied forms of
cultivation, such as fields versus kitchen gardens, orchards versus annual crops.
Starchy plants were added late in the sequence but occur earlier at the larger
settlements. These distributions suggest a cultural rather than a geographical
pattern of plant adoption.

There is no evidence for a single ethnicity or polity in the region(s) during the time periods discussed here. Rather, the irregular distribution of plants
in time and space suggest that the cultural adoption of nonlocal domesticates
was variable. Variations in crop combination exist between neighboring valleys.
These patterns imply that communities or families made local decisions about
what they would plant and nurture. Although limited, the evidence supports
the proposed social model for group identification through specific plant adoption during these phases. These foragers took up cultivation and later became
farmers in a selective, meaningful, patterned, and intentional way.

CONCLUSIONS

The human-plant interactions changed between 8000 and 1000 BC along
the western coast of South America. The 6,000-year story begins with early
inhabitants foraging marine resources in and along the sea and ends with their
descendents becoming farmers who performed rituals at large, complex monu-
ments and used massive networks of irrigation canals in the river valleys. This
is a slow and nonuniform history of culturally engaged plant movement and
adoption and of new culturally constructed relationships between plants and
people. The early plant adoptions shed light on how groups interacted with
nature and each other, using the plants and meals made from them to form
different identities within the various coastal valleys, as nonlocal foods were
brought into their diet, landscape and lives.

This chapter suggests that increased plant engagement paralleled the social
relations among families. In later phases, political use of starchy plants and
associated feasting at larger settlements for aggrandizement and power are
more evident. Full agriculture and the generation of surplus food only appears
thousands of years after the arrival and adoption of cultivated plants. These
two agricultural stages—selective adoption of domestic plants and full-fledged
farming—are not formed by the same cultural processes. I argue that the
social meanings that formed the earlier activities were framed by a curiosity

and interest in remembering previous experiences and loved ones. The practical implications of such plant-human relations were redefined through the repeated acts of planting and tending, each plant carrying its own significance for the farmer, remembered through the work of each harvest and consumption of each meal.

I proposed four cultural concepts as influential in this long trajectory of plant tending. These concepts are derived from cross-cultural and specific ethnographic analogy about social life and plants and from basic historical ecology concepts about the historical and cultural factors that create landscapes. The first concept focused on creating identity through one's food procurement and foodways. The second notion was the importance of developing a horticultural mindset before farming could take hold and how that might have occurred along the dry coast. Farming and tending plants all day would come about with a shift in daily practice that accompanies a new worldview. The third concept was the importance of ancestral and family maintenance, created through women's use of plants that activate memories about family and ancestors. Such memorializing through inherited plants and recipes selects core plants in the family garden that are to be nurtured, collected, smelled, and eaten; at the same time, however, these plants provide physical and physiological links to other family members. The last participating concept in the development of agriculture was the importance of gift giving in society and how most gifts are plant derived. I have applied these concepts to a historical presentation of the coastal plant use. Each of these ideas has a role in this long, complex history of movement and adoption of nonlocal plants. The limited native plant communities and excellent preservation of the arid western coast allow us to track the cultural and historical dimensions of this plant movement and adoption that eventually resulted in farming and cities.

The specific plants found along the coast during this early history helped create the cultures that developed on the coast. This plant history is also a history of gift giving, of women nurturing their gardens and their families, of the plants domesticating the residents. By applying this timescale, we gain a sense of the shift in worldview about the coastal environment from a foraging mindset to a horticultural mindset as decisions to focus on domesticated plants gradually overruled the centrality of a marine-based subsistence.

Detailed archaeological study of plant distributions through time and space provide excellent case studies for testing assumptions about the practical and psychological relationships between humans and plants, human perceptions of their environment, as well as landscape manipulation and change. Whether farming or foraging, people are continually constituting their cultural beliefs in their world through the food they eat and through the constructed landscape within which they live. Such an emphasis on people's practical engagement with their environment is the essence of historical ecology and, it is to be hoped, has contributed to our understanding of long-term plant use in South America.

ACKNOWLEDGMENTS

This paper was presented at the 2002 Symposium on Neotropical Historical Ecology at Tulane University organized by William Balée. All of the presentations at the conference were inspirational. I especially want to thank Charles Clement for his feedback and conversations. Both Clark Erickson and William Balée have been keen and perceptive in their editing. Bill Whitehead created the figures.

NOTES

1. In contrast to my use in this chapter, the Spanish term *loma* (plural, *lomas*) in chapters 7 and 8 refers to a hill or hillock on a flat plain, specifically to an artificial earthen mound of the Bolivian Amazon.
2. Although men are also involved in farming and at times the only people allowed into some fields, as discussed by Heckenberger, chapter 10, in this volume, I am focusing here on house gardens and the early stages of farming.
3. Early cotton and gourd uncovered at the Real Alto site and at the contemporary highland site of La Galgada lessen support for the thesis that these two crops were initially farmed together because of their benefit to marine fishing (Moseley 1975).

REFERENCES

Allen, C. 1988. *The Hold Life Has.* Washington, D.C.: Smithsonian Institution Press.

Anderson, E. 1952. *Plants, Man, and Life.* Berkeley: University of California Press.

Appadurai, A. 1981. Gastro-politics in Hindu South Asia. *American Ethnologist* 8:494–511.

Atkinson, J. 1989. *The Art and Politics of Wana Shamanship.* Berkeley: University of California Press.

Balée, W. 1989. The culture of Amazonian forests. *Advances in Economic Botany* 7: 1–21.

———. 1994. *Footprints of the Forest: Ka'apor Ethnobotany—the Historical Ecology of Plant Utilization by an Amazonian People.* New York: Columbia University Press.

Bender, B. 1978. Gatherer-hunter to farmer: A social perspective. *World Archaeology* 10 (2): 204–222.

Benfer, R. 1982. Proyecto Paloma de la Universidad de Missouri y el Centro de Investigaciones de Zonas Aridas. *Zonas Aridas* 2:34–73.

Bergh, B. O. 1976. Avocado *Persea americana* (Lauraceae). In N. W. Simmonds, ed. *Evolution of Crop Plants,* 148–151. London: Longman.

Bird, J., J. Hyslop, and M. D. Skinner. 1985. *The Preceramic Excavations at Huaca Prieta, Chicama Valley, Peru.* Anthropological Papers no. 62, Part 1. New York: American Museum of Natural History.

Bloch, M. 1998. Why trees, too, are good to think with: Towards an anthropology of the meaning of life. In L. Rival, ed. *The Social Life of Trees,* 39–55. Oxford: Berg.

Braidwood, R. J., J. D. Sauer, H. Helbaek, P. C. Mangelsdorf, H. C. Cutler, C. S. Coon, R. Linton, J. Steward, and A. L. Oppenheim. 1953. Did man once live by beer alone? *American Anthropologist* 55:515–526.

Brown, C. 1985. Mode of subsistence and folk biological taxonomy. *Current Anthropology* 26 (1): 43–64.

Burger, R. L. 1992. *Chavin and the Origins of Andean Civilization.* London : Thames and Hudson.

Cauvin, J. 2000. *The Birth of the Gods and the Origins of Agriculture.* Cambridge, U.K.: Cambridge University Press.

Clement, C. 1999a. 1492 and the loss of Amazonian crop genetic resources I: Relations between domestication and human population decline. *Economic Botany* 53 (2): 188–202.

———. 1999b. 1492 and the loss of Amazonian crop genetic resources II: Crop biogeography at contact. *Economic Botany* 53 (2): 203–216.

Connerton, P. 1989. *How Societies Remember.* Cambridge, U.K.: Cambridge University Press.

Counihan, C., and P. Van Esterik, eds. 1997. *Food and Culture.* London: Routledge.

Crumley, C., ed. 1994. *Historical Ecology.* Santa Fe, N.Mex.: School of American Research.

DeBoer, W. 2001. The big drink: Feast and forum in the upper Amazon. In M. Dietler and B. Hayden, eds., *Feasts,* 215–239. Washington, D.C.: Smithsonian Institution Press.

Denevan, W. M. 2002. *Cultivated Landscapes of Native Amazonia and the Andes.* Oxford: Oxford University Press.

Descola, P. 1992. Societies of nature and the nature of society. In A. Kuper, ed., *Conceptualizing Society,* 107–126. London: Routledge.

———. 1996. *In the Society of Nature.* Cambridge, U.K.: Cambridge University Press.

Dietler, M. 1996. Feasts and commensal politics in the political economy: Food, power, and status in prehistoric Europe. In P. Weissner and W Schiefenhövel, eds., *Food and the Status Quest,* 87–125. Berghahn, Germany: Providence.

Dietler, M., and B. Hayden, eds. 2001. *Feasts.* Washington, D.C.: Smithsonian Institution Press.

Donnan, C. B. 1976. *Moche Art and Iconography.* Los Angeles: UCLA Latin American Center.

Douglas, M.1973. The Bog Irish. In M. Douglas, ed., *Natural Symbols,* 59–76. New York: Vintage.

———. 1984. *Food in the Social Order: Studies of Food and Festivities in Three American Communities.* New York: Russell Sage Foundation.

———. [1972] 1997. Deciphering a meal. In C. Counihan and P. Van Esterik, eds., *Food and Culture,* 36–54. New York: Routledge.

Enfield, D. B. 1992. Historical and prehistorical overview of El Niño/Southern Oscillation. In H. F. Diaz and V. Markgraf, eds., *El Niño: Historical and Paleoclimatic Aspect of the Southern Oscillations,* 95–117. Cambridge, U.K.: Cambridge University Press.

Engel, F. 1970. *Las Lomas de Iguanil y el Complejo de Aldas.* Lima: Universidad Nacional Agraria.

Erickson, C. 2000. An artificial landscape-scale fishery in the Bolivian Amazon. *Nature* 408:190–193.

———. 2001. The Lake Titicaca basin: A pre-Columbian built landscape. In D. Lentz, ed., *Imperfect Balance: Landscape Transformations in the Precolumbian Americas,* 311–356. New York: Columbia University Press.

Falk, Pasi. 1991. Homo culinaris: Towards an historical anthropology of taste. *Social Science Information* 30 (4): 757–790.

Farrington, I., and J. Urry. 1985. Food and the early history of cultivation. *Journal of Ethnobiology* 5:143–157.

Fischler, C. 1988. Food, self, and identity. *Social Science Information* 27 (2): 275–292.

Gepts, P. 1990. Biochemical evidence bearing on the domestication of *Phaseolus* (Fabaceae) beans. *Economic Botany* 44 (3): 28–38.

Grieder, T., A. Bueno, A. B. Mendoza, C. E. Smith, and R. M. Malina. 1988. *La Galgada*. Austin: University of Texas Press.

Harlan, J. 1975. *Crops and Man*. Madison, Wisc.: American Society of Agronomy.

Hastorf, C. A. 1993. *Agriculture and the Onset of Political Inequality before the Inka*. Cambridge, U.K.: Cambridge University Press.

——. 1998. The cultural life of early plant use. *Antiquity* 72:773–782.

——. 1999. Cultural implications of crop introductions in Andean prehistory. In J. Hather and C. Gosden, eds., *Change in Agrarian Systems,* 35–58. London: Routledge.

——. 2001. Making the invisible visible: The hidden jewels of archaeology, In P. B. Drooker, ed., *Fleeting Identities: Perishable Material Culture in Archaeological Research,* 27–42. Center for Archaeological Investigations Occasional Paper no. 28. Carbondale: Southern Illinois University Press.

——. 2003. Community with the ancestors: Ceremonies and social memory in the Middle Formative at Chiripa, Bolivia. *Journal of Anthropological Archaeology* 22:305–332.

Hastorf, C. A., and S. Johannessen. 1993. Pre-Hispanic political change and the role of maize in the central Andes of Peru. *American Anthropologist* 95 (1): 115–138.

Hayden, B. 1990. Nimrods, piscators, pluckers, and planters: The emergence of food production. *Journal of Anthropological Archaeology* 9 (1): 31–69.

Heiser, C. B. 1965. Cultivated plants and cultural diffusions in nuclear America. *American Anthropologist* 67 (4): 930–949.

Hodder, I. 1991. *The Domestication of Europe*. Oxford: Basil Blackwell.

Hugh-Jones, C. 1979. *From the Milk River: Spatial and Temporal Processes in Northwest Amazonia*. Cambridge: Cambridge University Press.

Hugh-Jones, S. 1995. Inside-out and back-to-front: The androgynous house in northwest Amazonia. In. J. Carsten and S. Hugh-Jones, eds., *About the House,* 226–252. Cambridge, U.K.: Cambridge University Press.

——. 1996. Bonnes raisons ou mauvaise conscience? De l'ambivalence de ertains Amazoniens envers la consommation de viande. *Terrain* 26: 123–148.

——. 2001. The gender of some Amazonian gifts: An experiment with an experiment. In T. A. Gregor and D. Tuzin, eds., *Gender in Amazonia and Melanesia: An Exploration of the Comparative Method,* 245–278. Berkeley: University of California Press.

Iltis, H. H. 2000. Homeotic sexual translocations and the origin of maize (Zea mays, Poaceae): A new look at an old problem. *Economic Botany* 54 (1): 7–42.

Ingold, T. 1993. The temporality of the landscape. *World Archaeology* 25:152–174.

——. 1996. Growing plants and raising animals: An anthropological perspective on domestication. In D. Harris, ed., *The Origins and Spread of Agriculture and Pastoralism in Eurasia,* 12–24. London: University of London Press.

Kahn, M. 1986. *Always Hungry, Never Greedy*. Cambridge, U.K.: Cambridge University Press.

Kojan, D. 2002. Cultural identity and historical narratives of the Bolivian eastern Andes: An archaeological study., Ph.D. diss., University of California, Berkeley.

Lanning, E. 1967. *Peru before the Incas*. Englewood Cliffs, N.J.: Prentice-Hall.

Lathrap, D. W. 1973. The antiquity and importance of long distance trade relationships in the moist tropics of pre-Colombian South America. *World Archaeology* 5:170–186.

Lathrap, D. W. 1977. Our father the cayman, our mother the gourd: Spinden revisited, or a unitary model for the emergence of agriculture in the New World. In C. A. Reed, ed., *Origins of Agriculture*, 713–751. The Hague: Mouton.

Latour, B. 1999. *Pandora's Hope*. Cambridge, Mass.: Harvard University Press.

LeFebvre, H. 1991. *The Production of Space*. Oxford: Basil Blackwell.

Lentz, D. L., ed. 2001. *Imperfect Balance: Landscape Transformations in the Precolumbian Americas*. New York: Columbia University Press.

Lévi-Strauss, C. 1969. *The Raw and the Cooked*. Translated by John Weightman and Doreen Weightman. New York: Harper and Row.

Macbeth, H., ed. 1997. *Food Preferences and Taste: Continuity and Change*. Providence, R.I.: Berghahn.

Martin, L. M., P. Fournier, A. Mourguiart, A. Sifeddine, and B. Turcq. 1993. Southern oscillation signal in South American paleoclimatic data of the last 7000 years. *Quaternary Research* 39:338–346.

Mauss, M. 1990. *The Gift*. London: Routledge.

Meggers, B. J. 1971, *Amazonia: Land and Culture in a Counterfeit Paradise*. Chicago: Aldine.

Messer, E. 1997. Three centuries of changing European tastes for the potato. In H. Macbeth, ed., *Food Preferences and Taste: Continuity and Change*, 101–113. Oxford: Berghahn.

Milton, K. 1987. Primate diets and gut morphology: Implications for hominid evolution. In M. Harris and E. B. Ross, eds., *Food and Evolution: Toward a Theory of Human Food Habits*, 93–115. Philadelphia: Temple University Press.

Moseley, M. 1975. *The Maritime Foundations of Andean Civilization*. Menlo Park, Calif.: Cummings Archaeology Series.

——. 1983. The good old days were better: Agrarian collapse and tectonics. *American Anthropologist* 85:773–799.

——. 1999. *The Incas and Their Ancestors*. London: Thames and Hudson.

Ortiz, A. 1994. Some cultural meanings of corn in aboriginal North America. In S. Johannessen and C. A. Hastorf, eds., *Corn and Culture in the Prehistoric New World*, 527–544. Boulder, Colo.: Westview Press.

Oyuela-Caycedo, A. 1996. The study of collector variability in the transition to sedentary food producers in northern Colombia. *Journal of World Prehistory* 10:49–93.

Pauketat, T. R. 2000. The tragedy of the commoners. In M. A. Dobres and J. Robb, eds., *Agency in Archaeology*, 113–129. London: Routledge.

Pearsall, D. 1978. Paleoethnobotany in western South America: Progress and problems. In R. I. Ford, M. F. Brown, M. Hodge, and W. L. Merrill, eds., *The Nature and Status of Ethnobotany*, 389–418. Anthropological Papers no. 67, Museum of Anthropology. Ann Arbor: University of Michigan.

——. 1992. The origins of plant cultivation in South America. In C. W. Cowan and P. J. Watson, eds., *Origins of Agriculture in the World: An International Perspective*, 173–205. Washington, D.C.: Smithsonian Institution Press.

——. 2003. Plant food resources of the Ecuadorian Formative: An overview and comparison to the central Andes. In J. S. Raymond and R. L. Burger, eds., *Archaeology of Formative Ecuador*, 213–257. Washington, D.C.: Dumbarton Oaks.

Pearsall, D., and D. Piperno. 1990. Antiquity of maize cultivation in Ecuador: Summary and reevaluation of the evidence. *American Antiquity* 55:324–337.

Pickersgill, B. 1969. The archaeological record of chili peppers (*Capsicum* spp.) and the sequence of plant domestication in Peru. *American Antiquity* 34:54–61.

Pickersgill, B., and C. B. Heiser. 1977. Origins and distribution of plants domesticated in the New World Tropics. In C. A. Reed, ed., *Origins of Agriculture,* 803–835. The Hague: Mouton.

Piperno, D. R., and D. M. Pearsall, 1998. *The Origins of Agriculture in the Lowland Neotropics.* San Diego: Academic Press.

Pollan, M. 2001. *The Botany of Desire: A Plant's Eye View of the World.* New York: Random House.

Popper, V. S., and C. A. Hastorf. 1988. Introduction. In C. A. Hastorf and V. S. Popper, eds., *Current Paleoethnobotany: Analytical Methods and Cultural Interpretation of Archaeological Plant Remains,* 1–16. Chicago: University of Chicago Press.

Posey, D. A. 2002. *Kayapo Ethnoecology and Culture.* Edited by Kristina Plenderleith. London: Routledge.

Posey, D. A., and W. Balée, eds. 1989. *Resource Management in Amazonia: Indigenous and Folk Strategies.* Bronx: New York Botanical Garden.

Pullman, P. 2000. *The Amber Spyglass.* New York: Random House.

Quilter, J. 1989. *Life and Death at Paloma.* Iowa City: University of Iowa Press.

Quilter, J., B. Ojeda E., D. M. Pearsall, D. H. Sandweiss, J. G. Jones, and E. S. Wing. 1991. Subsistence economy of El Paraíso, an early Peruvian site. *Science* 251:277–283.

Rappaport, R. 1976. *Pigs for the Ancestors.* New York: Columbia University Press.

Rival, L., ed. 1998a. Domestication as a historical and symbolic process: Wild gardens and cultivated forests in the Ecuadorian Amazon. In W. Balée, ed., *Advances in Historical Ecology,* 232–250. New York: University of Columbia Press.

——, ed. 1998b. *The Social Life of Trees.* Oxford: Berg.

Rival, L., and N. L. Whitehead, eds. 2001. *Beyond the Visible and the Material: The Amerindianization of Society in the Work of Peter Rivière.* New York: Oxford University Press.

Roosevelt, A. C. 1984. Problems interpreting the diffusion of cultivated plants. In Doris Stone, ed., *Pre-Columbian Plant Migration,* 1–18. Papers of the Peabody Museum of Archaeology and Ethnology. Cambridge, Mass.: Harvard University Press.

Roosevelt, A. C., M. Lima da Costa, C. Lopes Machado, M. Michab, N. Mercier, H. Valladas, J. Feathers, W. Barnett, M. Imazio da Silveira, A Henderson, J. Sliva, B. Chernoff, D. S. Reese, J. A. Holman, N. Toth, and K. Schick. 1996. Paleoindian cave dwellers in the Amazon: The peopling of the Americas. *Science* 272:373–384.

Rossen, J., T. Dillehay, and D. Ugent. 1996. Ancient cultigens or modern intrusions?: Evaluating plant remains in an Andean case. *Journal of Archaeological Science* 23:391–407.

Rowe, J., and D. Menzel, eds. 1967. *Peruvian Archaeology: Selected Readings.* Palo Alto, Calif.: Peek.

Rozin, P. 1976. The selection of food by rats, humans, and other animals. In J. Rosenblatt, R. A. Hinde, C. Beer, and E. Shaw, eds., *Advances in the Study of Behavior,* 6:21–76. New York: Academic Press.

——. 1982. Human food selection: The interaction of biology, culture, and individual experience. In L. M. Barker, ed., *The Psychobiology of Human Food Selection,* 225–254. Chichester, U.K.: Ellis Horwood.

Sandweiss, D. H., J. B. Richardson III, E. J. Reitz, J. T. Hsu, and R. A. Feldman. 1989. Early maritime adaptations in the Andes: Preliminary studies at the Ring Site, Peru. In D. S. Rice, C. Stanish, and P. R. Scarr, eds., *Ecology, Settlement, and History of the Osmore Drainage, Peru,* 35–84. British Archaeological Reports International Series no. 545i. Oxford: British Archaeological Reports.

Sauer, C. O. 1952. *Agricultural Origins and Dispersals.* Bowman Memorial Lecture no. 5. New York: American Geographical Society.

Shady, R. 1997. *La Ciudad Sagrada de Caral.* Lima: Universidad Nacional Mayor de San Marcos.

Shady, R., J. Haas, and W. Creamer. 2001. Dating Caral, a Preceramic site in the Supe Valley on the central coast of Peru. *Science* 292:723–726.

Shipek, F. C. 1989. An example of intensive plant husbandry: The Kumeyaay of southern California. In D. R. Harris and G. C. Hillman, eds., *Foraging and Farming: The Evolution of Plant Exploitation,* 159–170. London: Unwin and Hyman.

Simoons, F. J. 1998. *Plants of Life, Plants of Death.* Madison: University of Wisconsin Press.

Smalley, J., and M. Blake. 2003. Stalk sugar and the domestication of maize. *Current Anthropology* 44 (5): 675–703.

Sperber, D. 1980. Is symbolic thought pre-rational? In M. Foster and S. Brandes, eds., *Symbol as Sense,* 25–44. New York: Academic Press.

Thomas, J. 2004. *Archaeology and Modernity.* London: Routledge.

Towle, M. 1961. *The Ethnobotany of Pre-Columbian Peru.* Viking Fund Publications no. 20. Chicago: Aldine.

Viveiros de Castro, E. 1992. *From the Enemy's Point of View: Humanity and Divinity in an Amazonian Society.* Chicago: University of Chicago Press.

Wagner, R. 1975. *The Invention of Culture.* New York: Prentice Hall.

Watson, P. J., and M. Kennedy. 1991. The development of horticulture in the eastern woodlands of North America: Women's role. In J. Gero and M. Conkey, eds., *Engendering Archaeology,* 255–275. Oxford: Basil Blackwell.

Weiner, A. 1992. *Inalienable Possessions.* Berkeley: University of California Press.

Weismantel, M. 1988. *Food, Gender, and Poverty in the Ecuadorian Andes.* Philadelphia: University of Pennsylvania Press.

Wiessner, P., and W. Schiefenhövel, eds. 1996. *Food and the Status Quest.* Oxford: Berghahn.

Wilson, D. 1988. *Prehispanic Settlement Patterns in the Lower Santa Valley, Peru.* Washington, D.C.: Smithsonian Institution Press.

Young, M. 1971. *Fighting with Food.* Cambridge, U.K.: Cambridge University Press.

Zimmerer, K. S. 1994. Human geography and the new ecology: The prospects and promise of integration. *Annals of the American Association of Geographers* 84 (1): 108–125.

4

MICROVERTEBRATE SYNECOLOGY AND ANTHROPOGENIC FOOTPRINTS IN THE FORESTED NEOTROPICS

PETER W. STAHL

HISTORICAL ECOLOGY AND archaeology share a mutual yet asymmetrical relationship, as the former must rely on the techniques and methodologies of the latter for generating inferences about a deep time that existed beyond human memory and before the advent of written documents. Humans impacted most of their biosphere in a dialectical relationship that produced culturally and historically determined landscapes (Balée 1998:16) at least since the onset of the Holocene. As expressions of cumulative human existence, ancient landscapes are cultural artifacts that archaeologists can recover and recognize. This is particularly important in the Western Hemisphere, where much human memory was tragically obliterated and about which too little was written too late. Like historians sorting out the uneven accounts of memory or writing, archaeologists are concerned with the vagaries of (1) how, where, and when the material they excavate originally accumulated; (2) what was deposited and what was not; and (3) what did and what did not survive through preservation.

Zooarchaeologists are those of us who identify and interpret preserved bone assemblages recovered from archaeological contexts. Although the bones of large animals are often our focus, we have had a long and cautious interest in the analysis of microvertebrate specimens from small animals whose live weights generally do not exceed 1 kilogram. Microvertebrates can be important in human subsistence, especially where populations of these organisms are locally abundant and relatively inexpensive to harvest. They can also be important for paleoecological inferences because their narrow niche requirements and restricted ranges can make them highly sensitive to local habitats. However, a frequent inability to identify the precise mechanisms responsible for the initial accumulation and deposition of their skeletons in archaeological contexts can compromise their value for interpreting human subsistence and paleoecology (Falk and Semken 1998; Stahl 1996).

In living tropical forests as well as in fossil assemblages originally deposited within the context of ancient forests, microvertebrate populations can be numerically abundant and taxonomically rich. Depending on context, fossil assemblages can serve as excellent evidence for interpreting prehistoric diet and paleoecology. Paleoecology traditionally emphasizes *stenotopic* taxa (specialists with narrow niche requirements) over *eurytopic* taxa (generalists with broad niche requirements) for analogical reconstruction. Eurytopic taxa are able to thrive under such diverse conditions, so that little is revealed about the environment at the time of deposition (Findley 1964:24). For reasons explored in this chapter, however, I have come to appreciate the potential importance of eurytopic taxa for studying forested neotropical lowlands before the arrival of Europeans, particularly because they can signal the anthropogenic footprint of vegetational clearance. In the following sections, I review theoretical literature on eurytopic and stenotopic taxa in the forested Neotropics. I then summarize recent field experiments in neotropical forest fragments, emphasizing microvertebrate synecology, which deals with the community of small animals associated together under certain ecological conditions. I conclude with archaeological data from the seasonally dry tropics of western lowland Ecuador (cf. Hastorf, chapter 3, this volume) and emphasize the potential significance of eurytopic microvertebrates in paleoecological reconstructions based on archaeofaunal analysis.

EURYTOPIC AND STENOTOPIC TAXA IN TROPICAL FORESTS: THEORETICAL CONSIDERATIONS

A well-known tendency is for diversity to increase toward the tropics, particularly in terrestrial ecosystems and especially in forests, where species richness and equitability are relatively higher (Fleming 1973; Keast 1972, but see Mares 1992; Mares and Willig 1994). Typically tropical environments have developed over a considerable period of time within the ideal growing conditions created by high and steady energy inputs, ample humidity, and decreased seasonal variation. Environmental heterogeneity supports a greater variety of diverse organisms with different requirements. Different species exploit environments in different ways based on their ability to partition niches broadly to narrowly along any number of dimensions. In general, temperate species tend to be generalists that do not distinguish subtle differences among niches as they adapt to greater temperature extremes over larger expanses of time and space. Relatively free from having to adapt to seasonal fluctuations, tropical species tend to specialize because they have the option to focus on microhabitats and to detect heterogeneity (Eldredge 1991).

Relative body size and systematic affiliation of tropical taxa are also important. Species of smaller-bodied rodents and insectivores tend to be stenotopic,
whereas larger-bodied carnivores, perissodactyls, and artiodactyls tend to
be increasingly eurytopic (Mares and Willig 1994:40). This difference makes
intuitive sense because small animals perceive their environment in a relatively
continuous or fine-grained manner, which enables them to partition the same
environment in more complex ways than larger animals. The latter need greater
foraging space in order to exploit temporally and spatially dispersed resources,
which appear to them in an increasingly patchy or coarse-grained distribution.

Other important considerations are the relative stability and level of resource
supply. Situations with fluctuating resources favor large, broadly tolerant populations capable of high reproduction, particularly where resource levels are elevated. Stable environments favor small populations with narrow tolerance and
low reproduction, particularly where resource levels are low (Valentine 1971).
Stable environments with low resource levels are theoretically characterized by
many different kinds of species (high richness), each with comparable population sizes (high equitability). Small populations of K-selected specialists
produce larger or fewer offspring, or both, that can effectively exploit these heterogeneous environments (Valentine 1971:57). Maximum population sizes have
been reached in these crowded conditions where fewer members of individual
species efficiently crop resources and convert food into offspring (MacArthur
and Wilson 1967:149). However, unstable environments with high resource
levels sustain low species richness and low equitability. Here, large populations of r-selected generalists with high reproduction rates have the advantage
(Valentine 1971:57), which is particularly important for colonizing taxa, whose
large and productive populations can harvest the most food and thus rear the
largest families (MacArthur and Wilson 1967:149).

Unstable conditions in the highly productive tropics interest me for a number of reasons. They are important for understanding the historic evolution
of tropical species richness. Museum models emphasize that ancient tropical
environments have accumulated species over time in the absence of regular
extinction. Engine or pump models emphasize that continuous but modest
change creates new niche opportunities (Colinvaux 1998:1; Erickson and Balée,
chapter 7, this volume). The fossil record attributes species diversification to
instances of unique evolutionary opportunities when inhabitable biospace
opens up through preceding extinction, adaptive innovation, or environmental
expansion (Valentine 1971:57). From a historical perspective, the ecological history of the forested Neotropics can be characterized as one of dynamic stability. Temporally and spatially localized instability can cull but not exterminate
species, while opening up new opportunities for others (Colinvaux 2001).

Unstable conditions in canopied forests can result from many sporadic and
localized natural perturbations such as fire, tree fall, disease, drought, and flood.

Animals routinely disturb forest conditions in their regular activities. Terborgh (2000) recently characterized a "beater syndrome" in which a commensal relationship is created between animals wherein one species unintentionally increases food availability for another through forest disturbance. The unstable conditions of most interest to me as an archaeologist are those that result from anthropogenic clearance and modification, a universal cultural footprint in the tropical forests. Of course, zooarchaeologists have long been aware of the relationship between humans and commensal or synanthropic species. Many of these animals do not show regular irruption cycles, but have a propensity for dramatic population increase under favorable anthropogenic conditions associated with agricultural production or storage (for references, see Stahl 1996:39). Certainly, humans actively manipulate their environments to create or take advantage of the increased productivity found in unstable forest clearings (e.g., "garden hunting," as in Linares 1976). In either case, tropical zooarchaeologists should be aware of the potential signature (in the form and nature of buried animal bones) of anthropogenic clearance left by the high abundance of eurytopic taxa in archaeological context. Numerous field studies, to which I now turn, substantiate this point.

EURYTOPIC TAXA IN NEOTROPICAL FOREST FRAGMENTS: FIELD STUDIES

The relationship between forest disturbance, microvertebrate populations, and potential signatures for anthropogenic clearance in the archaeological record is clarified in field-based studies of neotropical forest fragmentation that have been initiated over the past few decades for practical and theoretical reasons. As tropical forests are progressively cleared by growing human populations for their natural resources and agricultural space, data are urgently required to assess the effects of clearance and to gauge designs for effective future management of tropical ecosystems (Debinski and Holt 2000; Laurance and Bierregaard 1997; Lovejoy and Oren 1981; Lovejoy et al. 1984, 1986). The creation of forest edges that abut cleared areas is highly complex, so I summarize here only those observations most relevant to my argument.

Immediately obvious abiotic effects of forest fragmentation include light and wind penetration, increasing temperature, and reduction in soil moisture, with movement from the forest center toward the edge. Fragmentation involves elevated tree mortality, dramatic rates of leaf litter fall, and increased plant growth, particularly among pioneering heliophytes. Animals react according to their willingness or ability to cross nonforested gaps, or the minimum area of forest required for their normal maintenance, and according to the temporal/spatial patchiness of resource distribution (Dale et al. 1994:1028). Gap-crossing ability and area requirements are functions of the scale at which resources are utilized, which involve physiological and behavioral traits such as body size, mobility, and trophic status. For example, mobile birds or large carnivores may not be hindered by gaps as they exploit resources in continuous clusters (Dale et al. 1994:1033). Similarly, small nocturnal mammals that discriminate their immediate environment via short-distance visual cues may not distinguish between secondary habitats and forest gaps, as opposed to highly visual birds or obligate arboreal primates that exploit discretely isolated patches (Malcom 1997a:219).

Forest fragmentation dramatically affects arthropod biomass in that overstory insects move closer to the ground, and the increased vegetational biomass and productivity of new edges support more insect life (Malcolm 1997b:530–531). As a rule, invertebrate abundance increases from the forest interior toward the edge, and invasion by edge generalists changes community richness (Didham 1997). Increased light penetration encourages new understory plant growth and can increase the richness and abundance of butterflies through colonization by edge species and heliophytes, at the same time that large and mobile butterflies disappear (Brown and Hutchings 1997:107–108). Similarly, insect pollinators and detritivores also decline in abundance and richness (Didham et al. 1996).

Forest fragmentation has major consequences for vertebrate populations. Snakes appear in areas of greater habitat diversity, in particular ecological edges (Rand and Myers 1990:399). They are able to range over primary, secondary, agricultural, and other habitats; therefore, extensive deforestation, clearing, and secondary regrowth do not greatly alter their populations (Zimmerman and Rodrigues 1990:446). Similarly, few forest frogs are lost following forest fragmentation. As heterotherms with variable body temperature, many frogs can utilize matrix conditions because they demand less energetic and spatial requirements than homeothermic birds and mammals with uniform body temperatures (Tocher, Gascon, and Zimmerman 1997:137). Frog species richness increases, particularly when forest isolates are surrounded by dense undergrowth. These areas are invaded by taxa not normally associated with high forest, especially by opportunists of disturbed areas and by species capable of long-distance migration (Tocher, Gascon, and Zimmerman 1997). Compared to interior forests, edge communities may provide a greater resource base for insectivorous amphibian predators that exploit butterflies and insects (Brown and Hutchings 1997:108; Didham and Stork 1998:39).

Whereas many mobile organisms flee fragmented forest isolates, insectivorous populations may increase (Lovejoy et al. 1986:283). The immediate increase in phytophageous insects capitalizing on greater plant productivity along edges of recently isolated forest fragments may initially attract certain insectivorous birds (Lovejoy et al. 1986:265). As with frogs, edge communities may provide a greater resource base for avian predators that exploit butterflies and insects. Birds that exploit secondary growth may increase in abundance (Bierregaard and Stouffer 1997:147; Blake, Styles, and Loiselle 1990:162), as do nonforest and uncommon species (Stouffer and Bierregaard 1995b:2436). Canopy birds that readily cross forest gaps also continue to use fragments, including pigeons, parrots, oropendolas, some toucans, and tanagers (Stouffer and Bierregaard 1993:901). Obligate ant followers, which are among the first birds to disappear with the onset of fragmentation, are the first to reappear, followed by mixed flock species if the surrounding secondary growth is sufficiently developed (Bierregaard and Lovejoy 1989:228; Bierregaard and Stouffer 1997:146–147; Stouffer and Bierregaard 1995b:2433–2434). Hummingbird populations persist or increase after isolation because many canopy species are now forced to forage in the understory (Stouffer and Bierregaard 1995a:1091). Eurytopic nectivorous understory hummingbirds appear to be less vulnerable to forest fragmentation because they move back and forth between primary and secondary vegetation, taking advantage of spatial and temporal variability in flower production (Bierregaard and Stouffer 1997:153) and readily using edges and gaps with a high abundance of flowering bodies (Stouffer and Bierregaard 1995a:1091).

The proportion of small mammals close to the ground is markedly increased through fragmentation. Neither the biomass nor richness of small arboreal mammals is affected, whereas small terrestrial mammal populations tend to be richer and more abundant in recently isolated forest fragments (Malcolm 1997a:216). Edge communities may provide a greater resource base than forest interiors for insectivorous mammalian predators that exploit understory butterflies and insects (Brown and Hutchings 1997:108; Didham and Stork 1998:39). Increased light penetration from the fragment edge and decreased root competition from trees contribute to greater volume and productivity in understory vegetation, which result in an explosion of arthropods (Malcolm 1995:190). Not only is arthropod prey far more conspicuous, but destruction of the canopy also decreases competition from arboreal insectivores (Malcolm 1997b: 530–531). Reduced temporal and spatial variation in prey distribution prompts an abundant and predictable supply of understory insect food. More prey of the same type can be encountered on a regular basis, which contributes to a "superabundance" of small, insectivorous faunas (Malcolm 1995:191, 1997a:216, 1997b:530–531, 1998:48). Fragment edges also experience a net increase of small mammals due to an overflow of abundant and mobile individuals from surrounding secondary habitats (Malcolm 1997a:216, 1997b:531). Overall terrestrial microhabitat diversity is much higher in secondary forests because of the increased understory plant productivity and fallen timber (Malcolm 1997a:216). Many smaller terrestrial mammals exploit the tangled structure of gaps. For example, spiny rats (*Proechimys*) use fallen logs as conduits, and the short-tailed opossum (*Monodelphis*) forages under fallen wood and litter (Malcolm 1997a:219). The mature second-growth vegetation that can follow recent disturbance is often strongly dominated by one or two species, as are climax conditions in seasonal areas (Eisenberg 1983:273). Certain small mammals such as rice rats (*Oryzomys*) and spiny rats thrive in secondary vegetation, forest isolates, and gaps (Lovejoy at al. 1984:313, 1986:276; Malcolm 1995:192, 1998:47). Small mouse opossums (*Marmosa* spp.), which utilize early-successional stages, appear to be abundant in the secondary growth that gradually encroaches on newly formed forest fragments (Lovejoy at al. 1986:277).

Many larger terrestrial mammalian taxa that are normally eurytopic thrive in the edge conditions created through forest fragmentation, notably nine-banded armadillo (*Dasypus novemcinctus*), white-tailed (*Odocoileus virginianus*) and some Brocket (*Mazama* spp.) deer, collared peccary (*Tayassu tajacu*), and many carnivorous predators. Inside the newly created forest fragment, certain larger forest mammals are capable of surviving and thriving away from the edge. Howler monkeys (*Alouatta*), in particular, are the most commonly observed primate species in forest fragments and the only ones that appear to increase in numbers (Schwarzkopf and Rylands 1989). Folivory enables them to persist in small areas of isolated trees and to exploit the high productivity of second-growth forest, especially in the absence of frugivorous competition for large fruit (Lovejoy et al. 1986:276). Their persistence in fragments may also be attributed to flexible home range, ability to travel in small groups, and willingness to cross open ground and secondary growth (Schwarzkopf and Rylands 1989:9–10).

PREHISTORIC MICROVERTEBRATE ASSEMBLAGES FROM THE JAMA VALLEY, ECUADOR

I now examine the potential significance of eurytopic microvertebrates for reconstructing incidences of prehistoric deforestation by focusing on animal bone assemblages recovered from excavated archaeological sites in the Jama River drainage of northern Manabí Province in western Ecuador (figure 4.1). The assemblages were retrieved as part of the larger Jama Valley Archaeological/ Ethnobotanical Project, initiated by James A. Zeidler and Deborah M. Pearsall in 1988 as an interdisciplinary investigation of long-term sociopolitical change and agricultural intensification. Three seasons of regional survey and selective site testing revealed some 236 archaeological sites with occupations organized into a seven-part cultural sequence (table 4.1) spanning more than 3,500 years before the arrival of the first Europeans in the early sixteenth century (Zeidler 1995; Zeidler, Buck, and Litton 1998; Zeidler and Pearsall 1994). In the course of excavation, zooarchaeological data were recovered from 39 archaeological sites, representing several different temporal occupations throughout various portions of the valley.

The Jama River and its tributaries constitute the largest drainage basin in northern Manabí Province, forming a transverse valley system that courses

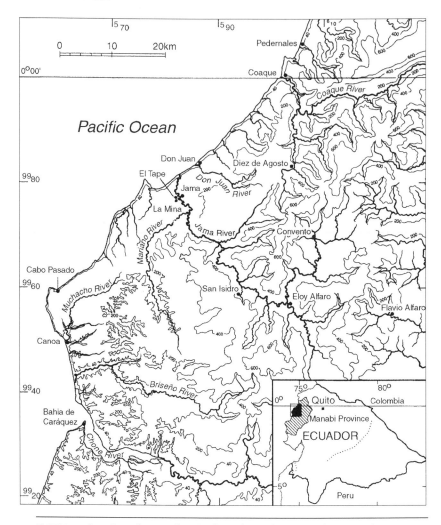

FIGURE 4.1 Jama River drainage (from Zeidler and Pearsall 1994:5 with permission).

through 1,612 square kilometers of rolling terrain. Interior portions of the 75-linear-kilometer river channel lie at 600 meters above sea level and eventually drop to sea level at its embouchure on the Pacific Coast. The valley is located just south of the equator between 0° 15' E and 0° 30' S. Lat. in a transitional littoral area between coastal dry and inland wet regimes. Lower reaches of the valley support a dry tropical climate with less than 1,000 millimeters of annual rainfall. Precipitation increases inland, where higher elevations are subject to a semihumid megathermic (continuous high temperatures and abundant rainfall) tropical climate, and seasonal rainfall supports dry and humid premontane forests, depending on elevation (Cañadas 1983; Zeidler and Kennedy 1994).

TABLE 4.1 Prehistoric Cultural Sequence in the Jama Valley

CERAMIC PHASE	CULTURAL COMPONENT	MODAL BEGINNING/ ENDING DATES
Piquigua	Valdivia 8	2030–1880 cal BC
Tabuchila	Chorrera	1300–750 cal BC
Muchique 1	Jama-Coaque I	240 cal BC–cal AD 90
Muchique 2	Jama-Coaque II	cal AD 420–790
Muchique 3	Jama-Coaque II	cal AD 880–1260
Muchique 4	Jama-Coaque II	cal AD 1290–1430
Muchique 5	Campace?	cal AD 1430–1640

Source: Zeidler, Stahl, and Sutliff 1998.

Recent settlement of the area dates from the late 1800s, with interior centers settled some 50 years later in wide expanses of alluvium. Since about 1920, logging, extensive cattle grazing, and intensive agriculture have removed most of the primary forest. Today, more than 93 percent of the valley is nonforested land utilized for agriculture and grazing. The remaining terrain is covered in a patchwork of primarily second-growth forest with high representation of exotic invader species (Pearsall 2004; Zeidler and Kennedy 1994).

The recovered archaeofaunal collection is somewhat unique. Comprising more than 85,000 preserved specimens, it is the largest assemblage from one area of prehistoric Ecuador, yet it weighs only 6.7 kilograms and fits easily inside a small suitcase. The collection is largely a by-product of intensive archaeobotanical recovery involving water separation and flotation of large constant volume and blanket soil samples extracted from each archaeological context (Pearsall 2004). The use of intensive flotation recovery strongly affects the size and composition of the assemblage, which is dominated by small fragments (Stahl 1995). Almost half the assemblage (figure 4.2) consists of small fish specimens (Osteichthyes); the other half is dominated by small mammals (Mammalia) and indeterminate fragments so small that they could not be reliably identified to zoological class. The remainder includes reptiles (Reptilia), amphibians (Amphibia), aquatic invertebrates (Crustacea), birds (Aves), cartilaginous marine taxa (Chondrichthyes), and marine shell (Mollusca). The highly fragmented sample is taxonomically rich, but unevenly dominated by small animals, some of which were deposited with higher skeletal completeness. Larger taxa, in contrast, appear as relatively isolated finds dispersed throughout the archaeological contexts. Factors contributing to the nature of the archaeofaunal sample include the natural composition and characterization of neotropical forest animal populations; prehistoric human activities associated with hunting, transport, distribution, consumption, and disposal; natural processes that affect animal bones before and after their deposition; and the way in which the bones were recovered from archaeological context (Stahl 1995).

Although our program of intensive recovery provides a highly representative sample of those archaeofaunal specimens preserved in buried contexts throughout the Jama River valley, the use of microvertebrate samples in archaeological inference is often plagued by potential ambiguity. In neotropical contexts, the taxonomy and natural histories of small animals tend to be poorly studied, and their diminutive and fragile elements are frequently difficult to identify to high taxonomic resolution. In particular, microvertebrate specimens can accumulate in many ways, including accidental entrapment and death; predation by humans and nonhumans alike; and postmortem scavenging and transport. The evidence we use to identify modes of accumulation is sometimes vague, frequently obscured, and often equivocal; however, it is essential to recognize how an assemblage originally accumulated in order to determine if the data are appropriate to the questions we ask (Stahl 1996).

In certain high-resolution contexts, these ambiguities can be surmounted. One such context is a large bell-shaped pit from the site of Pechichal (M3B4-011) in the middle section of the Jama River valley. The pit was likely first excavated and used for storage, but was subsequently filled with rubbish and abandoned. Twenty clearly visible depositional strata of garbage in the eventually sealed archaeological context were completely processed and sampled through water flotation. Separate accumulation histories were inferred from the relative numerical abundances and differential skeletal representation of different taxa throughout the pit's internal stratigraphy. On the basis of relative skeletal completeness, high frequency, and context within the pit, hardy generalists that thrive in forest fragment edges, such as frogs, snakes, and rodents, were

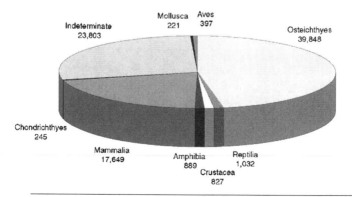

FIGURE 4.2 Taxonomic composition of the Jama archaeofaunas.

identified as local faunas that became entrapped in the pit while it was still open. Relatively small numbers of isolated bone specimens from neotropical taxa capable of persisting in or near forest fragments were stratigraphically dispersed throughout the pit. I suggested that the pit faunas included a mix of culturally and nonculturally or accidentally deposited faunas that implicated accumulation episodes while the pit was in or near the edge of a modified forest fragment, an inference corroborated by both micro- and macrobotanical data (Stahl 2000). We have as yet not completed assigning cultural phases to each archaeological context, nor have we similar resolution for other contexts; however, I think it significant that the pit's archaeofaunal inventory resembles the entire Jama assemblage.

Figure 4.3 illustrates the frequency of all recovered taxa identified to the ordinal level, excluding aquatic faunas, most of which are marine in origin. The entire Jama sample is heavily dominated by rodents (Rodentia, n = 4307), which outnumber the next highest taxonomic category by a factor of nearly seven. It is significant that the most abundant categories are also microvertebrates, an overwhelming number of whose specimens are represented by frogs/toads (Anura, n = 618), snakes (Serpentes, n = 484), salamanders (Caudata, n = 257), lizards (Sauria or Lacertilia, n = 139), many small opossums (Marsupialia, n = 77), and bats (Chiroptera, n = 28). Combined, microvertebrates comprise more than 86 percent of the faunal categories shown in figure 4.3. The remaining taxa include numerous edge generalists among the edentates (Xenarthra, n = 258), deer and peccary (Artiodactyla, n = 196), turtles (Chelonia, n = 188), rabbits (Lagomorpha, n = 54), and monkeys that persist in forest fragments (Primates, n = 18). The rest of the subsample includes mammalian carnivores (Carnivora, n = 51) and isolated specimens of tinamous (Tinamiformes, n = 6), pelicans (Pelicaniformes, n = 1), ducks (Anseriformes, n = 5), herons/egrets (Ciconiformes, n = 3), vultures/hawks (Falconiformes, n = 12), guans (Galiformes, n = 4), rails (Gruiformes, n = 2), snipes (Charadriiformes, n = 1), doves (Columbiformes, n = 4), parrots (Psittaciformes n = 1), cuckoos (Cuculiformes, n = 1), and tapirs (Perissodactyla, n = 2)—many of which include eurytopic edge species.

Figure 4.4 illustrates the same set of Jama taxa ranked according to their relative ubiquity, or the total number of different sampled archaeological contexts in which they were recovered. Figures 4.3 and 4.4 are virtually identical in their rank orders (r_s = .96, P = .01, where r_s is the Spearman's rho or rank order correlation and P is the confidence level). In other words, the taxa represented by the most numerous specimens tend to be found in more contexts. For example, rodent specimens were recovered in almost 80 percent of the sampled archaeological contexts. The only taxon that differs somewhat between the two figures is the order (Artiodactyla), which includes peccary and deer. Relatively low numbers of deer specimens are distributed throughout nearly one-half of

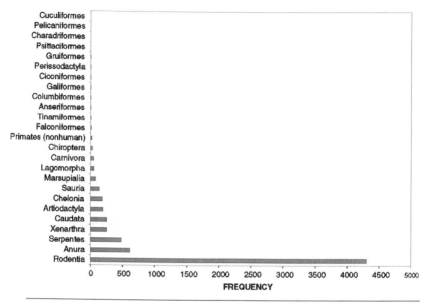

FIGURE 4.3 Frequency of nonaquatic orders (n = 6,817).

the sampled archaeological contexts, suggesting that deer specimens tend to be found in many temporally and spatially different deposits around the Jama River valley, but as isolated specimens.

Figure 4.5 illustrates the 13 most frequently represented taxa in the Jama assemblage, arranged in order of the average number of recovered specimens per sampled archaeological context. The rank order of these taxa is identical to the 13 most highly represented taxa in figure 4.3 and is almost identical to that in figure 4.4. In other words, the same taxa, virtually all of them microvertebrates, dominate the sample by frequency and tend to be found throughout the most sampled contexts, where they are also relatively abundant. As previously mentioned, deer specimens tend to be ubiquitous yet isolated fragments; the same might be said for all larger and, by definition, rarer taxa.

Figure 4.6 illustrates the same 13 most frequently represented taxa in the Jama assemblage, but arranges them in order of the average number of different kinds of skeletal elements per sampled archaeological context. Compared with figure 4.5, the rank order of these taxa in this figure differs somewhat ($r_s = .47$, $P = 0.1$) and likely reflects preservation, skeletal variation, and identification. Depressed averages in figure 4.6 are obviously a factor of few preserved skeletal elements that can be positively identified to taxon—for example, salamanders (mandibles, vertebrae) and snakes (mandibles, vertebrae, ribs). Identified bat

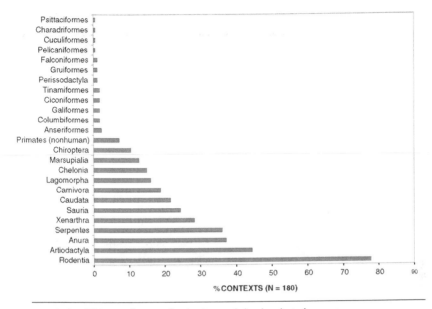

FIGURE 4.4 Ubiquity of recovered orders in sampled archaeological contexts.

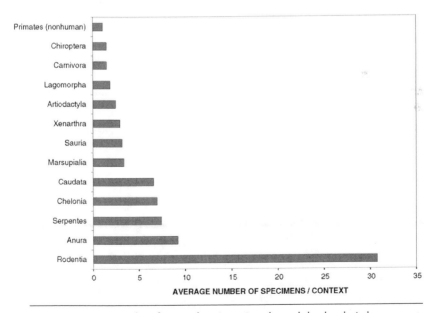

FIGURE 4.5 Average number of recovered specimens in each sampled archaeological context.

specimens are all dental or cranial fragments. In fact, most higher-resolution identifications are based on dental specimens that preserve well. Where relative skeletal completeness could be estimated in sampled archaeological contexts, rodents and anurans (frogs/toads) are characterized by high abundance and relatively more complete skeletons. I suspect the same could have been said of salamanders and snakes were it not for obvious problems of identification. Vertebral elements dominate the preserved elements of these taxa. Overall, the identified microfaunas within the Jama River valley assemblage include the kinds of animals we would expect to find in unstable forest edges, and the larger taxa include many that survive or thrive in fragmented forests.

It is often difficult to identify vertebrate specimens to genus or species based on the skeletal fragments that typically are preserved in humid neotropical forest settings (Stahl 1995). The majority of genus-level identifications, especially of microvertebrates, are based on durable dental fragments. Table 4.2 lists high-resolution identifications of nonaquatic taxa in the Jama assemblage and summarizes some typical ecological data (from Eisenberg and Redford 1999; Emmons and Feer 1990; Ridgely and Greenfield 2001; Savage 2002; Tirira 1999), along with the number of identified specimens. This subsample should be viewed with some caution because it is likely a biased representation of the originally deposited faunas. Nonetheless, the typical natural histories associated with this suite of identified faunas are revealing.

High-resolution microvertebrate taxa in the Jama assemblage consist of habitat generalists, most of which include naturally disturbed areas and

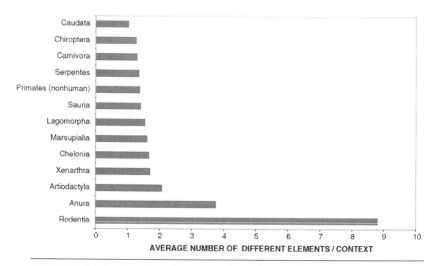

FIGURE 4.6 Relative skeletal completeness in each sample archaeological context.

TABLE 4.2 Indigenous Vertebrate Genera and Species Identified in Prehistoric Context

TAXON	NAME	TYPICAL NATURAL HISTORY
Didelphis	opossum	Variable, including multistratal forest, arboreal, scansorial, terrestrial, solitary, nocturnal, omnivore, nests in tree cavity or burrow (n = 59)
Marmosa	mouse opossum	Variable, undergrowth, arboreal, scansorial, terrestrial, nocturnal, omnivore, mainly insects, nests in cavities (n = 18)
Tamandua	northern anteater	Secondary, gallery to multistratal forests, arboreal, terrestrial, solitary, diurnal and nocturnal, insectivore, nests in tree hollows (n = 1)
Bradypus	sloth	Evergreen and dry forests, isolated trees in pastures, arboreal, nocturnal, diurnal, solitary, folivore (n = 2)
Dasypus	armadillo	Open areas to forests, terrestrial, fossorial, small groups, diurnal, nocturnal, omnivore (n = 252)
Phyllostomus	leaf-nosed bat	Variable, anthropogenic clearings, volant, nocturnal, frugivore, carnivore (n = 1)
Artibeus	fruit-eating bat	Mature and secondary forests, gardens, volant, nocturnal, frugivore (n = 5)
Alouatta	howler monkey	Dry deciduous to multistratal forest, arboreal, gregarious, diurnal, frugivore, folivore (n = 12)
Cebus	capuchin monkey	Mature and disturbed forest, arboreal, gregarious, diurnal, omnivore (n = 2)
Nasua	coati	Variable, forest to scrub, arboreal and terrestrial, solitary and groups, diurnal, omnivore (n = 1)
Felis pardalis	ocelot	Variable with cover, human settlements, terrestrial, solitary, nocturnal and diurnal, carnivore (n = 7)

TABLE 4.2 *(continued)*

TAXON	NAME	TYPICAL NATURAL HISTORY
Felis yagouaroundi	jaguarundi	Variable, secondary vegetation near human settlements, terrestrial and scansorial, solitary and pairs, nocturnal and diurnal, carnivore (n = 2)
Felis concolor	puma	Variable, terrestrial, solitary, nocturnal and diurnal, carnivore (n = 1)
Panthera onca	jaguar	Variable, terrestrial, solitary, nocturnal and diurnal, carnivore (n = 11)
Tapirus	tapir	Forest to grassland near water, terrestrial, usually solitary, nocturnal and diurnal, browser (n = 1)
Tayassu	peccary	Variable, including forest, savanna, cropland, terrestrial, gregarious, nocturnal, diurnal in rain forest, omnivore (n = 42)
Odocoileus	white-tailed deer	Mixed cover, cropland, terrestrial, solitary and small groups, diurnal, nocturnal, crepuscular, grazer and browser (n = 83)
Mazama	brocket deer	Variable, forest to savanna, terrestrial, solitary or pairs, diurnal, nocturnal, browser, frugivore (n = 13)
Sciurus	squirrel	Variable, forest, secondary, cropland, solitary or small groups, scansorial, diurnal, omnivore (n = 4)
Oryzomys	rice rat	Deciduous to evergreen forest, commensal, semiarboreal, scansorial, terrestrial, solitary, nocturnal, omnivore (n = 420)
Rhipidomys	climbing rat	Dry deciduous to multistratal forest, commensal, arboreal, scansorial, solitary, nocturnal, omnivore (n = 6)
Akodon	grass mouse	Open grass and scrubland, cropland, terrestrial, nocturnal, diurnal, omnivore (n = 19)

TABLE 4.2 (*continued*)

TAXON	NAME	TYPICAL NATURAL HISTORY
Sigmodon	cotton rat	Open grass and scrubland, cropland, commensal, terrestrial, nocturnal, diurnal, omnivore (n = 120)
Zygodontomys	cane mouse	Forest clearings, cropland, terrestrial, nocturnal, omnivore (n = 3)
Coendou	porcupine	Mature and disturbed forest, arboreal, pairs, nocturnal, herbivore (n = 1)
Agouti	paca	Mature to disturbed and secondary forest, cropland, terrestrial, solitary or pair, nocturnal, frugivore and browser (n = 10)
Dasyprocta	agouti	Humid forest, cropland, terrestrial, solitary and pair, diurnal, crepuscular, herbivore, frugivore (n = 52)
Proechimys	spiny rat	Forest, riverine, terrestrial, solitary, nocturnal, omnivore (n = 103)
Sylvilagus	rabbit	Variable, forest to semiarid brush, cropland, terrestrial, solitary, nocturnal, browser (n = 51)
Crypturellus	tinamou	Forest, woodland, regenerated scrub, numerous in fragments, commonly terrestrial, solitary (n = 2)
Bufo	toad	Variable, forest to disturbed, commensal, terrestrial, fossorial, nocturnal, diurnal, insectivore (n = 4)
Rana	frog	Forest, semiaquatic, near water's edge, nocturnal, diurnal, insectivore, carnivore (n = 2)
Iguana	iguana	Forest, dry forest, near water, arboreal, nest in burrows, solitary or group (n = 5)
Anolis	anoles	Variable, riparian, terrestrial to arboreal, insectivore, especially arthropods (n = 6)
Boa	boa	Variable, secondary growth, terrestrial to semiarboreal, nocturnal, carnivore (n = 27)
Rhinoclemmys	forest turtle	Forest, terrestrial, diurnal, herbivore (n = 27)

anthropogenic clearing in their range. Human disturbance favors "weedy" species of hardy, adaptable plants and animals such as rodents, frogs, snakes, and understory avian insectivores that become superabundant as specialized animals disappear (Malcolm 1998:48). As predicted from field studies, certain micromammals are dominant with respect to numerical abundance and ubiquity. Rice rats (*Oryzomys*) are found in a variety of habitats and tend to be particularly common in second-growth, disturbed habitats and in human clearings (Musser et al. 1998), as are cotton rats (*Sigmodon*) especially where rainfall is seasonal (Voss 1992). The spiny rat (*Proechimys*) is a dominant contributor to nonvolant biomass and is particularly abundant in areas of dense undergrowth, which provides preferred nesting sites (Emmons 1982). Various generalists that utilize disturbed habitats to a significant degree complete the list of identified small rodents, including: squirrel (*Sciurus*), climbing rat (*Rhipidomys*), grass mouse (*Akodon*), and cane mouse (*Zygodontomys*). Tiny mouse opossums (*Marmosa*) frequent openings in pursuit of insects. Leaf-nosed bats (*Phyllostomus*) and especially fruit-eating bats (*Artibeus*) frequent anthropogenic clearings and tend to be important seed dispersers in early-successional forests. The depositional history of volant bats is peculiar; however, many are likely accidental intrusions in a refuse pit (Stahl 2000) or in one instance may have been intentionally interred as a grave inclusion (Zeidler, Stahl, and Sutliff 1998). Small generalist herpetofaunas (*Bufo, Rana, Anolis*) round out the microvertebrate profile.

Larger rodent taxa are similarly implicated in anthropogenic clearings, including such neotropical staples as agouti (*Dasyprocta*) and paca (*Agouti* or *Cuniculus*). Opossum (*Didelphis*) is quite common, as is armadillo (*Dasypus*), whose high abundance is partially explained by the preservation of abundant and durable carapace fragments, especially in lower levels of a pit (Stahl 2000). In the Esmeraldas Province of Ecuador, twice as many smaller mammals, especially spiny rat (*Proechimys semispinosus*), rice rat (*Oryzomys caliginosus*), common opossum (*Didelphis marsupialis*), mouse opossum (*Marmosa robinsoni*), and wooly opossum (*Caluromys derbianus*) are captured around family garden plots than on forest trails (Suárez, Stallings, and Súarez 1995:39). Of course, active and abandoned gardens are attractive venues for wildlife that are effectively managed for human consumption, a point first raised by Linares (1976). Here, we must include many of the larger game animals, which are often edge generalists, or animals that thrive in forest fragments: anteater (*Tamandua*), sloth (*Bradypus*), howler monkey (*Alouatta*), peccary (*Tayassu*), deer (*Odocoileus, Mazama*), rabbit (*Sylvilagus*), tinamou (*Crypturellus*), and iguana (*Iguana*). The menagerie is rounded out by a host of opportunistic nonhuman predators, including felids (*Felis, Panthera*) and boa constrictors (*Boa*).

CONCLUSION

Much of the theoretical literature discussed in this chapter is already well known, and the results of pertinent field-based projects have been appearing in publications the past few decades. Certainly, zooarchaeologists have long understood the particular value of microvertebrates for establishing inferences regarding anthropogenic landscape modification. Nevertheless, I believe it is important to assemble these ideas to improve our understanding of a systematically and rigorously excavated archaeofaunal data set from the forested neotropical lowlands. It is particularly significant because these areas have not received their requisite share of archaeological attention. Also, for more than half a century anthropologists have produced radically conflicting paradigms to interpret neotropical lowland prehistory. Whereas one opinion projects only a very minor human contribution into past landscape management, an opposing viewpoint extols significant and profound changes prior to the arrival of Europeans in the sixteenth century (Stahl 2002).

Prehistoric microvertebrate assemblages from the Jama River valley strongly suggest a prevalence of unstable edge environments and forest fragments in the local environment. Of course, archaeologists have traditionally focused on excavating human settlements, which are by definition anthropogenic disturbances. Furthermore, ancient human settlement and farming in the valley was strongly focused on the riverine floodplain, especially in its earliest phases. These habitats may have originally been more open than inland forested areas or may have been initially cleared and maintained through repeated reoccupation and farming by humans. Nevertheless, a significant component of larger taxa within the assemblage further implicates nearby stands of fragmented forest, at varying distances from floodplain habitats. We have yet to complete our interpretation of prehistoric human settlement in the Jama River valley of western Ecuador, and until we do, I caution that some of the ideas offered in this chapter are subject to change. However, my overall impression of the prehistoric Jama environment so far includes a heavy anthropogenic involvement spanning at least 3,600 years.

ACKNOWLEDGMENTS

This research was made possible through grants from the National Science Foundation (BNS-8709649 and BNS-9108548) awarded to James A. Zeidler and Deborah M. Pearsall. I thank Jim and Debby for their continued support, encouragement, insight, and permission to reprint the project map in figure 4.1. I thank the various curators and staff of the American Museum of Natural

History who granted me access to comparative collections, in particular Rob Voss (mammalogy) and Allison Andors (formerly of ornithology). I appreciate the opportunity to contribute to this volume and thank William Balée and Clark Erickson for the invitation and for their excellent editorial advice. I remain responsible for the contents of this chapter.

REFERENCES

Balée, W. 1998. Historical ecology: Premises and postulates. In W. Balée, ed., *Advances in Historical Ecology*, 13–29. New York: Columbia University Press.

Bierregaard, R. O., Jr., and T. E. Lovejoy. 1989. Effects of forest fragmentation on Amazonian understory bird communities. *Acta Amazonica* 19:215–241.

Bierregaard, R. O., and P. C. Stouffer. 1997. Understory birds and dynamic habitat mosaics in Amazonian rain forests. In W. F. Laurance and R. O. Bierregaard Jr., eds., *Tropical Forest Remnants: Ecology, Management, and Conservation of Fragmented Communities*, 138–155. Chicago: University of Chicago Press.

Blake, J. G., F. G. Styles, and B. A. Loiselle. 1990. Birds of La Selva Biological Station: Habitat use, trophic composition, and migrants. In A. H. Gentry, ed., *Four Neotropical Rainforests*, 161–182. New Haven, Conn.: Yale University Press.

Brown, K. S., Jr., and R. W. Hutchings. 1997. Disturbance, fragmentation, and the dynamics of diversity in Amazonian forest butterflies. In W. F. Laurance and R. O. Bierregaard Jr., eds., *Tropical Forest Remnants: Ecology, Management, and Conservation of Fragmented Communities*, 91–110. Chicago: University of Chicago Press.

Cañadas Cruz, L. 1983. *El mapa bioclimático y ecológico del Ecuador*. Quito: Banco Central del Ecuador.

Colinvaux, P. 1998. The Ice-Age Amazon and the problem of diversity. *Review of Archaeology* 19:1–10.

———. 2001. Paradigm lost: Pleistocene environments of the Amazon basin (continued forest cover in perpetual flux, part 2). *Review of Archaeology* 22:20–31.

Dale, V., S. M. Pearson, H. L. Offerman, and R. V. O'Neill. 1994. Relating patterns of land-use change to faunal biodiversity in the central Amazon. *Conservation Biology* 8:1027–1036.

Debinski, D. M., and R. D. Holt. 2000. A survey and overview of habitat fragmentation experiments. *Conservation Biology* 14:342–355.

Didham, R. K. 1997. The influence of edge effects and forest fragmentation on leaf litter invertebrates in central Amazonia. In W. F. Laurance and R. O. Bierregaard Jr., eds., *Tropical Forest Remnants: Ecology, Management, and Conservation of Fragmented Communities*, 55–70. Chicago: University of Chicago Press.

Didham, R. K., J. Ghazoul, N. E. Stork, and A. J. Davis.1996. Insects in fragmented forests: A functional approach. *Trends in Ecology and Evolution* 11:255–260.

Didham, R. K., and N. E. Stork. 1998. Rise of the supertramp beetles. *Natural History* 107 (6): 38–39.

Eisenberg, J. F. 1983. Behavioral adaptations of higher vertebrates to tropical forests. In F. B. Golley, ed., *Tropical Rainforest Ecosystems, Structure, and Function*, 267–278. Amsterdam: Elsevier Scientific.

Eisenberg, J. F., and K. H. Redford. 1999. *Mammals of the Neotropics*. Vol. 3 of *The Central Neotropics*. Chicago: University of Chicago Press.

Eldredge, N. 1991. *The Miner's Canary: Unraveling the Mysteries of Extinction*. New York: Prentice Hall.

Emmons, L. H. 1982. Ecology of *Proechimys* in southeastern Peru. *Tropical Ecology* 23:280–290.

Emmons, L. H., and F. Feer. 1990. *Neotropical Rainforest Mammals: A Field Guide*. Chicago: University of Chicago Press.

Falk, C. R., and H. A. Semken Jr. 1998. Taphonomy of rodent and insectivore remains in North American archaeological sites: Selected examples and interpretations. In J. J. Saunders, B. W. Styles, and G. F. Baryshnikov, eds., *Quaternary Paleozoology in the Northern Hemisphere*, 285–321. Illinois State Museum Scientific Papers no. 27. Springfield: Illinois State Museum.

Findley, J. S. 1964. Paleoecological reconstruction: Vertebrate limitations. In J. J. Hester and J. Schoenwetter, comps., *The Reconstruction of Past Environments*, 23–25. Proceedings of the Fort Burgwin Conference on Paleoecology, 1962. Fort Burgwin, Taos, N.Mex.

Fleming, T. H. 1972. Numbers of mammal species in North and Central American forest communities. *Ecology* 54:555–563.

Keast, A. 1972. Comparisons of contemporary mammal faunas of southern continents. In A. Keast, F. C. Erk, and B. Glass, eds., *Evolution, Mammals, and Southern Continents*, 433–501. Albany: State University of New York Press.

Laurance, W. F., and R. O. Bierregaard Jr., eds. 1997. *Tropical Forest Remnants: Ecology, Management, and Conservation of Fragmented Communities*. Chicago: University of Chicago Press.

Linares, O. F. 1976. "Garden hunting" in the American tropics. *Human Ecology* 4:331–349.

Lovejoy, T. E., R. O. Bierregaard Jr., A. B. Rylands, J. R. Malcolm, Q. E. Quintelas, L. H. Harper, K. S. Brown Jr., A. H. Powell, G. V. N. Powell, H. O. R. Schubart, and M. B. Hays. 1986. Edge and other effects of isolation on Amazon forest fragments. In M. E. Soulé, ed., *Conservation Biology: The Science of Scarcity and Diversity*, 257–285. Sunderland, Mass.: A. Sinauer Associates.

Lovejoy, T. E., and D. C. Oren. 1981. The minimum critical size of ecosystems. In R. L. Burgess and D. M. Sharpe, eds., *Forest Island Dynamics in Man-Dominated Landscapes*, 7–12. New York: Springer-Verlag.

Lovejoy, T. E., J. M. Rankin, R. O. Bierregaard Jr., K. S. Brown Jr., L. H. Emmons, and E. Vander Voort. 1984. Ecosystem decay of Amazonian forest remnants. In M. H. Nitecki, ed., *Extinctions*, 295–325. Chicago: University of Chicago Press.

MacArthur, R. H., and E. O. Wilson. 1967. *The Theory of Island Biogeography*. Princeton, N.J.: Princeton University Press.

Malcolm, J. R. 1995. Forest structure and the abundance and diversity of neotropical small mammals. In M. D. Lowman and N. M. Nadkarni, eds., *Forest Canopies*, 179–197. San Diego: Academic Press.

——. 1997a. Biomass and diversity of small mammals in Amazonian forest fragments. In W. F. Laurance and R. O. Bierregaard Jr., eds., *Tropical Forest Remnants: Ecology, Management, and Conservation of Fragmented Communities*, 207–221. Chicago: University of Chicago Press.

——. 1997b. Insect biomass in Amazonian forest fragments. In N. E. Stork, J. Adis, and R. K. Didham, eds., *Canopy Arthropods*, 510–533. London: Chapman and Hall.

——. 1998. High roads to oblivion. *Natural History* 107 (6): 46–49.

Mares, M. A. 1992. Neotropical mammals and the myth of Amazonian biodiversity. *Science* 255:976–979.

Mares, M. A., and M. R. Willig. 1994. Inferring biome associations of recent mammals from samples of temperate and tropical faunas: Paleoecological considerations. *Historical Biology* 8:31–48.

Musser, G. G., M. D. Carleton, E. M. Brothers, and A. L. Gardner. 1998. Systematic studies of Oryzomyine rodents (Muridae, Sigmodontinae): Diagnoses and distributions of species formerly assigned to *Oryzomys* "*Capito.*" Bulletin no. 236. New York: American Museum of Natural History.

Pearsall, D. M. 2004. *Plants and People in Ancient Ecuador: The Ethnobotany of the Jama River Valley.* Belmont, Calif.: Wadsworth.

Rand, A. S., and C. W. Myers. 1990. The herpetofauna of Barro Colorado Island, Panama: An ecological summary. In A. H. Gentry, ed., *Four Neotropical Rainforests,* 386–409. New Haven, Conn.: Yale University Press.

Ridgely, R. S., and P. J. Greenfield. 2001. *Field Guide.* Vol. 2 of *The Birds of Ecuador.* Ithaca, N.Y.: Cornell University Press.

Savage, J. M. 2002. *The Amphibians and Reptiles of Costa Rica.* Chicago: University of Chicago Press.

Schwarzkopf, L., and A. B. Rylands. 1989. Primate species richness in relation to habitat structure in Amazonian rain forest fragments. *Biological Conservation* 48:1–12.

Stahl, P. W. 1995. Differential preservation histories affecting the mammalian zooarchaeological record from the forested neotropical lowlands. In P. W. Stahl, ed., *Archaeology in the Lowland American Tropics,* 154–180. Cambridge, U.K.: Cambridge University Press.

——. 1996. The recovery and interpretation of microvertebrate bone assemblages from archaeological contexts. *Journal of Archaeological Method and Theory* 3:31–75.

——. 2000. Archaeofaunal accumulation, fragmented forests, and anthropogenic landscape mosaics in the tropical lowlands of prehispanic Ecuador. *Latin American Antiquity* 11:241–257.

——. 2002. Paradigms in paradise: Revising standard Amazonian prehistory. *Review of Archaeology* 23:39–50.

Stouffer, P. C., and R. O. Bierregaard Jr. 1993. Spatial and temporal abundance patterns of ruddy quail-doves (*Geotrygon montana*) near Manaus, Brazil. *Condor* 95:896–903.

——. 1995a. Effects of forest fragmentation on understory hummingbirds in Amazonian Brazil. *Conservation Biology* 9:1085–1094.

——. 1995b. Use of Amazonian forest fragments by understory insectivorous birds. *Ecology* 76: 2429–2445.

Suárez, E. J., J. Stallings, and L. Súarez.1995. Small-mammal hunting by two ethnic groups in north-western Ecuador. *Oryx* 29:35–42.

Terborgh, John. 2000. In the company of humans. *Natural History* 109 (4): 54–62.

Tirira, Diego. 1999. *Mamíferos del Ecuador.* Museo Zoológico Publicación Especial no. 2. Quito: Pontificia Universidad Católica del Ecuador.

Tocher, M., C. Gascon, and B. L. Zimmerman. 1997. Fragmentation effects on a central Amazonian frog community: A ten-year study. In W. F. Laurance and R. O. Bierregaard Jr., eds., *Tropical Forest Remnants: Ecology, Management, and Conservation of Fragmented Communities,* 124–137. Chicago: University of Chicago Press.

Valentine, J. W. 1971. Resource supply and species diversity patterns. *Lethaia* 4:51–61.

Voss, R. S. 1992. A revision of the South American species of *Sigmodon* (Mammalia: Muridae) with notes on their natural history and biogeography. *American Museum Novitates* 3050:1–56.

Zeidler, J. A. 1995. Archaeological survey and site discovery in the forested Neotropics. In P. W. Stahl, ed., *Archaeology in the Lowland American Tropics,* 7–41. Cambridge, U.K.: Cambridge University Press.

Zeidler, J. A., C. E. Buck, and C. D. Litton. 1998. Integration of archaeological phase information and radiocarbon results from the Jama River valley, Ecuador: A Bayesian approach. *Latin American Antiquity* 9:160–179.

Zeidler, J. A., and R. C. Kennedy. 1994. Environmental setting. In J. A. Zeidler and D. M. Pearsall, eds, *Environment, Cultural Chronology, and Prehistoric Subsistence in the*

Jama River Valley, 13–41, vol. 1 of *Regional Archaeology in Northern Manabí.* Memoirs in Latin American Archaeology no. 8. Pittsburgh: University of Pittsburgh.

Zeidler, J. A., and D. M. Pearsall, eds. 1994. *Environment, Cultural Chronology, and Prehistoric Subsistence in the Jama River Valley.* Vol. 1 of *Regional Archaeology in Northern Manabí.* Memoirs in Latin American Archaeology no. 8. Pittsburgh: University of Pittsburgh.

Zeidler, J. A., P. W. Stahl, and M. J. Sutliff. 1998. Shamanistic elements in a terminal Valdivia burial, northern Manabí, Ecuador: Implications for mortuary symbolism and social ranking. In A. Oyuelo-Caycedo and J. S. Raymond, eds., *Recent Advances in the Archaeology of the Northern Andes: In Memory of Gerardo Reichel-Dolmatoff,* 101–112. Institute of Archaeology Monograph no. 39. Los Angeles: University of California.

Zimmerman, B. L., and M. T. Rodrigues. 1990. Frogs, snakes, and lizards of the INPA-WWF Reserves near Manaus, Brazil. In A. H. Gentry, ed., *Four Neotropical Rainforests,* 426–454. New Haven, Conn.: Yale University Press.

PART 2

5

PRE-EUROPEAN FOREST CULTIVATION IN AMAZONIA

WILLIAM M. DENEVAN

TRADITIONAL CULTIVATION IN Amazonian forests today is dominated by long-fallow shifting cultivation. However, there is little evidence for it prior to about AD 1600. A new model of pre-European agriculture has been suggested based on the inefficiency of stone axes, the evidence of anthropogenic soils, and archaeology. The productive landscape probably consisted of semi-intensive fields and managed bush fallows, surrounded by zones of modified forest manipulated by hunting and gathering and other human activities. Not only was soil fertility apparently maintained artificially by organic inputs and in-field burning, but also some soil was altered in the form of dark earth, which continues to be visible and fertile. Thus, pre-1492 cultivation practices and landscapes seem to have been considerably different from those of current traditional people, a perspective gained from a historical ecological approach to stability versus change in people-environment interactions.

Historical ecology is "the study of changing human-environmental relations" (Crumley 1996:560; also see Balée 1998; Balée and Erickson, introduction, this volume). It is an interdisciplinary subject involving anthropology (including archaeology), geography, history, sociology, demography, and the physical and biological sciences. The expression *historical ecology* became current in the 1990s, but it has been used from time to time since at least the 1970s. The topic subsumes some of the previous and ongoing research in cultural and human ecology, cultural and historical geography, environmental history, and landscape ecology. However, cultural ecology has often lacked a historical dimension; environmental history focuses on historical documents; and landscape ecology may minimize or ignore human agency.

Within my field of geography, people-environment research has been a major concern since at least the time of Alexander von Humboldt (1769–1859).

Three benchmark historical volumes dominated by geographers or geographical approaches are *Man and Nature* in 1864 (Marsh [1864] 2003), *Man's Role in Changing the Face of the Earth* (Thomas 1956), and *The Earth as Transformed by Human Action* (Turner et al. 1990). Particularly important have been Carl Sauer (1938, 1950, 1956, 1958) and his students at Berkeley. Sauer was instrumental in turning early American geography away from environmental determinism (1927:165–175). Today, "nature and society" is one of the four principal subfields of geography.

Historical ecology includes prehistory. Sauer's influence long ago turned me toward early human impacts on environment in Latin America (Denevan 1961) and to native cultivated landscapes (Denevan 1966). I have argued that indigenous alterations of nature were present to some degree wherever native peoples were present (Denevan 1992b, forthcoming). People cannot live on and from the environment without changing it. However, this perspective is not without controversy (Vale 2002). Regarding cultivation, I became more and more convinced that pre-European agricultural practices were sophisticated and productive. Many of these practices were lost after 1492, but their vestiges have persisted on the landscape in various places. Here I briefly present some of the evidence and thinking about native cultivation in Amazonian forests.[1]

The prevailing image of pre-European agriculture in upland Amazonian forests (*terra firme*) is that of shifting cultivation (swidden) with periods of short cropping and long fallowing. This is the dominant pattern today for both native peoples and settlers. However, a revisionist model has been suggested, a landscape of semi-intensively cultivated fields[2] intermingled with fruit orchards, managed fallows, house gardens, and brief bush fallows, with semipermanent settlements, some numbering thousands of people, surrounded by zones of modified forest manipulated by hunting and gathering activities (Denevan 2001:102–132).[3] This complex system of integrated land use both created and exploited fertile, anthropogenic soils (dark earths, anthrosols), known in Brazil as *terra preta do índio* (Indian black earth, or known more broadly as Amazonian Dark Earths).

SEMIPERMANENT CULTIVATION AND THE INEFFICIENCY OF STONE AXES

In Amazonian forests, the vegetation must first be cleared before forms of cultivation of annual crops are possible. Light materials could have been cut using machetes made of hard wood, but trees were chopped down mainly with stone axes. However, stone axes are so inefficient for removing large trees, compared to metal axes introduced by Europeans, that long-fallow shifting cultivation was probably difficult, even with the girdling and burning of tree trunks (Denevan 1992c). Experimental research with both types of axes indicates that up to 60 times more energy and time is required to clear forest with stone axes, with an average of about ten to one, depending on tree diameter and hardness, axe type, cutting technique, arm strength, and use of auxiliary methods (Carneiro 1974, 1979a, 1979b; Hill and Kaplan 1989:331). Stone axes not only cut poorly, but also dull and break, and the hafting comes undone, requiring frequent polishing, repair, and replacement. Moreover, in Amazonia, suitable stone may be hundreds of miles away.

The historical short-cropping/long-fallowing shifting cultivation system has been made possible by labor-efficient metal axes. In pre-European times, there must have been much less frequent forest clearing with stone axes. Once an opening was established by clearing or by a treefall, natural burn, or a blowdown from violent wind shear (Nelson et al. 1994), the opening could have been cultivated semipermanently, possibly with gradual enlargement at the edges (Denevan 2001:120–127). Fertility could have been maintained by organic inputs of household garbage, ash and charcoal, and mulches and composts, and by the frequent in-field burning of weeds, crop residues, logs and branches, leaves and palm fronds from both within a field and carried from adjacent forest. Labor inputs would have been high, especially for controlling weeds, which are more aggressive with intensive cultivation compared to clearings from mature fallows or primary forests. However, even short fallows of a year or two can reduce weeds; hence, the likelihood of semipermanent rather than permanent cultivation—a few years of crops rotating with a few years of bush fallow. Dark-earth farmers in Brazil today do this to reduce the labor costs of weeding (German 2003a, 2003b:326).

Ethnographic examples of semi-intensive cultivation can be found in Amazonia. The best studies are for the Kayapó in central Brazil (Hecht 2003; Posey 2002:165–218). Cropping for up to 5 or 6 years and fallows as short as 8 to 10 years are possible because of site-specific plantings, soil-fertility management, polycropping, and crop zonation. Organic material is collected and burned periodically within fields. These small burns are managed for frequency, location, extent, and biomass, as well as for seasonal and diurnal timing and thus burn temperature. These burning techniques are important for specific crops, crop clustering, field architecture, and fertility characteristics. Small mounds are created from compost and are planted (*apête*), and rich top soil is placed in cracks in rocky sites and then planted.

Pre-European cultivation was probably often, if not usually, more intensive and more productive than postcontact cultivation. After 1492, metal axes became available, a technological revolution in terms of tools and forest-clearing efficiency. However, there was actually an agricultural de-evolution toward long-fallow shifting cultivation, a simplification and reduction in productivity both per unit of land and over time, which has continued to the present. In addition, some interior Amazonian societies abandoned cultivation completely for a foraging economy (Balée 1992:37–41, 1995:98–102). Both changes can be termed *agricultural regression*. Such regression also resulted from forced migration away from the attractive riverine zones onto the poor upland soils in the interior (Lathrap 1970:186–190).

Iron and, later, steel axes became the primary trade items for native people in colonial times and in some remote areas until recently because of their extraordinary value for clearing forest (Métraux 1959). Descriptions of shifting cultivation are rare throughout the Americas prior to about AD 1600 (Denevan 2001:115). However, with little or no evidence scholars frequently have assumed that long-fallow shifting cultivation was the dominant form of pre-European agriculture in neotropical forests (Meggers 1957:80–83; still unchanged 40 years later, see Meggers [1971] 1996:19–23) and that "intensive cultivation is an impossibility" (Roosevelt 1980:87, later reversed in Roosevelt 1999:381–382). Historical ecological and archaeological research and analysis cannot be too strongly emphasized in order to counter such assumptions, which are based primarily on ethnographic (recent or present-day) analogy (Erickson and Balée, chapter 7, this volume).

DARK EARTHS

Anthropogenic dark earths are widespread in Amazonia, probably covering at least 0.1 to 0.3 percent (15,500–20,700 square kilometers) of the forested area (Sombroek et al. 2003:130). Black dark-earth soil, or terra preta proper, usually contains high concentrations of ceramics, kitchen waste, and bones, which identify the soil as a former settlement site.[4] Soil scientists now consistently explain the soil color and fertility as the result of sustained accumulation of organic refuse in middens, including ash and charcoal from domestic fires, over a long period of time. The lighter or brownish form of anthropogenic dark earth, usually called *terra mulata,* is much more extensive and usually surrounds patches of darker terra preta. A section of both soils on the bluff of the Tapajós River near Belterra south of the city of Santarém was mapped long ago by the late Wim Sombroek (1966:175), a pioneer of Amazonian soil research. He was apparently the first to maintain that terra mulata was produced by "long-lasting cultivation."

The key to terra mulata formation and persistence seems to have been a burning practice that leaves intact charcoal, which is not degradable, in contrast to ash. This "cool" burning could have been a form of "slash and char," with incomplete combustion, in which moist slash is burned, in contrast to the slash and burn today, in which a "hot" burn is a more complete burn, accomplished after a long period of drying out. The resulting soil carbon from cool burning, along with high levels of soil microorganisms, apparently created the self-sustaining high fertility of these soils. The stone axe thesis helps explain why long-term, semipermanent cultivation could have taken place instead of long-fallow shifting cultivation, which does not produce dark earth. Frequent organic inputs and in-field burning could have made semipermanent cultivation possible as well as contributing to the creation of terra mulata.

Terra preta contains up to 70 times more carbon than does surrounding soil. Carbon itself is not a direct nutrient, but it retains nutrients and makes them available, stabilizes soil organic matter, raises pH level, raises soil microbial activity, maintains soil moisture, helps repel insects, reduces nutrient leaching, and thus maintains and improves soil fertility and hence crop-production level and sustainability (Lehmann, Kern, German et al. 2003; Steiner, Teixeira, and Zech 2004).

The largest known extents of dark earths are on the forested bluffs along the main Amazonian rivers (Denevan 1996, 2001:104–110). It is often believed that riverine settlement in Amazonia was located primarily below the bluffs in the floodplains. For example, archaeologist Anna Roosevelt states that late prehistoric people "were very densely settled ... along the banks, levees, and deltas of the major floodplains" (1987:154–155). And historian John Hemming says that "In the sixteenth century the native population was very dense in the flood plain or várzea" ([1978] 1995:191). However, the floodplains are a high-risk habitat for both settlements and crops because of periodic extreme flooding of even the highest natural levees. Houses can be built on pilings, but crops will be destroyed.

EXPLORATION, CULTIVATION, SETTLEMENT, AND DARK EARTHS

The first descriptions of the Amazon River were in the mid–sixteenth century by Gaspar de Carvajal ([1542] 1934) from the Orellana voyage and by the surviving men of the Pedro de Ursúa-Lope de Aguirre disaster (Mampel González and Escandell Tur 1981). These accounts clearly indicate that most of the large Amazonian settlements were located on the bluff edges, not within the floodplain. Long, linear settlements were reported, which extended continuously for several leagues (a sixteenth-century league was 5–6 kilometers). For example, Captain Altamirano and Francisco Vázquez, who were with Aguirre in 1561, mentioned a settlement that was two or three leagues long or more, with the houses touching one another (Mampel González and Escandell Tur 1981:225; Vázquez de Espinosa [1628] 1948:387). Carvajal gave similar information for 1541 ([1542] 1934:198–212).

How credible are these early reports (see discussion in Denevan 1996: 661–664)? First of all, the descriptions from the two expeditions are similar, and they were only 20 years apart. Such accounts of large bluff settlements are absent by the mid–seventeenth century, by which time there had been massive population reductions.[5] And although Carvajal and the men of Aguirre told some fanciful stories, these stories seem to have come mostly from poorly understood natives, not from direct observations by the Spaniards. Exaggeration is possible. Regardless, dark-earth soils full of potsherds—evidence for large, linear bluff settlements—cover bluffs for estimated distances of as much as 2–6 kilometers (Myers 1973:240; Petersen, Neves, and Heckenberger 2001:97; Smith 1980:560). However, these stretches of dark earth were not necessarily fully occupied at a point in time given probable local (short-distance) shifting of houses and fields.

Large bluff settlements could not have been supported by seasonal playa and natural-levee cultivation. The bluff soils are the same poor Oxisols that dominate the interior upland forests. There was likely a complimentary system of bluff cultivation and hunting combined with seasonal floodplain cultivation and fishing (Denevan 1996:671–672). Bluff cultivation for large, semipermanent settlements would have had to have been productive and sustainable, as I have suggested. Evidence for this assumption comes from the Araracuara sites on the bluff of the Caquetá River in the Colombian Amazon (Herrera et al. 1992). Radiocarbon dating and analyses of soils, pollen, phytoliths, plant remains, and ceramics indicate nearly continuous human occupation of one site for 800 years; the creation of fertile, brown anthropogenic soils; and intensive agroforestry systems with maize, manioc, and fruit trees.

Most large archaeological sites along the major rivers are located where the primary river channels, navigable year round, impinge against the bluffs, rather than where the channels are in the middle of the floodplain and less accessible from the bluffs. These bluff-channel junctures are the locations of pre-European dark earths and settlements, colonial missions, and most towns and cities today, with relatively empty lands in between (Denevan 1996).

If semi-intensive cultivation and dense settlement were located on the bluffs, then both were certainly also possible in the interior forests. Indeed, there is dark-earth, archaeological, and historical evidence for large permanent settlements in the interior forests where soils are similar. Oitavo Bec, a dark-earth site south of the city of Santarém, covers more than 120 hectares (Woods and McCann 1999:12), and the Comunidade Terra Preta site located between the lower Rio Tapajós and the Rio Arapiuns covers 200 hectares (Smith 1999:26). However, there are numerous small dark-earth sites of only 1 hectare or so, which must have been created by a few people (Smith 1980:563). Historical settlements in the interior mostly had 100 people or less, but there are reports of some with several thousand inhabitants.[6]

CONCLUSIONS

Thus, several lines of evidence and reasoning suggest that in pre-European Amazonia cultivation was often semipermanent rather than long fallow. Such cultivation could well have created terra mulata, and, if so, then terra mulata soil, wherever it now occurs, may be indicative of former semipermanent cultivation. Once established, self-perpetuating dark earths could have made possible ongoing semi-intensive cultivation to the present day, with further organic additives being unnecessary or minimal, although other factors would have caused periodic field abandonment (Petersen, Neves, and Heckenberger 2001:92; Woods and McCann 1999:10–12). The implications for understanding both the Amazonian past and future agricultural development in Amazonia and elsewhere are dramatic. Uncertainties remain, but the necessary field and laboratory research is now under way (see, e.g., Lehmann, Kern, Glaser et al. 2003).

A historical ecological approach to people-environment interactions involves examination of change over time: change in society, change in technology, change in the environment (both natural and human induced), and consequently changes in the interactions involved. For pre-European Amazonia, I suggest a very different form of indigenous cultivation from that of today, one often (not always) based on the human formation of fertile soils and their utilization, with a concentration of land use and associated settlement rather than dispersal, as is characteristic of long-fallow extensive cultivation in recent times.

The present is not necessarily an extension of the past. Although there are continuities, there are also dramatic disjunctures. For Amazonian cultivation and the associated landscape, such a disjuncture seems to have occurred after 1492. Just as some agriculturalists regressed to foragers, others regressed from semi-intensive cultivators to long-fallow shifting cultivators.

Finally, if pre-European cultivation was often semipermanent or permanent, then the disturbance patterns of distribution, kind, quality, and diversity of natural resources would have been different from those produced through post-1492 long-fallow swidden; that is, patches of intensive modification were interspersed with less-intensive disturbance based on hunting and gathering activity, in contrast to a more uniform disturbance based on the constant shifting of cultivation, foraging, and settlement.

NOTES

1. For indigenous cultivation in floodplains and savannas in Amazonia, see Denevan 2001.

2. *Semipermanent cultivation, semi-intensive cultivation,* and *intensive shifting cultivation* are more or less synonymous terms for *crop/fallow rotations* in which both the cropping and fallow periods are brief (from one to ten years or so). In contrast, a long-fallow system can be considered extensive, and permanent (annual harvests) or nearly permanent cultivation can be considered intensive.

3. This view of semi-intensive cultivation is part of a revised (or new) paradigm (or model or synthesis) of Amazonian prehistory, which argues for complex societies (see Denevan 2001:130–131; Heckenberger et al. 1999; Neves 1999; Roosevelt 1994;

Stahl 2002; Viveiros de Castro 1996). The older, standard paradigm of egalitarian societies, low populations, small temporary settlements, and long-fallow shifting cultivation (reviewed in Stahl 2002:39–42) is still persistent (Meggers 1957, 1996, 2001).

4. For recent studies and discussions of anthropogenic dark earths in Amazonia, see Glaser et al. 2001; Glaser and Woods 2004; Lehmann, Kern, Glaser et al. 2003; Mann 2002; McCann 2004; McCann et al. 2001; Petersen et al. 2001; Woods and McCann 1999.

5. Populations in the lowland Amazon basin proper circa 1492 have been roughly estimated at between 2.5 and 5 million and may have been considerably higher (Denevan 1992a, 1996:656, 672–674, 2003:186–187; Hemming 1995:505–521). By 1900, the total indigenous population was reduced by disease, war, and slavery to a few hundred thousand; rose to about 500,000 in 1972; and subsequently has risen somewhat above that despite near extinction for some groups (Denevan 1992a:232). For Brazilian Amazonia only, Hemming estimates about 350,000 in 1910, decreasing to 100,000 in the 1950s, and then increasing again to 350,000 still living in tribal communities at the end of the twentieth century (2003:636–637). Recent genetic study indicated 45 million people in Brazil with some Indian ancestry (Hemming 2003:812).

6. Examples of large archaeological settlements (estimated) occur in the upper Xingu River basin (1,000–1,500) and in Goiás (1,043–1,738). Historical examples reported (varying accuracy) include settlements of the coastal Tupinambá in the sixteenth century (up to 8,000); and, in central Brazil, the Bororo in the early twentieth century (1,500), the Kayapó in 1896 (1,250 or 5,000 in four settlements) and in 1900 (3,500–5,000), the Paresi in the early eighteenth century (1,200), the Xarae in the sixteenth century (7,500), and the Apinayé in 1824 (1,400) (for sources, see Denevan 2003:182–183).

REFERENCES

Balée, W. 1992. People of the fallow: A historical ecology of foraging in lowland South America. In K. H. Redford and C. Padoch, eds., *Conservation of Neotropical Forests: Working from Traditional Resource Use*, 35–57. New York: Columbia University Press.

——. 1995. Historical ecology of Amazonia. In L. E. Sponsel, ed., *Indigenous Peoples and the Future of Amazonia: An Ecological Anthropology of an Endangered World*, 97–110. Tucson: University of Arizona Press.

——. 1998. Historical ecology: Premises and postulates. In W. Balée, ed., *Advances in Historical Ecology*, 13–29. New York: Columbia University Press.

Carneiro, R. L. 1974. On the use of the stone axe by the Amahuaca Indians of eastern Peru. *Ethnologische Zeitschrift Zürich* 1:107–122.

——. 1979a. Forest clearance among the Yanomamö: Observations and implications. *Antropológica* 52:39–76.

——. 1979b. Tree felling with the stone axe: An experiment carried out among the Yanomamö Indians of southern Venezuela. In C. Kramer, ed., *Ethnoarchaeology: Implications for Archaeology*, 21–58. New York: Columbia University Press.

Carvajal, G. de. [1542] 1934. *The Discovery of the Amazon*. Compiled by J. Toribio Medina. Edited by H. Heaton. Special Publications no. 17. New York: American Geographical Society.

Crumley, C. L. 1996. Historical ecology. In D. Levinson and M. Ember, eds., *Encyclopedia of Cultural Anthropology*, 2:558–560. New York: Henry Holt.

Denevan, W. M. 1961. The upland pine forests of Nicaragua: A study in cultural plant geography. *University of California Publications in Geography* 12 (4): 251–320.

Denevan, W.M. 1966. *The Aboriginal Cultural Geography of the Llanos de Mojos of Bolivia.* Ibero Americana no. 48. Berkeley: University of California Press.

——. [1976] 1992a. The aboriginal population of Amazonia. In W.M. Denevan, ed., *The Native Population of the Americas in 1492*, 2d ed., 205–234. Madison: University of Wisconsin Press.

——. 1992b. The pristine myth: The landscape of the Americas in 1492. *Annals of the Association of American Geographers* 82:369–385.

——. 1992c. Stone vs. metal axes: The ambiguity of shifting cultivation in prehistoric Amazonia. *Journal of the Steward Anthropological Society* 20:153–165.

——. 1996. A bluff model of riverine settlement in prehistoric Amazonia. *Annals of the Association of American Geographers* 86:654–681.

——. 2001. *Cultivated Landscapes of Native Amazonia and the Andes.* Oxford: Oxford University Press.

——. 2003. The native population of Amazonia in 1492 reconsidered. *Revista de Indias* (Madrid) 63:175–187.

——. Forthcoming. Pre-European human impacts on tropical lowland environments. In T. Veblen, K. Young, and A. Orme, eds., *The Physical Geography of South America.* Oxford: Oxford University Press.

German, L. 2003a. Ethnoscientific understandings of Amazonian Dark Earths. In J. Lehmann, D.C. Kern, B. Glaser, and W.I. Woods, eds., *Amazonian Dark Earths: Origin, Properties, Management*, 179–204. Dordrecht, Netherlands: Kluwer Academic.

——. 2003b. Historical contingencies in the coevolution of environment and livelihood: Contributions to the debate on Amazonian black earth. *Geoderma* 111:307–331.

Glaser, B., L. Haumaier, G. Guggenberger, and W. Zech. 2001. The "terra preta" phenomenon: A model for sustainable agriculture in the humid tropics. *Naturwissenschaften* 88:37–41.

Glaser, B., and W.I. Woods, eds. 2004. *Amazonian Dark Earths: Explorations in Space and Time.* Berlin: Springer-Verlag.

Hecht, S.B. 2003. Indigenous soil management and the creation of *terra mulata* and *terra preta* in the Amazon basin. In J. Lehmann, D.C. Kern, B. Glaser, and W.I. Woods, eds., *Amazonian Dark Earths: Origin, Properties, Management*, 355–372. Dordrecht, Netherlands: Kluwer Academic.

Heckenberger, M.J., J.B. Petersen, and E.G. Neves. 1999. Village size and permanence in Amazonia: Two archaeological examples from Brazil. *Latin American Antiquity* 10:353–376.

Hemming, J. [1978] 1995. *Red Gold: The Conquest of the Brazilian Indians.* Rev. ed. London: Papermac.

——. 2003. *Die If You Must: Brazilian Indians in the Twentieth Century.* London: Macmillan.

Herrera, L.F., I. Cavelier, C. Rodríguez, and S. Mora. 1992. The technical transformation of an agricultural system in the Colombian Amazon. *World Archaeology* 24:98–113.

Hill, K., and H. Kaplan. 1989. Population and dry-season subsistence strategies of the recently contacted Yora of Peru. *National Geographic Research* 5:317–334.

Lathrap, D.W. 1970. *The Upper Amazon.* New York: Praeger.

Lehmann, J., D.C. Kern, L.A. German, J.M. McCann, G.C. Martins, and A. Moreira. 2003. Soil fertility and production potential. In J. Lehmann, D.C. Kern, B. Glaser, and W.I. Woods, eds., *Amazonian Dark Earths: Origins, Properties, Management*, 105–124. Dordrecht, Netherlands: Kluwer Academic.

Lehmann, J., D.C. Kern, B. Glaser, and W.I. Woods, eds. 2003. *Amazonian Dark Earths: Origin, Properties, Management.* Dordrecht, Netherlands: Kluwer Academic.

Mampel González, E., and N. Escandell Tur, eds. 1981. *Lope de Aguirre Crónicas: 1559–1561.* Barcelona: Ediciones Universidad de Barcelona.

Mann, C. C. 2002. The real dirt on rainforest fertility. *Science* 297:920–923.

Marsh, G. P. [1864] 2003. *Man and Nature.* Edited by D. Lowenthal. Seattle: University of Washington Press.

McCann, J. M. 2004. Subsidy from culture: Anthropogenic soils and vegetation in Tapajônia, Brazilian Amazonia. Ph.D. diss., University of Wisconsin, Madison.

McCann, J. M., W. I. Woods, and D. W. Meyer. 2001. Organic matter and anthrosols in Amazonia: Interpreting the Amerindian legacy. In R. M. Rees, B. C. Ball, C. D. Campbell, and C. A. Watson, eds., *Sustainable Management of Soil Organic Matter,* 180–189. New York: CAB International.

Meggers, B. J. 1957. Environment and culture in the Amazon basin: An appraisal of the theory of environmental determinism. Studies in Human Ecology. *Social Science Monographs* 3:71–89.

——. [1971] 1996. *Amazonia: Man and Culture in a Counterfeit Paradise.* Rev. ed. Washington, D.C.: Smithsonian Institution Press.

——. 2001. The continuing quest for El Dorado: Round two. *Latin American Antiquity* 12: 304–325.

Métraux, A. 1959. The revolution of the ax. *Diogenes* 25:28–40.

Myers, T. P. 1973. Toward the reconstruction of prehistoric community patterns in the Amazon basin. In D. W. Lathrap and J. Douglas, eds., *Variation in Anthropology: Essays in Honor of John C. McGregor,* 233–252. Urbana: Illinois Archaeological Survey.

Nelson, B. W., V. Kapos, J. B. Adams, W. J. Oliveira, O. P. G. Braun, and I. L. do Amaral. 1994. Forest disturbance by large blowdowns in the Brazilian Amazon. *Ecology* 75:853–858.

Neves, E. G. 1999. Changing perspectives in Amazonian archaeology. In G. G. Politis and B. Alberti, eds., *Archaeology in Latin America,* 216–243. London: Routledge.

Petersen, J. B., E. Neves, and M. J. Heckenberger. 2001. Gift from the past: *Terra preta* and prehistoric Amerindian occupation in Amazonia. In C. McEwan, C. Barreto, and E. Neves, eds., *Unknown Amazon: Culture in Nature in Ancient Brazil,* 86–105. London: British Museum Press.

Posey, D. A. 2002. *Kayapó Ethnoecology and Culture.* Edited by K. Plenderleith. London: Routledge.

Roosevelt, A. C. 1980. *Parmana: Prehistoric Maize and Manioc Subsistence along the Amazon and Orinoco.* New York: Academic Press.

——. 1987. Chiefdoms in the Amazon and Orinoco. In R. D. Drennan and C. A. Uribe, eds., *Chiefdoms in the Americas,* 153–185. Lanham, Md.: University Press of America.

——. 1994. Amazonian anthropology: Strategy for a new synthesis. In A. C. Roosevelt, ed., *Amazonian Indians from Prehistory to the Present: Anthropological Perspectives,* 1–29. Tucson: University of Arizona Press.

——. 1999. Twelve thousand years of human-environment interaction in the Amazon floodplain. In C. Padoch, J. Márcio Ayres, M. Pinedo-Vasquez, and A. Henderson, eds., *Várzea: Diversity, Development, and Conservation of Amazonia's Whitewater Floodplains,* 371–392. Bronx: New York Botanical Garden Press.

Sauer, C. O. 1927. Recent developments in cultural geography. In E. C. Hayes, ed., *Recent Developments in the Social Sciences,* 154–212. New York: Lippincott.

——. 1938. Theme of plant and animal destruction in economic history. *Journal of Farm Economics* 20:765–775.

——. 1950. Grassland climax, fire, and man. *Journal of Range Management* 3:16–21.

——. 1956. The agency of man on the earth. In W. L. Thomas Jr., ed., *Man's Role in Changing the Face of the Earth,* 49–69. Chicago: University of Chicago Press.

Sauer, C. O. 1958. Man in the ecology of tropical America. *Proceedings of the Ninth Pacific Science Congress* 20:104–110.

Smith, N. J. H. 1980. Anthrosols and human carrying capacity in Amazonia. *Annals of the Association of American Geographers* 70:533–566.

——. 1999. *The Amazon River Forest: A Natural History of Plants, Animals, and People.* New York: Oxford University Press.

Sombroek, W. G. 1966. *Amazon Soils: A Reconnaissance of the Soils of the Brazilian Amazon Region.* Wageningen, Netherlands: Centre for Agricultural Publications and Documentation.

Sombroek, W. G., M. L. Ruivo, P. M. Fearnside, B. Glaswer, and J. Lehmann. 2003. Amazonian Dark Earths as carbon stores and sinks. In J. Lehmann, D. C. Kern, B. Glaser, and W. I. Woods, eds., *Amazonian Dark Earths: Origins, Properties, Management,* 125–139. Dordrecht, Netherlands: Kluwer Academic.

Stahl, P. W. 2002. Paradigms in paradise: Revising standard Amazonian prehistory. *Review of Archaeology* 23:39–51.

Steiner, C., W. G. Teixeira, and W. Zech. 2004. Slash and char: An alternativce to slash and burn practiced in the Amazon basin. In B. Glaser and W. I. Woods, eds., *Amazonian Dark Earths: Explorations in Space and Time,* 183–193. Berlin: Springer-Verlag.

Thomas, W. L., Jr., ed. 1956. *Man's Role in Changing the Face of the Earth.* Chicago: University of Chicago Press.

Turner, B. L., II, W. C. Clark, R. W. Kates, J. F. Richards, J. T. Mathews, and W. B. Meyer, eds. 1990. *The Earth as Transformed by Human Action: Global and Regional Changes in the Biosphere over the Past 300 Years.* Cambridge, U.K.: Cambridge University Press.

Vale, T. R., ed. 2002. *Fire, Native Peoples, and the Natural Landscape.* Washington, D.C.: Island Press.

Vázquez de Espinosa, A. [1628] 1948. *Compendio y descripción de las Indias Occidentales.* Edited by C. U. Clark. Smithsonian Miscellaneous Collections no. 108. Washington, D.C.: Smithsonian Institution.

Vivieros de Castro, E. 1996. Images of nature and society in Amazonian ethnology. *Annual Review of Anthropology* 25:179–200.

Woods, W. I., and J. M. McCann. 1999. The anthropogenic origin and persistence of Amazonian Dark Earths. *Yearbook, Conference of Latin Americanist Geographers* 25:7–14.

6

FRUIT TREES AND THE TRANSITION TO FOOD PRODUCTION IN AMAZONIA

CHARLES R. CLEMENT

THE GENESIS CHAPTER of the Bible and the archaeological record agree: fruit trees were important subsistence resources during the Pleistocene/Holocene transition to food production in the Middle East (Tudge 1999), the Americas in general (Lentz 2000), Amazonia (Roosevelt 1999), and other tropical and subtropical areas. Hence, fruit trees must have contributed significantly to human carrying capacity: the ability of the landscape to supply sufficient nutritious food and other materials to guarantee human reproduction. As food production became dominant, numerous fruit trees were domesticated, many being modified by selection as much as were domesticated annual crops, but over time they became less important in subsistence in many areas. Although this is the general trend, in the humid tropics food production often included arboriculture, as in Southeast Asia and Oceania (Latinis 2000), or anthropogenic forests, as in Amazonia (Balée 1989), both systems being essentially fruit forests and reminiscent of the Garden of Eden.

The term *food production* includes all human activities involved in growing food plants (Piperno and Pearsall 1998), such as fruit trees. In this sense, it includes such widely used concepts as cultivation, domestication, horticulture, and agriculture, phenomena that always involve a mixture of land preparation, plant selection, and plant propagation. More generally, I have defined *landscape domestication* as "a conscious process by which human manipulation of the landscape results in changes in landscape ecology and in the demographics of its plant and animal populations, resulting in a landscape more productive and congenial for humans" (Clement 1999a:190), and *plant population domestication* as "a co-evolutionary process by which human selection on the phenotypes of promoted, managed or cultivated plant populations results in changes in the population's genotypes that make them more useful to humans and better adapted to human intervention in the landscape" (189).

Domestication is a process; therefore, its results form a continuum from least to most modified, and sections of that continuum can be categorized (landscapes can be pristine, promoted, managed, or cultivated; plant populations can be wild, incipiently domesticated, semidomesticated, and domesticated—see Clement 1999a for full details). Even though defined separately, landscape and plant population domestication are inseparable in practice, as Wiersum (1997) has emphasized with respect to the fruit trees of Southeast Asia. Both types of domestication generally focus on food production, although fibers, medicinals, stimulants, and other types of technological products are often domesticated. Landscape and plant population domestication are central to historical ecology, as they form the most important relationship between humans and their landscape, designed to guarantee social and biological reproduction, the main tenets of historical ecology (Balée 1998).

In this chapter, I examine three factors that might explain why fruit trees became relatively less important during the transition to food production and especially as it intensified during the mid-Holocene. I focus on Amazonian fruit trees in various stages of domestication to understand this change of importance. What fruit tree characteristics might explain fruit tree importance and then the subsequent relative loss of its importance?

I hypothesize that nutritional quality, fruit-production phenology (the seasonal variation in fruit harvests), and water content are important limitations to a predominantly frugivorous subsistence or food-production strategy. Gross (1975) postulated that animal protein limited human carrying capacity in Amazonia, supporting Meggers's (1971) argument that Amazonia was a counterfeit paradise due to ecological limits on resource abundance. Beckerman (1979) countered that vegetable protein was abundant in Amazonia, so much so that protein was probably not a limiting factor to population expansion. Additional information is provided here in support of Beckerman, while showing that fruit-production phenology limits the availability of vegetable protein during certain periods. Carrying capacity is always limited by the availability of nutritious foods during the period when these are least available.

Fruit-production phenology has not previously been studied in relation to why the transition to food production favored annuals over perennials, though mention is often made of the availability of starch reserves in annual root crops during the dry season when other resources are scarce (Harlan 1992, 1995). The humid to dry tropical transition zones of South America, Africa, Oceania, and Australia are prime examples (Harlan 1995), where *Manihot, Ipomoea, Xanthosoma, Dioscorea, Colocasia,* and other roots, tubers, and corms were and often still are important. A quick survey suggests that fruit phenology should be an important consideration, perhaps a primary one, especially when food processing and storage techniques were rudimentary.

Food processing is an important way to manage the water content of foods so that they can be stored for future use (Harlan 1995). Abundant water, when combined with starches, oils, and proteins, makes excellent substrates for bacteria and fungi that start degrading foods immediately after their harvest. Modifying water content and oxygen levels during processing controls microorganisms in that they have difficulty reproducing when water content is below 10 percent and in the absence of free oxygen (except anaerobic bacteria, which require other processing controls). Although processing has been discussed in other respects during the transition to food production, its relationship with the decision (certainly unconscious) to favor annuals over perennials has not been examined.

Fruit trees can usefully be classified as producing nuts or seeds, on the one hand, or as producing starchy, oily, or juicy pulps, on the other, each with different nutritional qualities. Other useful classifications are tree growth habit (especially palms and dicotyledons), with plant-management implications; tree population density (oligarchic[1] and dispersed), with fruit abundance implications; and tree adaptation to the floodplain or to uplands, with phenological implications. These distinctions are useful because domestication advanced more rapidly in some of these classes than in others, as well as in some species more than in others, and in some parts of Amazonia more than in others. Some of these contrasts are highlighted here.

Patiño (1964) suggested that some pre-Columbian human groups in certain locations of the lowland humid Neotropics may have relied more heavily on fruit crops than did groups in other areas. Northwestern Amazonia was one area he identified. I have identified a center of crop genetic diversity in that region, based largely on the number and diversity of fruit trees (Clement 1989, 1999b). The importance of arboriculture, especially of fruits, has long been recognized in Southeast Asia and Oceania (Kirch 1989; Latinis 2000; Lepofsky 1992; Yen 1974), and similar work has identified anthropogenic forests in Amazonia (Balée 1989; Frikel 1978), many of which are fruit forests. This work in other regions and with other conceptual orientations highlights the importance of arboriculture in the humid tropics and the intraregional distribution of the origins of Amazonian fruit domesticates indirectly supports Patiño's hypothesis. I think that generalizations about the importance of fruit trees to foraging and food-production strategies are either site specific (because of an abundance of candidate species), ethnospecific (as appears to be the case in Southeast Asia and perhaps in northwestern South America), or perhaps even humid tropic specific (as the seasonally dry tropics followed other routes to intensify food production).

This chapter evaluates the importance of fruit trees to subsistence during the transition from foraging to food production and why fruit trees became relatively less important as the transition advanced in many areas. Because the data are modern biological data, the inferences drawn about the Holocene are designed to stimulate hypotheses rather than test them. The data set is derived from Clement 1999a, where I identified 80 fruit crops in Amazonia at the time of European conquest, of which 45 are presented here. I selected these 45 based on the availability of seed or pulp chemical composition and yield data (principally Aguiar 1996; FAO 1986; Villachica et al. 1996; Wu Leung and Flores 1961). Thirty-four of the forty-five also had phenological information from the Belém, Pará, market (Cavalcante 1976), which can be thought of as a modern "foraging" territory in eastern Amazonia that offers a model for examining phenology. Although this data set is far from complete for Amazonia as a whole, it allows a preliminary evaluation of the reasons why fruit trees became relatively less important during the transition to food production in most areas.

NUTRITIONAL QUALITY

Nuts and seeds tend to be high in oil or starch, hence in energy, and have good protein levels (table 6.1; appendix 6.1). Because seeds are propagules, they must contain sufficient energy to guarantee germination and initial development of the seedling (Rosengarten 1984). Germination and subsequent differentiation are all enzyme-mediated activities, as is transformation of storage products (oil or starch) into new tissues, which helps explain the seed's abundant proteins. Nut trees often yield relatively well also because population survival depends on propagule abundance and dispersal efficiency. Nuts and seeds are high-quality resources, but the trees are generally widely dispersed in the landscape (except some palms and anthropogenic, oligarchic dicots) and tend to have long generation spans. Hence, they are generally not the major sources of protein for human groups, nor do their populations show significant modifications due to domestication.

TABLE 6.1 Comparison of the Mean Chemical Compositions* of Contrasting Fruit Groups

Group	water	protein	fats	carbs.	fiber	energy
			g / 100 g			kcal
Nuts and seeds	3.9	14.1	57.4	18.1	4.8	621
Palm fruits	45.3	3.5	21.8	16.0	12.2	310
Starchy/oily fruits	51.1	2.5	8.3	32.4	9.0	231
Juicy fruits	82.8	0.9	0.8	11.9	2.9	63

Note: The mesocarp is generally the most important part of the palm, starchy/oily, and juicy fruit groups.

* Fresh weights; the difference between the sum of these means and 100 is due to ash content; see appendixes for species-specific compositions and standard deviations of these means.

Starchy/oily fruits have a preponderance of one of these energy-storage forms, but tend not to have much protein because the pulp is an attractant for the animals that disperse the species. Their seeds, however, are occasionally used and possess the same characteristics mentioned earlier. This group usefully can be divided into monocotyledons (palms) versus dicotyledons (table 6.1; appendixes 6.2 and 6.3, respectively). The palms tend to have more oil in their mesocarps, whereas the dicots tend to be starchier, but this distinction is not uniform. Some species contribute significant amounts of beta-carotene, a precursor of vitamin A (data not presented, but examples are buriti, tucumã, pupunha [Wu Leung and Flores 1961]). Both starchy and oily fruit trees often yield relatively well. Because some of these species were domesticated to a greater or lesser degree, they must have provided a significant amount of protein, vitamins, and energy for both foraging and food-producing subsistence diets.

Juicy fruits tend to be low in energy and low in protein (table 6.1; appendix 6.4), hence they contribute variety and pleasure to diets rather than energy or protein. Some contribute significant amounts of vitamin C, other vitamins, and minerals. Most yield abundantly because juicy fruits contain little energy. A reasonable number of juicy fruits were domesticated at one time or another, suggesting the importance of variety and pleasure in diets. As an Australian horticulturist once commented to me, "Agriculture makes modern human civilization possible; horticulture makes it bearable!" A fruit-based food-production system would theoretically be more bearable than other types given the wide variability in flavors and nutrient contents in fruits.

PHENOLOGY

Most fruit trees are moderately to highly seasonal (figure 6.1; appendix 6.5), which means that their often high-quality or desirable resources are abundant only for a single, short period each year or, rarely, for several short periods. The partial exception are the floodplain oligarchic palms, which have longer populational fruiting seasons but not year-round fruiting. The great majority of fruit trees, including the floodplain oligarchic palms, do not fruit during the dry season (probably a Pleistocene adaptation[2]). Hence, foraging groups' carrying capacity would have been limited by the lack of fruit during the dry season, and their dependence on the availability of other foods would have increased.

Fruit-production phenology may be the most important of the three characteristics examined here. It may also explain why so many species with root and tuber storage organs were domesticated in greater Amazonia,[3] given that these storage organs fill during the late rainy season and attain maximum storage

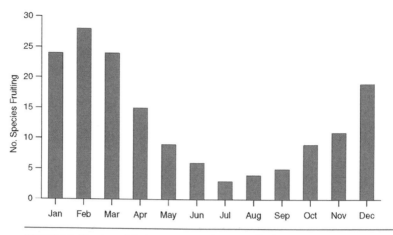

FIGURE 6.1 Number of fruit tree species that generally fruit in eastern Amazonia in each month of the year. The rainy season generally extends from October to May in eastern Amazonia.

capacity (equivalent to maturity for human use) during the dry season (Harlan 1992; Piperno and Pearsall 1998), when fruit resources are less abundant. These species have an additional advantage over fruit species: they can be stored in the field for long periods (at least for the dry season) before harvesting. Without food technologies, fruits must be used when they are ripe; they cannot be stored in the field with no effort.

WATER CONTENT AND FOOD TECHNOLOGY

Storage of fruit products is a common way to overcome the limitations imposed by phenology, but high water contents are often a barrier to efficient preparation for storage. Harlan (1995) emphasizes that both drying and fermentation were early strategies for food storage. Fruit pulps can be stored dry or wet, and nuts and seeds must be stored whole or, if shelled, well dried, in order to avoid rancification (Rosengarten 1984; most consumers interpret as spoilage the oxidation of fats that causes flavor changes) or contamination by aflatoxin-producing fungi. For dry storage, the pulp can be sun dried (difficult in the rainy season, when most species fruit!) or transformed into a coarse meal and dried over fire. For wet storage, the pulp can be fermented and stored for months, if isolated from the atmosphere. In both cases, starchy pulps preserve better and longer than oily pulps, nuts or seeds, which rancify unless well protected from the atmosphere.

Fruit pulp fermentation has uses above and beyond food storage. Adequately managed fermentation produces alcohol used widely to animate group festivities (Patiño 1965). All of the predominantly starchy fruit pulps, as well as roots and tubers, provide excellent substrates for yeasts, and beverages made from both fruits and roots are still common today among the remaining native populations (Patiño 1965). These beverages were likely to have been early reasons to experiment with fruit pulp fermentation because alcohol is a common by-product of fermentation if appropriately managed.

Both storage technologies would have inhibited travel because dry meal must be kept dry (easier when kept in one place) and fermented pulp must be kept well isolated from the atmosphere (easiest when buried in the ground or sealed in ceramic pots). Hence, the benefit-cost ratio of using these technologies before permanent habitation was established may have been considered unacceptable. These technological limitations would have been less severe for affluent foragers because they are hypothesized to have developed sedentary habits before developing food production (Price and Gebauer 1995). The number of low-fat/high-starch pulps among the Amazonian starchy/oily fruits (appendixes 6.2 and 6.3), however, suggests that oil may have constrained processing of these pulps, again helping to explain the number of root/tuber species that were domesticated because the latter could be harvested, processed, and used immediately during the dry season, as well as stored before or after processing. As food production became more important, however, both wet and dry storage certainly became more common.

HABITAT AND DEMOGRAPHY

Many oligarchic species are floodplain palms that have starchy/oily fruits and produce significant quantities of energy and some protein, on both a per tree and per hectare basis, making these important foraging resources (Beckerman 1979). Nonetheless, they are only incipient domesticates at most because there was little need to manage or cultivate them to enhance their yields and use. Peters and colleagues (1989) highlighted oligarchic forest species' potential yields, some of which exceed 10 tons per hectare per year with no management or fertilizer inputs. The clearest examples are the palms babaçu (appendix 6.1), açaí-do-Pará, burití, and patauá (appendix 6.2), and the Myrtaceous camu-camu (appendix 6.4). The few populations that suggest incipient domestication appear to be range extensions, and some selection was probably practiced when seeds were taken from the species' natural range to the human expanded range. Peters (2000) discusses modern management of these oligarchic forests and hypothesizes that incipient domestication may be more common than meets the eye. Morcote-Rios and Bernal (2001) review the archaeological evidence for several of the range expansions of palms, with macauba (appendix 6.2) the most dramatic example—its natural range in south-central Amazonia and the Brazilian Cerrados, and its human expanded range up to Mexico.

At the end of the Pleistocene and during the Early Holocene, the importance of oligarchic palms in the floodplains may have been less than during the later Holocene because the floodplains *(várzeas)* were still in formation (Ab'Saber 1982; Piperno and Pearsall 1998). Açaí-do-Pará, burití, and pataúa were probably gallery forest elements during the Pleistocene, although probably oligarchic even at that time, and only later invaded the expanding *várzeas* and creating the frequently enormous populations seen today in the Amazon River estuary (açaí-do-Pará) and upper Amazon near Iquitos (burití).

Most dispersed species are upland dicots, including all nuts listed here, and they would have been targeted for promotion and management to increase resource availability (Peters 2000), in the process becoming domesticated to some degree. As domestication progressed, they became more useful and more fruitful, potentially increasing their contribution to carrying capacity. Some species also have disease and insect resistance enough to allow them to be promoted, managed, or cultivated to significantly enhance their abundance in the landscape, some even approaching oligarchy in small areas that Balée (1989) calls anthropogenic forests (see also Campbell et al., chapter 1, this volume; Peters 2000). The clearest examples are the dicots Brazil nut and cacao (appendix 6.1); the palms caiaué, macaúba, and tucumã (both species; appendix 6.2); the dicots mari and piquiá (appendix 6.3), bacuri and taperebá (appendix 6.4)—all of which are occasionally found in small high-density populations (Balée 1989; Frikel 1978), similar to the Southeast Asian arboriculture systems (Latinis 2000). The Amazonian anthropogenic forests and Southeast Asian arboriculture systems probably have similar origins in horticulturally based agroforestry systems, the principal differences being the degrees of species diversity and plant management after establishment. The Amazonian anthropogenic forests are often dominated by a single species or a small number of species (an *oligarchy*, although Balée [1989] did not use that term originally) and receive little management today (generally just rudimentary clearing of weeds and brush to facilitate harvest) (Peters 2000). In contrast, the Southeast Asian arboricultural systems tend to be more diverse and are more intensively managed, probably because of the numerous harvests spread through the year for diversity.

A majority of the species that present evidence of some degree of domestication appear to be pioneers or successional species rather than climax species, following the trend of "weeds" becoming crops (Ford-Lloyd and Jackson 1986; Harlan 1992; Lentz 2000; Stahl, chapter 4, this volume). Ecologically, weeds are pioneer species adapted to disturbed and transitional environments and so are "preadapted" to colonize landscapes disturbed by humans and "offer" themselves to humans for whatever use may be made of them (Ford-Lloyd and Jackson 1986). Even Brazil nut is more of a successional than a climax species, although its size and occurrence in apparently climax forest today may suggest otherwise.

DOMESTICATION AND ITS CONSEQUENCES

Both landscape and plant population domestication can change nutritional quality, extend the fruiting season, and increase overall yield. An example is the palm pupunha (also known as peach palm, pejibaye, and chontaduro), the only Amazonian fruit species well studied to provide useful information for this discussion (table 6.2). Its small oily fruit was transformed into a large starchy fruit (1 gram fruit with 40 percent water and 27 percent fats up to 100 grams fruit with 60 percent water and 3 percent fats), making it a much better fermentation substrate. Its small racemes were transformed into large ones (from 0.5 kilograms up to 6 kilograms/bunch), increasing the benefit-cost ratio involved in planting and tending it. Enhanced management probably contributed to extended harvest seasons (although no comparative information is available at present), but not enough to overcome seasonality.

In the other starchy domesticates, for example cutite and mari, and juicy domesticates, such as abiu, biribá, cubiu, and graviola, we can postulate that fruit size, pulp/fruit ratio, and harvest index increased due to domestication, as did starch or sugar content or both. These changes are the types expected in crops during the domestication process (Harlan 1995; Hawkes 1983) and are clearly evident in pupunha (Clement 1988) and other Amazonian fruits (Clement 1989).

With these changes, domestication provided greater yields of more desirable fruit products. In the case of pupunha, the more advanced domesticates are starchy rather than oily, allowing for both more efficient fermentation and greater ease of storage. There is no current evidence that the starchy cutite or oily mari were processed by either fermentation or drying, though the fermentation of cutite would have been perfectly feasible (appendix 6.3).

DOMESTICATION AND ORIGIN

The various regions of Amazonia have not contributed uniformly to fruit tree domestication. I identified two centers, two minor centers, and four regions of

TABLE 6.2 Individual Fruit Weight, Edible Mesocarp, Chemical Composition of the Mesocarp, and Individual Tree Yield of Two Populations of Pupunha Brava (*Bactris gasipaes* var. *chichagui*, Acre and Ucayali) and of Various Landraces of Pupunha (*B. gasipaes* var. *gasipaes*) along the Domestication Continuum

Name	Location	fruit g	edible %	water	prot.	fats	carbo.	fiber	energy kcal	yield kg
						g / 100 g				
Acre	Acre / Rondônia, Brazil	1.0	60	41.1	5.4	27.1	0.3	24.6	365	2
Ucayali	Pucallpa, Peru	1.4	66	42.8	3.4	27.2	12.0	13.8	362	3
Juruá	Cruzeiro do Sul, Acre, Brazil	20	80	54.4	3.1	13.8	19.6	8.4	249	12
Solimões	Coari, Amazonas, Brazil	60	91	42.7	4.1	12.0	31.2	9.3	286	36
Putumayo	Benjamin Constant, Amazonas, Brazil	70	96	52.6	1.9	3.5	38.0	3.2	204	30
Vaupés	São Gabriel, Amazonas, Brazil	80	96	60.9	2.4	2.6	31.6	1.8	167	30

Source: Mora Urpí, Weber, and Clement 1997.

crop genetic diversity in Amazonia at the time of European conquest, but specifically stated that these concentrations of diversity were not necessarily centers of origin (Clement 1999b). Given our still rudimentary knowledge about the origins of native Amazonian crops, a simple division into central and cardinal quadrants allows an appreciation of where and how much domestication may have occurred (table 6.3).

IMPLICATIONS

As Beckerman (1979) argued, the right combination of fruit could provide an adequate diet, even in terms of protein. The additional data supplied here (appendixes 6.1–6.4) support this argument and explain the potential importance of fruit trees before and during the transition to food production, even though seasonally constrained.

Patiño (1964) affirmed that western Amazonia, northwestern South America, and southern Mesoamerica were home to ethnic groups that relied heavily on fruit trees for subsistence, but this reliance appears to have been rare in other neotropical regions—for example, eastern Amazonia—since the number of domesticated fruit trees is small and change due to selection is minor (table 6.3). Although the caloric or subsistence importance of fruit trees in Amazonia, especially in western Amazonia, has not been examined yet, it is unlikely that humans would have domesticated numerous fruit species unless they used them

TABLE 6.3 Probable Origin and Degree of Domestication of the 43 Fruit Species (Appendixes 6.1–6.4) That Have at Least One Population Showing Some Degree of Domestication in Amazonia

Group	n	North	East	South	West	Central	All?
Nuts & seeds	8	–	2 i / 1 s	1 i	1 s	2 i	1 i
Palm pulps	8	1 i	2 i	1 i / 1 s	2 i / 1 d	–	–
Starchy pulps	6	–	1 i	2 i	2 d	1 i	–
Juicy pulps	20	2 s / 1 d	1 i / 3 s	2 i / 1 s	1 i / 4 s / 4 d	1 i	–
Total species	42	4	10	8	15	4	1
Mean dom.*		8	14	10	34	4	1

* *Mean domestication* is the summation of the number of species by each species' degree of domestication, where incipient domestication (i) = 1, semidomestication (s) = 2, domestication (d) = 3. This function is designed exclusively to highlight differences among Amazonian origins in terms of number of species and their degree of domestication.

Source: Probable origin following Clement 1999b. Degree of domestication following Clement 1999a.

extensively for subsistence or pleasure. Hence, it is possible that fruit-based food-production systems were developed in the regions that Patiño identified, but that their users were supplanted by annual food-production systems used before or after European conquest. The abundance of fruit tree domesticates in the northwestern Amazonia center of crop genetic resources (Clement 1989, 1999b) supports Patiño's (1964) hypothesis. Even if fruit trees were extremely important in this area, root and tuber crops were additional staples, and maize became a staple during the past two millennia (Piperno and Pearsall 1998; Roosevelt 1980). Why is this the case?

The implications of fruit-production phenology appear to be most important in explaining the relative loss of importance of fruit trees to human subsistence in Amazonia because energy crops that are available precisely when fruits are unavailable were domesticated and became major sources of food. Fruit nutritional quality certainly contributed to subsistence, but only during harvest season. Water content was an important factor until human populations became more sedentary, when various food-preservation technologies overcame this limitation.

The relative importance of fruit trees, however, certainly varied among localities because each locale would have access to different wild populations for promotion, management, and domestication, and later to different domesticated populations. As the vast majority of crops appear to have originated from single domestication events (Blumler 1992), differential access to these wild populations determined which were domesticated and which not. Lentz (2000) emphasizes that plant seeds were also widely traded throughout the Americas. As the transition to food production progressed and root/tuber resources became more important, succulent fruit trees became the principal domesticates because they contributed variety and pleasure to the diet. The few starchy fruits that were domesticated, such as pupunha, must have been maintained and continued being domesticated for reasons beyond mere subsistence, such as fermentation for festivities, which can be equated to pleasure (intoxication! see Pollan 2001).

ACKNOWLEDGMENTS

I thank William Balée and the Department of Anthropology of Tulane University for the invitation and wherewithal to participate in the stimulating meeting that gave rise to this volume, as well as various colleagues present at the meeting who improved this essay, especially William Balée and Clark Erickson. Special thanks to Jaime P. L. Aguiar, INPA, for unpublished chemical composition data on *Astrocaryum aculeatum*.

APPENDIX 6.1 Individual Fruit Weight, Edible Portion, Chemical Composition of the Edible Portion, and Individual Tree Yield of Several Amazonian Nuts and Seeds in Various Stages of Domestication

Names		fruit	edible	water	protein	fats	carbo.	fiber	energy	yield
Common*	Species	g	%			g / 100 g			kcal	kg
Brazil nut[I]	*Bertholletia excelsa*	1000	15	3.3	16.3	64.0	11.3	1.7	671	100
Cashew nut[S]	*Anacardium occidentale*	30	10	2.7	15.2	37.0	42.0	1.4	533	70
Cutia nut[I]	*Couepia edulis*	45	30	3.6	16.6	74.1	2.5	1.75	743	100
Pendula nut[I]	*Couepia longipendula*	20	25	5.0	10.8	74.0	4.8	3.0	729	50
Sapucaia nut[I]	*Lecythis pisonis*	2000	10	3.5	15.4	49.2	20.3	7.2	586	100
Babaçu[I] – seed	*Attalea speciosa*	110	6	4.2	7.2	66.1	14.5	6.0	682	100
Cacao[S] – seed	*Theobroma cacao*	500	20	3.6	12.0	46.3	34.7	8.6	456	20
Cupuaçu[I] – seed	*Theobroma grandiflorum*	1000	23	5.0	18.9	48.2	15.0	9.1	569	20
Mean				3.9	14.1	57.4	18.1	4.8	621	
St. Deviation				0.8	3.5	13.1	12.9	3.0	95	

* W = wild; I = incipient; S = semi; D = domesticate

Source: Aguiar 1996; FAO 1986; Villachica et al. 1996; Wu Leung and Flores 1961.

APPENDIX 6.2 Individual Fruit Weight, Edible Mesocarp, Chemical Composition of the Mesocarp, and Individual Tree Yield of Eight Amazonian Palms in Various Stages of Domestication

Names		fruit	edible	water	protein	fats	carbo.	fiber	energy	yield
Common*	Species	g	%			g / 100 g			kcal	kg
Açaí-do-Pará[I]	Euterpe oleracea	2	7	41.0	3.6	8.8	22.5	22.9	258	24
Buriti[I]	Mauritia flexuosa	75	23	70.8	2.7	10.5	4.5	10.4	139	150
Caiaué[I]	Elaeis oleifera	7	26	33.5	3.0	16.2	39.1	6.8	341	50
Macaúba[I]	Acrocomia aculeata	27	27	51.8	4.4	13.8	14.5	13.4	231	40
Patauá[I]	Oenocarpus bataua	8	42	35.6	3.3	12.8	15.7	31.5	317	40
Pupunha[D]	Bactris gasipaes	20	80	45.0	3.5	27.0	19.8	3.8	351	16
Tucumã[S]	Astrocaryum aculeatum	47	22	43.0	4.2	39.5	5.9	6.4	396	22
Tucumã[I]	Astrocaryum vulgare	30	22	41.7	3.6	45.5	5.6	2.0	450	20
Mean				45.3	3.5	21.8	16.0	12.2	310	
St. Deviation				11.0	0.5	13.1	10.8	9.5	92	

* W = wild; I = incipient; S = semi; D = domesticate

Source: Aguiar 1996; FAO 1986; Villachica et al. 1996; Wu Leung and Flores 1961.

APPENDIX 6.3 Individual Fruit Weight, Edible Mesocarp, Chemical Composition of the Mesocarp, and Individual Tree Yield of Several Starchy Amazonian Fruits in Various Stages of Domestication

Names		fruit	edible	water	protein	fats	carbo.	fiber	energy	yield
Common*	Species	g	%			g / 100 g			kcal	kg
Cutite[D]	Pouteria macrophylla	100	70	75.6	1.7	0.5	19.3	1.0	92	20
Cutite grande[I]	Pouteria macrocarpa	150	70	62.4	2.1	0.9	32.7	1.0	143	20
Jatobá[I]	Hymenaea courbaril	30	10	14.6	5.9	2.2	75.3	13.4	309	30
Mari[D]	Poraqueiba sericea	70	25	55.6	2.3	17.6	16.7	7.4	281	100
Pajurá[I]	Couepia bracteosa	100	10	67.6	1.8	0.1	27.8	2.0	127	10
Piquiá[I]	Caryocar villosum	300	13	41.9	1.4	22.7	26.9	6.7	358	120
Uxi[W]	Endopleura uchi	50	40	48.9	2.2	10.1	38.5	20.5	252	300
Uxi coroa[W]	Duckesia verrucosa	70	25	42.5	2.7	12.1	21.7	20.0	287	100
Mean				51.1	2.5	8.3	32.4	9.0	231	
St. Deviation				17.7	1.3	8.2	17.5	7.6	91	

* W = wild; I = incipient; S = semi; D = domesticate

Source: Aguiar 1996; FAO 1986; Villachica et al. 1996; Wu Leung and Flores 1961.

APPENDIX 6.4 Edible Portion, Chemical Composition of the Edible Portion, and Individual Tree Yield of Various Amazonian Fruits in Various Stages of Domestication

Names		fruit	edible	water	protein	fats	carbo.	fiber	energy	yield
Common*	Species	g	%			g / 100 g			kcal	kg
Abiu[D]	Pouteria caimito	150	70	82.0	0.8	1.6	12.0	2.5	66	75
Araçá-boi[S]	Eugenia stipitata	200	80	90.0	0.6	0.2	8.3	0.6	40	25
Araticum[I]	Annona montana	1000	50	86.8	0.4	1.6	6.5	3.8	52	35
Bacuri[S]	Platonia insignis	400	10	70.0	1.8	2.0	19.7	5.3	125	200
Biribá[D]	Rollinia mucosa	1000	60	85.0	1.1	0.4	11.7	1.2	53	40
Camu–camu[W]	Myrciaria dubia	10	50	93.1	0.5	0.1	5.5	0.6	24	10
Cashew apple[S]	Anacardium occidentale	30	55	87.5	0.8	0.3	9.6	1.4	46	70
Cubiu[D]	Solanum sessiliflorum	40	60	92.0	0.6	1.4	4.7	0.4	35	4
Cupuaçu[I]	Theobroma grandiflorum	1000	35	85.3	1.2	0.4	10.5	1.8	58	20
Graviola[D]	Annona muricata	1000	56	83.1	1.0	0.4	13.8	1.1	60	30
Guava[S]	Psidium guayava	100	70	80.8	0.9	0.4	12.0	5.3	69	30
Inga[S]	Inga edulis	450	20	84.7	0.8	0.1	7.1	7.0	60	100
Genipapo[S]	Genipa americana	300	60	83.9	1.2	0.1	12.4	1.6	55	90
Mapati[S]	Pourouma cecropiifolia	4	26	87.2	0.3	0.2	10.0	1.9	47	20
Murici[S]	Byrsonima crassifolia	4	40	82.8	0.9	1.3	12.2	2.2	66	15

APPENDIX 6.4 (*continued*)

Names		fruit	edible	water	protein	fats	carbo.	fiber	energy	yield
Common*	Species	g	%			g / 100 g			kcal	kg
Pitanga[I]	*Eugenia uniflora*	4	66	85.8	0.8	0.4	11.9	0.6	51	3
Pitomba[I]	*Talisia esculenta*	5	20	78.4	1.1	0.2	18.5	1.4	59	20
Puruí[S]	*Borojoa sorbilis*	500	80	64.7	1.1	0.1	24.6	8.3	104	5
Sapota[S]	*Quararibea cordata*	300	34	81.1	1.0	0.4	11.3	5.5	71	100
Sorva[I]	*Couma utilis*	15	80	72.5	1.2	2.9	14.5	8.4	123	35
Taperiba[S]	*Spondias mombim*	10	36	82.7	0.8	2.1	12.8	1.0	70	100
Mean				82.8	0.9	0.8	11.9	2.9	63	
St. Deviation				6.7	0.3	0.8	4.6	2.5	25	

* W = wild; I = incipient; S = semi; D = domesticate

Source: Aguiar 1996; FAO 1986; Villachica et al. 1996; Wu Leung and Flores 1961.

	Jan	Feb	Mar	Apr	May	Jun	Jul	Aug	Sep	Oct	Nov	Dec
PALMS												
Açaí-do-Pará	–	–	–	–	–	–	–	x	x	x	x	x
Bacaba	–	x	x	x	–					–	–	–
Buriti	x	x	x	x	x	x	–			–	–	–
Patauá	x	x	x	–					–	x	x	x
Pupunha	x	x	x	–					x	x	–	x
Tucumá (A.a.)	–	x	x	x	x	–					–	–
Tucumá (A.v.)	x	x	x	x	x	x	–				–	–
NUTS												
Brazil nut	x	x	x	x	–							x
Caju	–	–			–	x	x	x	x	x	–	–
Pendula nut	x	x	x	–								
Sapucaia	–	x	x	x	–							–
STARCHY FRUITS												
Jatobá	x	x	x	–								–
Mari	x	x	x	x	–							
Piquia		–	x	x	x	x	–					
Uxi	x	x	x	x	x	–					–	x
Uxi coroa	x	x	x	–								–
JUICY FRUITS												
Abiu	x	x	–							–	x	x
Araçá-boi	–	x	x	x	x	–					–	–
Bacuri	x	x	x	x	x							–
Biribá	x	x	x	x	–	–	–	–	–	–	–	x
Cacao	x	x	x	x	x	x	–	–	–	–	x	x
Cupuaçu	x	x	x	x	–						–	x
Cutite	x	x	–						–	x	x	x
Cutite grande	–						x	x	x	x	x	x
Genipapo	x	x	x	–					–	x	x	x
Graviola	–	–	–	–	–	–	x	x	x	x	–	–
Guava	x	–		–	x	x	–			–	x	x
Ingá	x	x	x	–	–				–	x	x	x
Mapati	x	x	–								–	x

APPENDIX 6.5 (*continued*)

	Jan	Feb	Mar	Apr	May	Jun	Jul	Aug	Sep	Oct	Nov	Dec
Murici	x	x	x	–							–	x
Pitomba	x	x	–								–	x
Sapota		x	x	x								
Sorva	x	x	–							–	x	x
Tarepebá	x	x	x	–	–						x	x
FREQUENCY	24	28	24	15	9	6	3	4	5	9	11	19

Note: See appendixes 1–4 for scientific names.
Source: Cavalcante 1976.

NOTES

1. From *oligarchy*—dominance by a few. In this case, it means forests that are dominated by one or a few species (Peters et al. 1989). See also Campbell et al., chapter 1, this volume.
2. The Pleistocene adaptation is hypothesized because pronounced dry seasons occurred throughout the Pleistocene when the floodplains were minimal or nonexistent, whereas the Holocene floodplains appeared only in the past few millennia (Ab'Saber 1982). Because evolutionary change for ecological adaptation may take millennia when pressures are not significant, this adaptation appears to be a Pleistocene remnant.
3. Examples are *Xanthosoma sagittifolium* (L.) Schott, Araceae; *Canna edulis* Ker., Cannaceae; *Ipomoea batatas* (L.) Lam., Convolvulaceae; *Dioscorea trifida* L. f. and *D. dodecaneura* Steud., Dioscoreaceae; *Manihot esculenta* Crantz, Euphorbiaceae; *Pachyrhizus tuberosus* Spreng., Leguminosae Papilionoideae; *Calathea allouia* (Aubl.) Lindl., *Maranta arundinacea* L., *M. ruiziana* Korn., Marantaceae; *Heliconia hirsuta* L. f., Heliconiaceae (Clement 1999a).

REFERENCES

Ab'Saber, A. N. 1982. The paleoclimate and paleoecology of Brazilian Amazonia. In G. T. Prance, ed., *Biological Diversification in the Tropics,* 41–59. New York: Columbia University Press.

Aguiar, J. P. L. 1996. Tabela de composição de alimentos da Amazônia. *Acta Amazonica* 26:121–126.

Balée, W. 1989. The culture of Amazonian forests. In D. A. Posey and W. Balée, eds., *Resource Management in Amazonia: Indigenous and Folk Strategies,* 1–21. Advances in Economic Botany no. 7. Bronx: New York Botanical Garden.

——. 1998. Historical ecology: Premises and postulates. In W. Balée, ed., *Advances in Historical Ecology,* 13–29. New York: Columbia University Press.

Beckerman, S. 1979. The abundance of protein in Amazonia: A reply to Gross. *American Anthropologist* 81:533–560.

Blumler, M. A. 1992. Independent inventionism and recent genetic evidence on plant domestication. *Economic Botany* 46:98–111.

Cavalcante, P. B. 1976. *Frutas comestíveis da Amazônia.* 3rd ed. Manaus: Instituto Nacional de Pesquisas da Amazônia.

Clement, C. R. 1988. Domestication of the pejibaye palm *(Bactris gasipaes):* Past and present. In M. J. Balick, ed., *The Palm—Tree of Life: Biology, Utilization, and Conservation,* 155–174. Advances in Economic Botany no. 6. Bronx: New York Botanical Garden.

——. 1989. A center of crop genetic diversity in western Amazonia. *BioScience* 39:624–631.

——. 1999a. 1492 and the loss of Amazonian crop genetic resources. I. The relation between domestication and human population decline. *Economic Botany* 53:188–202.

——. 1999b. 1492 and the loss of Amazonian crop genetic resources. II. Crop biogeography at contact. *Economic Botany* 53:203–216.

——. 2004. Fruits. In G. T. Prance and M. Nesbitt, eds., *The Cultural History of Plants,* 77–95. London: Routledge.

Food and Agriculture Organization (FAO). 1986. *Examples from Latin America.* Vol. 3 of *Food and Fruit-Bearing Forest Species.* FAO Forestry Paper 44/3. Rome: FAO of the United Nations, Forestry Department.

Ford-Lloyd, B., and M. Jackson. 1986. *Plant Genetic Resources: An Introduction to Their Conservation and Use.* London: Edward Arnold.

Frikel, P. 1978. Áreas de arboricultura pré-agrícola na Amazônia: Notas preliminares. *Revista Antropológica* 21:45–52.

Gross, D. R. 1975. Protein capture and cultural development in the Amazon basin. *American Anthropologist* 77:526–549.

Harlan, J. R. 1992. *Crops and Man.* 2d ed. Madison, Wisc.: American Society of Agronomy, Crop Science Society of America.

——. 1995. *The Living Fields: Our Agricultural Heritage.* Cambridge, U.K.: Cambridge University Press.

Hawkes, J. G. 1983. *The Diversity of Crop Plants.* Cambridge, Mass.: Harvard University Press.

Kirch, P. V. 1989. 2nd millennium BC arboriculture in Melanesia : Archaeological evidence from the Mussau Islands. *Economic Botany* 43:225–240.

Latinis, K. D. 2000. The development of subsistence system models for island Southeast Asia and Near Oceania: The nature and role of arboriculture and arboreal-based economies. *World Archaeology* 32:41–67.

Lentz, D. L. 2000. Anthropocentric food webs in the Precolumbian Americas. In D. L. Lentz, ed., *Imperfect Balance: Landscape Transformations in the Precolumbian Americas,* 89–119. New York: Columbia University Press.

Lepofsky, D. 1992. Arboriculture in the Mussau Islands, Bismarck Archipelago. *Economic Botany* 46:192–211.

Meggers, B. J. 1971. *Amazonia: Man and Culture in a Counterfeit Paradise.* Chicago: Aldine.

Mora Urpí, J., J. C. Weber, and C. R. Clement. 1997. *Peach Palm: Bactris gasipaes Kunth.* Promoting the Conservation and Use of Underutilized and Neglected Crops no. 20. Gatersleben and Rome: Institute of Plant Genetics and Crop Plant Research and International Plant Genetic Resources Institute.

Morcote-Rios, G., and R. Bernal. 2001. Remains of palms (Palmae) at archaeological sites in the New World: A review. *Botanical Review* 67:309–350.

Patiño, V. M. 1964. *Frutales.* Vol. 1 of *Plantas cultivadas y animales domésticos en América equinoccial.* Cali, Colombia: Imprenta Departamental.

——. 1965. *Historia de la actividad agropecuaria en América equinoccial.* Cali, Colombia: Imprenta Departamental.

Peters, C. M. 2000. Precolumbian silviculture and indigenous management of neotropical forests. In D. L. Lentz, ed., *Imperfect Balance: Landscape Transformations in the Precolumbian Americas*, 203–223. New York: Columbia University Press.

Peters, C. M., M. J. Balick, F. Kahn, and A. B. Anderson. 1989. Oligarchic forests of economic plants in Amazonia: Utilization and conservation of an important tropical resource. *Conservation Biology* 3:341–349.

Piperno, D. R., and D. M. Pearsall. 1998. *The Origins of Agriculture in the Lowland Neotropics*. San Diego: Academic Press.

Pollan, M. 2001. *The Botany of Desire: A Plant's-Eye View of the World*. New York: Random House.

Price, T. D., and A. B. Gebauer. 1995. New perspectives on the transition to agriculture. In T. D. Price and A. B. Gebauer, eds., *Last Hunters, First Farmers: New Perspectives on the Prehistoric Transition to Agriculture*, 3–19. Santa Fe, N.Mex.: School of American Research Press.

Roosevelt, A. C. 1980. *Parmana: Prehistoric Maize and Manioc Subsistence along the Amazon and Orinoco*. San Diego: Academic Press.

Roosevelt, A. C. 1999. Twelve thousand years of human-environment interaction in the Amazon floodplain. In C. Padoch, J. M. Ayres, M. Pinedo-Vasquez, and A. Henderson, eds., *Várzea: Diversity, Development, and Conservation of Amazonia's Whitewater Floodplains*, 371–392. Advances in Economic Botany no. 13. Bronx: New York Botanical Garden Press.

Rosengarten, F., Jr. 1984. *The Book of Edible Nuts*. New York: Walker.

Tudge, C. 1999. *Neanderthals, Bandits, and Farmers: How Agriculture Really Began*. New Haven. Conn.: Yale University Press.

Villachica, H., J. E. Urano Carvalho, C. H. Müller, C. Díaz S., and M. Almanza. 1996. *Frutales y hortalizas promisorios de la Amazonia*. Lima, Peru: Tratado de Cooperación Amazonica.

Wiersum, K. F. 1997. From natural forest to tree crops: Co-domestication of forests and tree species, an overview. *Netherlands Journal of Agricultural Science* 15:425–438.

Wu Leung, W.-T., and M. Flores. 1961. *Tabla de composición de alimentos para uso en América Latina*. Guatemala City, Guatemala: INCAP-ICNND.

Yen, D. E. 1974. Arboriculture in subsistence of Santa Cruz, Solomon Islands. *Economic Botany* 28:247–284.

7

THE HISTORICAL ECOLOGY OF A COMPLEX LANDSCAPE IN BOLIVIA

CLARK L. ERICKSON AND WILLIAM BALÉE

IDEAL STUDIES OF the interface between people and the environment involve collaboration among scholars from diverse backgrounds to understand how complex landscapes came to be over periods of hundreds, if not thousands, of years. Historical ecological research displays such interdisciplinary underpinnings because it is situated at the interface and overlap of the social sciences, natural sciences, and humanities in the programmatic study of how the typically complex interactions of nature and culture have become historical and cultural landscapes. Such diversity is in evidence with regard to the scholars who have contributed to this volume, from archaeology, ethnography, ethnobotany, genetics, geography, biology, and ecology.

The historical ecology applied in this chapter underscores how to read physical signatures and material patterning, including similarity, disjuncture, and anomaly, embedded in a cultural landscape in the Bolivian Amazon. On heterogeneous landscapes of considerable age, these signatures may form a complex palimpsest, a vertical or horizontal layering or stratigraphy of signatures and patterning etched on the surface of the earth, deep into the soil (Erickson, chapter 8, this volume; or, as in the case of Amazonian Dark Earth, Glaser and Woods 2004; Lehmann et al. 2003), or above the surface fixed in the layers of vegetation extending up through the emergent canopy (Balée 1989, 1994; Campbell et al., chapter 1, this volume). The multiple scales of study range from microorganisms and soil chemistry (in the case of Amazonian Dark Earth), stratigraphic profiles of anthropogenic soil horizons, to rearrangements of local hydrology, drainage, and water tables to the structure and physiognomy of the living forest and landscape more generally—that is, to the terrestrial and aquatic plants and animals living below, on, and above the earth's surface (chapters 2, 4, and 6, this volume).

This particular historical ecological study is of a relatively medium-scale cultural landscape of pre-Columbian earthworks that consists of mounds, causeways, canals, ponds, and the present vegetation growing on them in the Llanos de Mojos region of the Bolivian Amazon. The investigation involved an ethnobotanist/ethnographer (Balée), a team of archaeologists (Erickson, Bolivian coinvestigator Wilma Winkler, ethnographer Kay Candler, and then-students John Walker and Marcello Canuto), together with a group of indigenous Sirionó. It soon became apparent to us that the ethnobotanist, archaeologist, and native person approach the same cultural landscape from different perspectives. In reading and interpreting an anthropogenic landscape, the ethnobotanist tends to look upward and then sees a cornucopia of plants and their myriad structures, together with plant parts: bole, branch, crown, bark, leaves, fruits, flowers, and seeds. In contrast, the archaeologist has a knack for staring at the ground, kicking leaf litter to expose bare ground, evaluating the topography of the surface, and rooting around under the roots of tree falls searching for pottery fragments, stone tool debris, charcoal, and any other signs of previous human activity. Although both are practicing science in the present, the archaeologist is thinking of the distant past and the ethnobotanist is contemplating the present and somewhat more recent past. When an ethnobotanist and archaeologist work together—seeking to merge their interpretations of the same landscape into a holistic framework that will betoken an enriched and deeper, more complex notion of the past—sparks fly, debate ensues, and significant new insights gradually emerge as one scholar learns from the other. Their hosts, the Sirionó, however, seem to be scanning the ground and sky in search of spoor left by the animals that they prize as game, traces of overgrown trails, and evidence of past campsites and farmsteads marking good agricultural soils and memories that link the anthropogenic past and present in a practical way into their remarkable landscape mosaic of forest, savanna, and wetland.

Erickson and Balée met on the streets of the tropical Bolivian city of Trinidad in 1994. Erickson and his Bolivian–U.S. team had been studying the cultural landscapes of earthworks and settlements since 1990 (Erickson, chapter 8, this volume). Balée had been investigating the economic plants, soils, languages, and management practices of Tupí-Guaraní-speaking peoples in the Amazon basin for more than ten years (Balée 1994). After finding out that Erickson was an archaeologist, Balée invited Erickson and his team to meet authorities of the Sirionó Indigenous Territory, where Balée had been collecting data on a plot deep under the canopy forest intermittently over the previous two years. What made this invitation interesting for Erickson the archaeologist was that the forest thrived on a *loma* or large, earthen, pre-Columbian settlement mound.[1] Balée, the ethnobotanist, needed long-term historical context in order to understand the human activities that structured the plant communities growing on the mound. A meeting was arranged with the Sirionó community in Ibiato (also located on a large pre-Columbian settlement mound). Some 80 people attended the meeting, and after four hours of public presentation, debate, and deliberation the Sirionó gave them permission to map and study the mound. Aside from children and those not directly involved in the study, 10 adult Sirionó worked with us at the Ibibate Mound Complex.

What makes the Ibibate Mound Complex an excellent case study in historical ecology and for collaboration between an archaeologist and ethnographer/ ethnobotanist is the medium-scale context of a heterogeneous cultural landscape well defined in time and space. We focus on comparing the biodiversity of two distinct environments (mound versus pampa) formed by different regimes of human activities. Differences in species composition and diversity demonstrate different degrees and scales of past human transformation and management of the Ibibate Mound Complex and surrounding anthropogenic landscape. Mapping and chronological dating provide high-definition spatial boundaries of the complex and heterogeneous topography of anthropogenic earthworks and natural geomorphological features of the landscape.

MOUNDS AND OTHER PRE-COLUMBIAN EARTHWORKS OF THE BOLIVIAN AMAZON

The Llanos de Mojos (Department of the Beni, Bolivia) is a prominent example of an anthropogenic landscape covered with earthworks such as raised fields, causeways, canals, reservoirs, fish weirs, and mounds (Erickson, chapter 8, this volume). Large artificial earthen mounds, locally referred to as lomas, are common in the Bolivian Amazon, especially near the city of Trinidad and along the Mamoré, Ibare, and Apere rivers and their tributaries (Denevan 1966; Erickson 2000c). In addition, smaller mounds and related forest islands cover the savannas and gallery forests of the Llanos de Mojos (Department of the Beni) (Erickson 2000c; Langstroth 1996). Archaeological studies show that mounds were originally used as settlements and for the burial of the dead. Some may have had important ritual functions as ceremonial and political centers. Mounds are important motifs in indigenous mythology of the region (Riester 1976) and local public imagination (Pinto Parada 1987).[2] Although most are neglected and overgrown with vegetation, some mounds support modern ranches, homesteads, fields, and even entire communities and a military base. Mounds have been the traditional focus of archaeologists since the first excavations of Erland Nordenskiöld (1910, 1913, 1924) at the turn of the twentieth century. Since that time and until the present, numerous national and international projects have explored mounds through survey, mapping, and excavation. Stig Rydén (1941, 1964) and Wanda Hanke (1957) reported pottery found on mounds in the Casarabe area. Geographer William Denevan (1966, 2001) was the first to provide a detailed inventory of mounds and other earthworks of the region. Research by local scholars such as Kenneth Lee (1979), Rodolfo Pinto Parada (1987), Mario Vilca, Ricardo Bottega, and others has raised public awareness of these features. The Argentinean-Bolivian team of Bernardo Dougherty, Horacio Calandra, Victor Bustos, and Juan Faldín surveyed, mapped, and excavated mounds in the Trinidad region during the late 1970s and 1980s (Bustos 1978a, 1978b, 1978c, 1978d; Dougherty and Calandra 1981, 1981–82, 1984, 1985; Faldín 1984), along with other scholars (Fernandez Distel 1987a; Pia 1983; Vejarano Carranza 1991). More recently, archaeologist Heiko Prümers and his German-Bolivian team has mapped and meticulously excavated the Mendoza Mound near Casarabe (Prümers 2000, 2001, 2002a, 2002b). Robert Langstroth's (1996) study of the vegetation, soils, and formation of mounds and forest islands is a landmark study. The U.S.–Bolivian team directed by Clark Erickson and Wilma Winkler has also reported mounds in addition to raised fields, fish weirs, causeways, and canals (Erickson 1980, 1995, 2000a, 2000b, 2000c, 2001, and chapter 8, this volume; Erickson et al. 1991; Erickson, Candler, Walker et al. 1993; Erickson, Candler, Winkler et al. 1993; Erickson, Winkler, and Candler 1997; Walker 2004).

ARCHAEOLOGICAL INVESTIGATIONS OF THE IBIBATE MOUND COMPLEX

The Ibibate Mound Complex is located in the Sirionó Indigenous Territory (Territorio Indígena de los Sirionó), Canton of San Javier, Cercado Province, Department of the Beni, Bolivia (14°48'14.47" S/64° 24' 30.53" W) (figure 7.1). The complex, made up of two large earthen mounds (lomas) referred to here as Mound 1 and Mound 2, are major anthropogenic topographic features on the local landscape. Today, the mounds are covered with old fallow forest. Community members have periodically used the mounds for swidden fields, gardens, temporary hunting camps, agroforestry, and ephemeral farmsteads over the past 60 years, although the evidence of these activities is subtle, and recent disturbance was minimal in the area of the vegetation studies, partly because certain long-lived, economic trees grow on the mound and only there, according to Sirionó informants themselves, so that to practice slash and burn at such a locale would destroy the resource. Associated earthworks include barrow pits, ponds (*pozas*), causeways (*terraplenes, calzadas*), and canals (*canales, zanjas*). Two impressive straight causeways with adjacent canals cross the savanna (pampa) and wetlands to the west of the forested mound complex. In the Sirionó language, *ibi* means "earth" (*junto al pie, tierra, planta*) (Coimbra Sanz 1980:8) and *ebate* means "high" or "tall" (*alto*) (Monje Roca 1981:51), so that, combined, Ibibate translates as "high earth mound."[3] Erickson's research

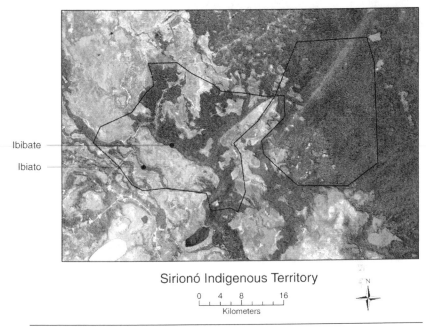

Ibibate

Ibiato

Sirionó Indigenous Territory

0   4   8   16
Kilometers

N

FIGURE 7.1 Location of the Ibibate Mound Complex, Bolivia.

team mapped the Mound Complex in July 1994,[4] while ethnobotanist Balée continued his multiyear study of its anthropogenic vegetation, which began in 1993. Due to dense vegetation and time limitations, we were unable to make a detailed topographic map of the entire Ibibate Mound Complex.

Sirionó informants identified more than 30 large mounds in the immediate area of the Ibibate Mound Complex and report many more within the nearby San Pablo Forest to the east. The mounds of the Sirionó Indigenous Territory are related in size, shape, and general artifact types to the numerous large mounds of the Casarabe, Loreto, and Trinidad regions to the north, west, and south. This zone has the highest concentration of large and medium-size mounds in the Bolivian Amazon The similar large mound, Loma Alta de Casarabe (16 meters tall and covering more than 10 hectares), is located less than 20 kilometers from the Ibibate Mound Complex (Dougherty and Calandra 1981–82; Rydén 1941).

MOUND 1

Mound 1 is located in the northern half of the Ibibate Mound Complex (figures 7.2 and 7.3). The base of Mound 1 covers approximately 2.25 hectares and has a volume of 75,294 cubic meters (table 7.2). A large marsh

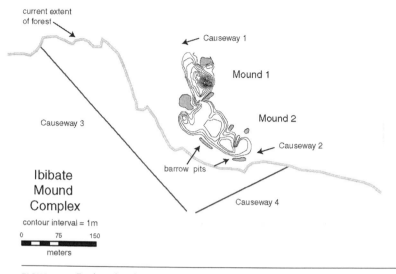

current extent
of forest

Causeway 1

Mound 1

Causeway 3

Mound 2

Causeway 2

Ibibate
Mound
Complex

barrow pits

Causeway 4

contour interval = 1m

0   75   150

meters

FIGURE 7.2 Earthworks of the Ibibate Mound Complex and surrounding cultural landscape: Mound 1, Mound 2, Causeway 1, Causeway/Canal 2, Causeway/Canal 3, Causeway/Canal 4, barrow pits, and ramps.

(*bajio*, or *curiche*) in the savanna to the west abuts the western edge of Mound 1. The orientation of the long axes of the Mound 1 and Mound 2 is northwest to southeast. Mound 1 has a gentle slope from the southeast to northwest and a steep slope on the northeast- and east-facing sides.

Mound 1 consists of at least two superimposed platforms. The 85- and 89-meter contours approximate the base and upper surface of the large lower platform. The footprint of the lower platform is "teardrop" in shape. The upper platform is located on the surface of the southeast end of the lower platform. The base of the upper platform is kidney bean shaped in footprint and follows roughly the 91-meter contour. The upper platform rises 9 meters above the surface of the lower platform. The highest point of the upper platform is 18 meters above the base of the barrow pits surrounding the mound, 15 meters above the present surface of the gallery forest, or 16 meters above the original surface of the gallery forest (based on posthole transects). Mound 1 has two "ramplike" features of earth (figure 7.2). The first begins at the northwest corner of the mound and ends at the highest point of the upper platform (figure 7.3). The other begins at Causeway 1 between Mound 1 and Mound 2 (described later) and ends at the highest point of the upper platform (figures 7.2 and 7.3).

A small extension of the lower platform (or possibly a separate mound that later merged with Mound 1) is located on the northeast side of Mound 1. This oval platform measures 30 meters long by 20 meters wide by 2.5 meters tall.

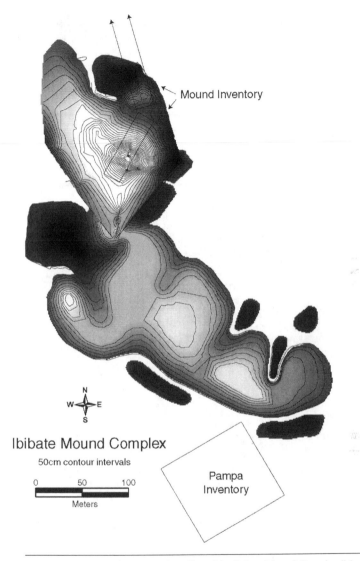

Mound Inventory

N
W — E
S

Ibibate Mound Complex

50cm contour intervals

0   50   100

Meters

Pampa
Inventory

FIGURE 7.3 Topographic map and profiles of the Ibibate Mound Complex (Mound 1, Mound 2, and associated features). Mound 1 was accurately mapped with a Topcon EDM transit; thus, contour lines are precise. Mound 2 was mapped with compass and tape, and contour lines were estimated. Contour interval is 0.5 meter.

TABLE 7.1 Description of Soil from the Ibibate Mound Complex (0–20 Centimeters Below Surface)

Inventory	pH	EC	Free Cations	Ca	Mg	Na	K	Sum Bases	CEC	Base Sat	Acidity	Al	P	O.M.	Total N	Sand	Silt	Clay	Texture
1.05		[MS cm-1]		Exchangable Cations [cmol kg-1]				[cmol kg-1]	[cmol kg-1]	%	[cmol kg-1]	[cmol kg-1]	[mg kg-1]	%	%	%	%	%	
Mound	6.3	195	absent	6.7	2.4	0.13	1	10.2	10.4	98	0.2		59	2.9	0.22	28	47	25	Silt
Pampa	4.5	144	absent	1.4	2	0.33	0.4	4.1	6.3	65	2.2	0.9	7	2.2	0.17	26	49	25	Silt

MOUND 2

Mound 2 is located in the southern part of the Ibibate Mound Complex (figures 7.2 and 7.3). It was mapped with compass and tape, so the map is not as detailed or accurate as that of Mound 1. The long axis of Mound 2 is oriented northwest to southeast. Mound 2 is composed of two or more superimposed platforms covering an estimated 9.33 hectares and having an estimated volume of 177,357 cubic meters (table 7.2). The edge of the lower platform of Mound 2 is irregular, with many projecting ridges from the long northwest–southeast axis, some surrounded by ponds, remnants of the abandoned river channel and artificial barrow pits dug for earthen fill to construct the mound. The lower platform is approximately 3 meters tall (85-meter to 88-meter contours). The broad flat surface of the lower platform of Mound 2 would have been a better location for houses than the steeper and more irregular surface of Mound 1. The upper platform is found in two separate locations: a 2-meter-tall cone-shaped extension to the northeast and a 1–2 meter high ridgelike spin along the long axis of Mound 2. Both platforms grade steeply toward the forest and savanna to the east and west. The highest location on the northeast part of the mound was cleared for house site and garden approximately 20 years ago, but now is abandoned and overgrown. The Sirionó regularly use a campsite *(pascana)* located at the southeastern tip of Mound 2 while hunting and gathering in the forest. The irregular surface and shape of Mound 2 may represent the aggregation of three to four individual mounds that grew into a single large mound over time.

BARROW PITS AND PONDS

Seven to eight water-filled depressions *(pozas, lagunas)* and the scars of one or two abandoned river channels or meander scars *(arroyos, cañadas)* surround the Ibibate Mound Complex. Topographic survey suggests that for several weeks during the height of the rainy season, most of the terrain adjacent to the Ibibate Mound Complex would be inundated with shallow water or waterlogged. The deepest depressions are to the west, northeast, and east of Mound 1 and to the west and southwest of Mound 2. Several of these depressions are circular and cover 0.25 hectares in area. The majority of the smaller ponds are curvilinear or elliptical in form. The depths of these water bodies in July 1994 (the dry season) varied from 0.5 to 2 meters deep. These features are clearly the barrow pits for the soil used to construct the massive mound complex. These depressions were made in the base of the channel of a long-abandoned river that meanders around the northeast and east sides of Mound 1 and the east side of Mound 2. Traces of a second abandoned river channel are found on the west and southwest side of

Mound 1 and on the west side of Mound 2. Water originally may have flowed between Mound 1 and Mound 2 before Causeway 1 was constructed, which blocked the channel. The unmodified abandoned river channels and the modified barrow pits probably provided year-round sources of drinking and bathing water for the residents of the Ibibate Mound Complex, in addition to a web of water networks for canoe transportation during the rainy season and a source of fish and other aquatic resources. These artificial water bodies would have been connected to natural water bodies in the forest and savanna during typical annual flooding during the rainy season.

The barrow pits do not have sufficient volume to account for the volume of earth in the mounds of the Ibibate Mound Complex. They may have originally been much deeper. Over time, they became dumps for garbage generated by the mound's inhabitants and collected soil and detritus that eroded off the steep sides of the two mounds. After settlement abandonment, natural processes of erosion and sedimentation and use of the mounds for slash-and-burn fields may have also contributed to reducing the size of the barrow pits. Despite these factors, large quantities of earth fill for the construction of the mounds must have been brought in from beyond the barrow pits and other adjacent low spots (although not necessarily from long distances). Topsoil removed as mound fill from the now forested landscape around the complex may have increased risk of flooding of house sites, gardens, and fields during the rainy season and reduced fertility.

CAUSEWAYS AND CANALS

We documented four sets of pre-Columbian causeways (*terraplenes, calzadas*) and canals (*canales, zanjas*) at the Ibibate Mound Complex (figures 7.2 and 7.4). Earthen causeways in the Bolivian Amazon are usually accompanied by one or two parallel adjacent canals from which earth was removed to construct the raised platform (Erickson 2000b, 2001). Causeways were used for pedestrian traffic, and canals provided channels for canoes. In addition to use for transportation and communication, causeways may have also controlled and managed water as dikes, diversion dams, fish weirs, and reservoir walls; delineated family and community fields or territories; and served ritual and political functions.

A deep, steep-sided ditch oriented northeast–southwest and measuring approximately 15 meters wide and 4 meters deep separates Mound 1 and Mound 2 (figure 7.2). This depression may originally have been the channel of a river or stream. Although this gap between the mounds could easily have been filled in, the builders of the mounds intentionally maintained the division between the two mounds that tower above it. Causeway 1, an artificial earthen structure now used as a trail between the mounds, bridges this ditch. The causeway measures

FIGURE 7.4 Oblique aerial photograph of Causeway/Canal 4 in the savanna west of Mound 1 and Mound 2. Note the line of motacú palms growing on the causeway.

3 meters wide at the top and 6 meters wide at the base, 15 meters long, and 3 meters high. The causeway orientation is northwest–southeast (108°). Causeway 1 is linked directly to the summit of Mound 1 by a ramplike ridge (figure 7.2).

Causeway/Canal 2, a short segment of a large well-preserved causeway and canal, is located between the north edge of Mound 1 and the abandoned river channel to the northeast (figure 7.2). This causeway measures 4 meters wide and 0.5 meters high. The adjacent canal measures 4 meters wide and 0.5 meter deep. Because of dense vegetation cover, the length and destination of Causeway/Canal 2 is unknown. It is not oriented toward Mound 1.

Causeway/Canal 3 is located in the savanna (pampa) and roughly parallel to the savanna-forest boundary west of Mound 1 and Mound 2 (figure 7.2). This causeway/canal is approximately 300 meters long, intersects with Causeway/Canal 4, and disappears into a wetland to the southwest of the mound complex. Its azimuth (135° or 315°) is oriented toward the Santa Fe Mound to the south.

Causeway/Canal 4 is located in the savanna to the southwest of Mound 2 (figures 7.2, 7.4). This badly eroded causeway (3 meters wide and 0.3 meter tall) is clearly defined by vegetation-filled canals along both sides and supports a line of motacú palms (*Attalea phalerata* Mart. ex. Spreng) and woody shrubs. The causeway is straight and oriented 245°. The causeway/canal is visible in the savanna for a distance of 100 meters and disappears into a wet land. Causeway/Canals 4 is not aligned to the Ibibate Mound Complex and enters the gallery forest approximately 50 meters from the southwest edge of Mound 2, where it disappears. The east end of the causeway aligns with the Ibiato Mound 10 kilometers across the pampa. The pioneer ethnographer of the Sirionó, Allan Holmberg, may have been referring to this causeway or to similar causeways when he mentioned that "rows of palm are sometimes encountered in the pampa" ([1950] 1985:7).

Causeway/Canal 4 also defines the straight southern boundary of a shallow wetland (*curiche* or *yomomo*) (figures 7.2, 7.4). The construction of the causeway may have intentionally or unintentionally impounded water and artificially created the wetland by blocking the natural flow of water from southeast to northwest in the savanna. Similar cases of water management have been documented in the Bolivian Amazon (Erickson 2000b, 2001, and chapter 8, this volume).

SURFACE COLLECTIONS OF ARTIFACTS

Cultural material, primarily fragments of pottery, is found on all exposed surfaces of Mound 1. Few potsherds are found on the vegetation-covered and less-disturbed surface of Mound 2. A number of potsherds were found on the foot trail across Mound 2 and at the edge of a small borrow pit on the east side of the mound. The densest concentration of surface remains was associated with the upper platform of Mound 1. The steep sides of the upper platform are conducive to soil erosion and slope wash. In addition, the upper platform is riddled with the burrows of armadillos, agoutis, and other animals that have made themselves at home in the well-drained areas of the mound. The spoil from this burrowing activity is filled with pottery, which over time has washed down the slope and collects on the surface. The pottery samples collected here are large fragments of vessels and, in one case, a complete vessel (figure 7.5), which suggests that the pottery is eroding from primary deposits within the upper platform of Mound 1 rather than from secondary fill taken from a previous midden filled with pottery and used to construct the mound. The high frequency of complete or near complete fine wares on the surface of the upper platform of Mound 1 suggests that the vessels were placed there as offerings in caches or burials.

The pottery is a mix of fine decorated wares and simple domestic wares.[5] Most of the pottery collected from the surface is from decorated serving vessels. Forms include simple open-mouth bowls and complex incurving-mouth

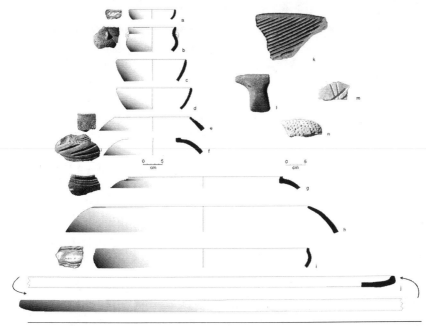

FIGURE 7.5 Examples of artifacts from surface collections at the Ibibate Mound Complex: (*a–j*) pottery vessels based on fragments, (*d*) a fragment of a ceramic hand-held grinder, (*k*) a fragment of a large ceramic "grater," (*l*) a broken ground stone ax of a T shape, (*m*) an arrow or spear shaft straightener or projectile point grinder.

bowls (figure 7.5). Based on ethnographic analogy with the lowland tropics, these vessels were probably used for serving and drinking of liquids, possibly a mild alcoholic beer *(chicha)* of manioc or maize. Decorations consist of fine and broad-line incision and geometric and curvilinear designs painted on slipped surfaces. Pottery "graters bowls" and hand "grinders," ubiquitous finds in pre-Columbian mounds of the Beni, are also present (figure 7.5). Archaeologists have described similar artifacts from mounds near Casarabe (Dougherty and Calandra 1981–82; Fernandez Distal 1987a; Prümers 2000, 2001, 2002a; Rydén 1941, 1964). The functions attributed to graters and grinders include grating manioc for preparation of *chivé* (a toasted, lightly fermented coarse flour) (Nordenskiöld 1913; Rydén 1964), producing barkcloth and distilling salt.

Fragments of eroded bone are common on the surface of Mound 1. The bones are of animals (species not yet determined) and of humans. Although some of the animal bones may be from animal burrows in the upper platform of the mound, the majority probably represents discarded food remains that were incorporated into the artificial fill of the mound. The human bone is probably from primary context burials, possibly in urns, eroded out of the upper platform; thus, one use of the upper platform of Mound 1 was apparently as a cemetery.

Another common surface find at Ibibate Mound Complex are the remains of large apple snail *(turos)* (*Pomacea* sp. or *Ampularia* sp.). These molluscs are common in the fill of pre-Columbian mounds in the Beni, often forming thick stratigraphic layers (Erickson 2000c; Pinto Parada 1987), and associated with fish weirs and artificial ponds in Baures (Erickson 2000a and chapter 8, this volume). Although rarely consumed by locals today, the highly productive snails are edible. Because of their low nutritional value and intact state, it is not clear if the molluscs in the mounds are food remains or served other purposes.

SOIL ANALYSES

Soil samples were taken from elevated areas of Mound 1 (Mound Inventory) and from the low-lying forest to the southwest of Mound 1 (Pampa Inventory; inventories are discussed in detail later) (figure 7.2). Each sample was based on 10 subsamples taken with a soil probe to depth of 20 centimeters at random locations within each inventory in July 1999. Subsamples were mixed thoroughly together, sun dried, and analyzed at Laboratorio del Centro de Investigación Agrícola Tropical, Santa Cruz, Bolivia (table 7.1).

The two samples have identical soil texture, reinforcing the idea that the mounds were constructed using local fill from nearby barrow pits. The soils from both inventories are relatively fertile by local standards, although the Mound Inventory obviously has better agricultural soil than the Pampa Inventory. The differences in pH, saturation of exchangeable bases, phosphorus, calcium, and total nitrogen are striking. The pH of the Mound Inventory is ideal for plant uptake of the cations of calcium, manganese, potassium, and phosphorus in contrast to the Pampa Inventory's low pH, which would limit nutrient availability for plants (Johannes Lehmann, personal communication, 2004) The phosphorus level is much higher in the Mound Inventory, sufficient for good yields of maize and other crops (Lehmann, personal communication, 2004). The low levels of organic matter in both samples is surprising due to the evidence of domestic activities or debris or both in the two inventories.

Although some of the differences in fertility may be due to the better drainage of the Mound Inventory, we believe that human activities best account for the soil characteristics. The Mound Inventory soils have maintained their fertility after 400 years since the settlement was densely occupied. With the exception of low levels of organic matter and charcoal, Johannes Lehmann (personal communication, 2004) favorably compares the Mound Inventory sample to those of anthropogenic Amazonian Dark Earths found in Brazil (Glaser and Woods 2004; Lehmann et al. 2003; Neves and Peterson, chapter 9, this volume). We had expected the soils of the Pampa Inventory to have been more fertile due to the presence of domestic debris in the subsurface soil probe transects. The recovery of pottery, charcoal, and bone at depths of up to 85 centimeters below surface suggests that occupation was not limited to the well-drained mound surfaces or that domestic midden was regularly spread over adjacent landscape as an organic soil amendment.

COMPARATIVE DATING OF THE IBIBATE MOUND COMPLEX

Because no excavations were done and no radiocarbon samples were collected for dating, we have no absolute dates for the construction, use, and abandonment of the Ibibate Mound Complex. Comparison of pottery styles to that collected at better-dated mounds provides some chronological control. The pottery we collected is almost identical to that described from surface collections at the Ibiato Mound (Fernandez Distel 1987a) and the Santa Fe Mound (Hanke 1957), and from stratigraphic excavations at Loma Alta de Casarabe ("High Mound of Casarabe") (Dougherty and Calandra 1981–82, 1984:191, table 2) and the Mendoza Mound (Prümers 2000, 2001, 2002a, 2002b). The later mounds are located within easy walking distance (10 kilometers) of the Ibibate Mound Complex.[6]

Radiocarbon dates (a total of 14) obtained from excavations of the Loma Alta de Casarabe range from 735–145 BC (2685–145 BP) to AD 1050–70 (900–70 BP; all uncorrected dates) (Dougherty and Calandra 1984:191, table 2). In a recent summary of the chronology of the Mendoza Mound, Prümers (2002b) states that the initial mound construction and occupation began around AD 400, followed by additional construction phases in AD 700 and AD 1200, with occupation ending around AD 1400. Based on the geographic proximity and strong stylistic similarity between the pottery from these better-dated mounds and pottery that we recovered, we are confident that the chronology of construction, occupation, and abandonment for the Ibibate Mound complex is comparable (1,000 to 1,500 years of use ending 500 years ago with the arrival of Europeans).

MOUND-FORMATION PROCESSES

The Ibibate Mound Complex is unquestionably an artificial, human creation. Archaeological investigations since the initial excavations of Erland Nordenskiöld in 1910 have clearly demonstrated that mounds are anthropogenic (Denevan 1966;

Erickson 2000c; Kenneth Lee in Pinto Parada 1987; Nordenskiöld 1913), despite unsubstantiated claims that the mounds are accumulation of flood sediments and were periodically occupied by nomadic peoples (see, e.g., Dougherty and Calandra 1981–82, 1984; Faldín 1984). Pre-Columbian mounds are quite distinct from natural rock formations (upwelling of the Brazilian Shield known locally as *cerros* or *cerritos*) (Denevan 1966; Hanagarth 1993), the topography created by dynamic fluvial processes (the levees of active and inactive rivers) (Hanagarth 1993), and certain forest islands formed by natural processes (Langstroth 1996).

Mounds are clearly associated with active and abandoned river channels. Most mounds were established on river levees, natural elevations formed by accumulations of sediments dropped as floodwater velocity is slowed when rivers overflow their banks. Levees on active rivers are often continuous along both banks and covered with gallery forests (Hanagarth 1993). Older levees are often fragmented, and forest growing on them forms "islands" surrounded by savanna and wetlands (Langstroth 1996). The mound builders first colonized natural better-drained levees for settlement, gardens, and fields with direct access to drinking water, fish, and game attracted to the water; for protection from fire; and for open water for their canoes. The levees of smaller rivers and abandoned river channels as in the case of the Ibibate Mound Complex are low and narrow, probably less than 0.5 meters above the surrounding landscape. Thus, river geomorphology accounts for the location of mounds, but not for the processes that created them.

The colonization of levees may have occurred when the rivers were more active. As rivers meandered and changed courses over time, some mounds were left on abandoned river channels and bends (oxbox lakes, cañadas). Abandoned channels still provide access to active channels of the rivers during the rainy season when most of the flat landscape is covered with shallow floodwaters.

Once established as a settlement, the mounds grew through a variety of cultural formation processes that can be inferred from mapping, surface collections, road cuts, excavations of mounds, and observations of houses and settlements today (see Erickson 2000c for a summary of this literature). House floors were raised 20 centimeters or more by constructing a clay platform to prevent waterlogging. Garbage (food and food-processing debris, charcoal, and ash) was used to fill depressions around houses and to level activity-area surfaces. Traditional houses in the Bolivian Amazon had wattle-and-daub walls roofed with wooden poles and palm thatch (figure 7.6). The technique involves creating a basketry-like web of wood and cane between upright house posts and covering it with thick clay mixed with grass. Houses tended to last 5–10 years in the humid environment. When the house was

FIGURE 7.6 The present-day Sirionó community on the Ibiato Mound. Similar, although more densely packed, thatched houses, work areas, plazas, and compact house gardens probably covered the Ibibate Mound Complex before it was abandoned.

abandoned, up to a ton of soil was added to the surface when its walls finally collapsed. These locations were leveled, and new houses were built on the raised surfaces. Thus, settlements surfaces were raised through a slow, but continuous, gradual accretion process as garbage was disposed and houses were abandoned and rebuilt by generations of inhabitants, which eventually resulted in the formation of a mound.

Stratigraphic profiles of some mounds often show wide strata of soil of uniform color and texture. In these cases, internal layers of some mounds were formed more rapidly through thick additions of construction fill over a short period of time, possibly a single episode. These construction layers are often associated with large pottery urns for human burials where the earth may have been added around a layer of urns or capped them to prevent damage. In other cases, layers of earth may have been added to mounds to increase their height and size or the area of flat surface for house and activities or to improve aesthetic appearance, or to achieve all three purposes. Clusters of large mounds such as those in the Sirionó Indigenous Territory and Casarabe may have resulted from competition between communities for the most impressive monumental landscape feature.

Most mounds such as the Ibibate Mound Complex are surrounded by deep depressions that were created as construction fill was removed. Soil could be easily excavated from these barrow pits using simple digging sticks and carried in baskets or carrying cloths to the mounds. These depressions also served as a year-round source of drinking water for the mound's inhabitants. In the case of the Ibibate Mound Complex, sections of the old river channel were mined for soil, thus creating the ponds. At the height of the rainy season and flooding, these depressions formed a near continuous moat around the mounds. During the dry season, they provided the only source of water.

Prümers (2000, 2001, 2002a, 2002b) has documented both slow accretion processes and mass additions of construction fill in the Mendoza Mound in Casarabe. Excavations in the exposed profile of a road cut through the center of the mound show that sections of the mound were built and abandoned as the settlement shifted horizontally over time. At various periods, thick layers of construction fill were added to raise and level the surface.

People have continued to return to these locations, once established, because of the dry surfaces and the rich soils they provide for settlement and gardens, even if humans occasionally abandoned them for short periods. Native peoples, ranchers, and colonists continue to seek out mounds for house sites and gardens. Mounds are permanent, valuable features on the local landscape.

Human activities are also a factor in the erosion and destruction of mounds. In the past, population densities were high on the mounds and surrounding landscape. Human disturbance such as deforestation, cultivation, burning, pathways, and hydraulic works would have impacted mounds. Mounds were eroding at the same time that they were being formed. The steep sides of the large mounds and barrow pits were easily eroded by the humans activities and heavy rains.

Our posthole testing transects demonstrated that buried pre-Columbian settlement debris (pottery sherds and charcoal) to a depth of 85 centimeters extended beyond the edges of Mound 1 and Mound 2 into the low ground to the north (the extension of the Mound Inventory to the northeast in figure 7.2). It is unlikely that these remains eroded from the mound located at a distance of 50–100 meters. Houses may have been established off mound, or garbage was tossed in this area for gardens and fields. As the mounds grew over time, flat space on the platforms and slopes for construction of houses would have been at a premium; thus, off-mound settlement may have been required. However, the low fertility of these soils suggests that anthropogenic activities were minimal.

The total amount of earth used in the construction of the Ibibate Mound Complex is impressive (table 7.2). Based on the topographic maps, an estimated 252,651 cubic meters of earth was moved to create the structures. Of course, the mounds grew through accretion, and this labor was spread out over many hundreds of years.

TABLE 7.2 Mound Descriptions and Labor Estimates

	Planimetric Area ha	Surface Area ha[a]	Volume m³	Labor minimum person–days[b]	Labor maximum person–days
Mound 1	2.21	2.25	75,294	30,117	65,473
Mound 2	9.27	9.33	177,357	70,942	154,223
Total	11.48	11.58	252,651	101,060	219,697

[a] Based on base at the 85-meter contour.
[b] Based on experimental raised-field construction: 1.15 to 2.5 cubic meters of earth moved per day based on a 5–hour workday (Erickson 1994; Erickson et al. 1991).

Continuous occupation of the Ibibate Mound Complex probably ended with the arrival of the Spanish conquistadors, missionaries, and colonists in the Bolivian Amazon. Native populations collapsed as the people died from Old World diseases to which they had no resistance, were enslaved, or perished from civil and international wars. Forest gradually became established in what had been houses, communities, gardens, orchards, and fields.

SOME COMPARISONS TO OTHER MOUNDS IN THE BOLIVIAN AMAZON

At 16–18 meters tall, with a volume of 252,651 cubic meters and a combined base covering approximately 11.5 hectares, the Ibibate Mound Complex is one of the largest and probably the tallest recorded mound groups in the Bolivian Amazon (table 7.2). Based on manual earth moving to construct experimental raised fields, we estimate that between 101,000 and 220,000 person-days of labor were required to construct the Ibibate Mound Complex. Because the mounds were constructed piecemeal over hundreds of years, the labor costs would have been spread out. For example, 100 people working 30 days a year could have constructed the Mound Complex in 33.7 to 73.3 years. Accurate comparisons of size, volume of fill, and form of pre-Columbian mounds are difficult because few mounds have been adequately mapped. The Loma Alta of Casarabe may be the Ibibate Mound Complex's closest rival (discussed earlier).

Two structural principles reflected in the Ibibate Mound Complex are notable. First, the mounds are made up of two distinct structures—a large flattened platform covering several hectares, with a smaller cone- or pyramid-shaped earthen structure on top (figure 7.7). This form has also been recorded for the Ibiato Mound, Loma Alta de Casarabe, Mendoza Mound, Suárez Mound (Mamoré River), Monte Sinai Mound (Apere River), and Cayalo Mound (Apere River). Mound 1 of the Ibibate Mound Complex, the Ibiato Mound, the Mendoza Mound, and Loma Alta de Casarabe are remarkably similar in form: all are

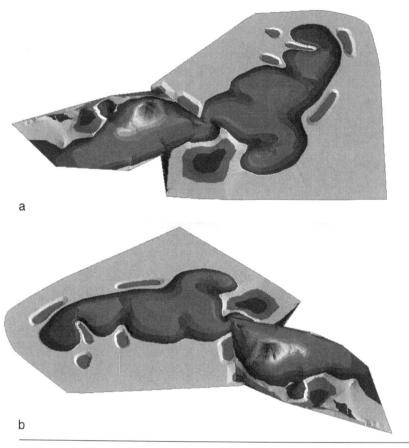

a

b

FIGURE 7.7 Perspectives of the Ibibate Mound Complex from different directions.

oriented northwest–southeast, and the highest elevation on the mounds is off center in the southeast of the lower platform (although the location is reversed in the case of Ibiato Mound) (compare figure 7.3 with Dougherty and Calandra 1981–82:17, lámina III; Prümers 2002a, figure 2; Stearman 1987:76, figure 3). The second characteristic is the pairing of mounds. In most cases, one mound is clearly larger or taller, or both, than the other. Dual mound groups have also been recorded for Dos Islas (Secure River), Cayalo Mound (Apere River), and Esperanza II Mound (Apere River). Concepts of duality embedded in settlement plans are often recorded in the ethnographic and historical Amazonian societies (Lévi-Strauss 1944). It is our central argument that these structures and the human intentionality that indisputably went into making them, perhaps principally for social and ritual purposes, have had a positive impact on the vegetation of this landscape.

Similar pre-Columbian networks of causeways and canals integrated with settlements and fields are common throughout the Bolivian Amazon (Denevan 1966; Erickson 2000b; Nordenskiöld 1916; Pinto Parada 1987).

FOREST INVENTORY METHODS

In order to understand similarities and differences of vegetation on the mounds and in the forest near the savanna (pampa) edge, two 1-hectare inventories associated with Ibibate Mound Complex and its immediate environs were carried out. These inventories are referred to as the Mound Inventory and the Pampa Inventory. The methods employed in this study follow largely those explained in Boom 1986; Campbell and colleagues 1986 (and Campbell et al., chapter 1, this volume); Gentry 1993; Mori and colleagues 1983; and Salomão, Silva, and Rosa 1988. The Mound Inventory was carried out in August 1993. Further collections and re-collections of trees in the Mound Inventory were made in July 1994 and September 1997. The Pampa Inventory began in September 1997. Revisits without plant collections being conducted were made to both inventory sites in July 1999 and August 2002, and soil samples were taken at both inventories in July 1999.

The Mound Inventory is a narrow rectangle, measuring 20 by 500 meters and includes the earthen pinnacle of Mound 1 located at 14°48'25" S/64°24'36" W. The inventory was divided into 40 subplots measuring 10 by 25 meters in order to calculate species' densities, ecological importance values (IV), species/area curves, and alpha diversity. The Mound Inventory is basically on *terra firme,* land that does not flood at any time of the year (as indicated in the term /ibi-bate/, "true land" or "high ground," in the Sirionó language).

The Pampa Inventory, at Subplot 1, is located at 14°48'23" S/64°24'30" W. The Pampa Inventory measures 100 by 100 meters. It also consists of 40 subplots of 10 by 25 meters each. The Pampa Inventory is located southwest of Mound 2. The area between the depression and the open savanna is a narrow ecotone about 200 meters wide at its widest point. The ecotone of forest and savanna, called savanna edge (*orilla de la pampa* in Spanish, *ibéra* in Sirionó), is zigzag in shape. It was not feasible to lay out a narrow rectangular inventory inside the single forest type in this area, but the 100-by-100-meter inventory did cover and stay within the boundaries of this single forest type.

On both inventories, all trees in the size class greater than or equal to 10 centimeters diameter at breast height (DBH) were tagged, and, with a few exceptions—as when the field identification indicated that the same species had been already collected numerous times—each was collected. Also included were woody vines with diameters greater than or equal to 10 centimeters DBH. Trees were tagged with consecutively numbered aluminum tree tags starting from the baseline of 10 meters in width at the zero point from each subplot. All inventory data were recorded in a large foldout notebook; a matrix was constructed whereby each row represented a numbered tree in the inventory, and these rows were listed consecutively in order of tree tag number, starting with number 1. Columns were (1) date of plant collection; (2) collection number (on Balée series); (3) subplot number (0–39); (4) location in relation to the transect (left or right of it when facing the heading of the inventory); (5) taxon (to genus and species, where possible); DBH; comments (for information such as the number of specimens made of each plant, whether it was fertile or not, whether it had leaves or was deciduous at the moment of collection, whether it had buttresses, sap, and other dendrological characteristics); (6) indigenous (Sirionó) name; and (7) indigenous uses. Items 6 and 7 were also repeatedly recorded in a separate notebook of ethnobiological responses by individual Sirionó consultants for the purposes of a different study. Vouchers of all specimens were deposited at the National Herbarium of Bolivia (Herbario Nacional de Bolivia) in La Paz.

Specimens that could not be determined to species, but were distinguished as being different species of the same genus, were treated as morphospecies and thus numbered differently; specimens that could not be identified to genus or, in rare cases, to family were treated likewise as undetermined (to genus or family) morphospecies and numbered accordingly. (The validity of morphospecies and their use in various aspects of understanding inventory data are discussed in Campbell 1989; Campbell et al. 1986; Campbell et al., chapter 1, this volume.) Identifications from the Mound Inventory are somewhat more reliable than those from the Pampa Inventory because (1) many trees at the time of the Pampa Inventory and collection were deciduous (in September 1997), so only dendrological collections in these cases could be made; (2) the plants were identified by more specialists; and (3) the vouchers were better distributed to determination specialists . Nevertheless, several of the individuals on the Pampa Inventory could be identified in the field with use of the keys given by Killeen, Garcia, and Beck (1993) and by Gentry (1993), as well as based on Balée's prior field experience and knowledge of species in the habitat, especially those shared at the Mound Inventory.

Two kinds of diversity, alpha and beta, are examined here. *Alpha diversity* is diversity of species in a restricted, given locale on the same substrate, as with the Mound Inventory or the Pampa Inventory. *Beta diversity* is the diversity of species sampled along an environmental gradient in a particular area across locales on different substrates; beta diversity in this case is a measure of change in species composition along the environmental gradient of slope (as one descends from the Mound Inventory, which is well drained, to the Pampa Inventory, which is flood prone).

The analysis of the inventory data was conducted using a program composed in DBASE Plus. Each raw tree database consists of the following structure (per tree): family, species, subplot, tree tag number, DBH, basal area (calculated from the DBH by dividing the DBH by 2 to get the radius and then applying the formula for the area of a circle, as each tree is conceived mathematically, πr^2). Measure of similarity between the inventory sites was calculated based on the Jaccard coefficient (given later). The DBASE Plus© program calculated the number of individuals of each species; the number of subplots on which each species is found; the total basal area of each species; the relative density of each species (the number of subplots in which each species is found is divided by the total number of all such subplots for all species multiplied by 100); the relative frequency of each species (the number of individuals in a species divided by the total number of individuals in the inventory multiplied by 100); the relative dominance of each species (the total basal area of a species divided by the total basal area of all species multiplied by 100); and the ecological importance value, which is the sum of relative density, relative frequency, and relative dominance for each species.

ANTHROPOGENIC FOREST

The total number of trees on the Mound Inventory was 448, and the total basal area was 27.730 square meters (appendix 7.1). The total number of trees on the Pampa Inventory was 425, and the total basal area was 26.158 square meters (appendix 7.2). These figures are similar both in tree numbers and basal areas and suggest maturity, possibly even great age, of both forests on these sites. That is because these basal areas, which are indirect measures or proxies for biomass, are higher than open forests (with or without palms) and vine forests (which are usually between 18 and 24 square meters [Pires and Prance 1985:112, 119]), though somewhat lower than or classical, mature forests of equatorial Amazonia (which can reach 40 square meters and greater [Pires and Prance 1985:112]). The basal areas of these forests, which are heavily represented by palms, especially *Astrocaryum murumuru* var. *murumuru* and *Attalea phalerata,* are roughly comparable to a set of four old-fallow forest (anthropogenic forest) inventories Balée conducted in Maranhão, Brazil (extreme eastern Amazonia), which had an average basal area, using essentially the same sampling techniques as here, of 26.25 square meters (Balée 1994:124–129). The forests of the Mound Inventory and adjoining Pampa Inventory seem slightly higher in biomass than two anthropogenic liana forests (with many palms present also) sampled in the middle Xingu River basin at 22.10 square meters and 21.90 square meters, respectively (Balée and Campbell 1990:47).

The two inventories are similar in terms of their respective alpha diversities, with 55 species in the Mound Inventory and 53 species in the Pampa Inventory (appendixes 7.1 and 7.2). They share 24 species (table 7.3), meaning that the

TABLE 7.3 Shared Species Between Mound Inventory and
Pampa Inventory with Ecological Importance Values (IV)
Shown at Each Site

Species	IV at Mound	IV at Pampa
Spondias mombin L.	3.93	3.04
Aspidosperma sp. 1	1.30	3.88
Dendropanax arboreus (L.) Decne. & Planchon	3.86	0.71
Astrocaryum murumuru Mart. var. *murumuru*	46.21	9.21
Attalea phalerata Mart. ex Spreng (= *Scheelea princeps*)	41.70	83.27
Syagrus sancona H. Karst	4.50	18.10
Clytostoma ulegnum Kanzlin	2.80	0.72
Ceiba pentandra Gaertn.	5.61	4.43
Cecropia concolor Willd.	2.08	1.15
Calophyllum brasiliense Camb.	6.38	2.82
Terminalia sp. 1	6.93	5.32
Hura crepitans L.	15.38	12.10
Sapium glandulosum (L.) Morong	5.51	5.32
Swartzia jororii Harms	2.44	2.24
Caesaria gossypiosperma Briq.	1.90	15.78
Lauraceae sp. 1	3.52	0.72
Acacia sp. 1	0.64	0.70
Gallesia integrifolia (Spreng.) Harms	23.47	23.15
Triplaris Americana Fisch & Meyer ex C. Meyer	4.47	4.34
Calycophyllum sp. 1	0.79	4.23
Genipa americana L.	0.94	1.45
Sapindus saponaria L.	0.97	0.72
Talisia hexaphylla M. Vahl.	3.93	6.24
Ampelocera ruizii Klotzsch	22.43	10.88

total number of species in both inventories is 84 (55 + 53 = 108; 108 − 24 = 84). The current alpha diversity at the Mound Inventory would not exist without human intervention, specifically the making of the built environment in prehistory. Langstroth (1996) has referred to the Ibibate Mound Complex and other forested mounds like it as "ceramic" forests because of the vast quantity of potsherds found inside them. The forest of the Pampa Inventory is anthropogenic as well, but for different reasons.

The analysis of shared species (table 7.3) yields significant circumstantial evidence as to anthropogenic origins of the forests on these two inventories. Both forests are, in a strictly vegetative sense, palm forests. They are also oligarchic, being heavily dominated by just a few species (Campbell et al., chapter 1, this volume; Peters et al. 1989). In terms of ecological importance values, the top 10 species in the Mound Inventory account for 65.4 percent of all importance values of all trees summed together (196.2/300 = 0.654). The top 10 species at Pampa Inventory account for 70.9 percent of all importance values of all trees summed together (212.7/300 = 0.709). The two ecologically most importance species on each inventory account for 23.12 percent of the Mound Inventory's total importance value and 41.65 percent of the Pampa Inventory's total importance value. However similar as oligarchies, though, the two forests are *different* types of oligarchic forests.

Many palm forests in Amazonia are anthropogenic (Balée 1988; Balée and Campbell 1990). In referring to forests in six 0.1-hectare sampled subplots in a vegetation study south of the Ibibate Mound Complex, Wendy Townsend noted, "Palm trees predominate in much of the forest area. The mounds and the canals of this landscape were built by humans before the arrival of the Spanish. The high proportion of palm trees in the forest is another indication of past human occupancy" (1995:26, see also 1996:16–19). Specifically, the motacú palm (*Attalea phalerata* Mart ex. Spreng) accounted for *at least* 50 percent of trees on four of six plots (Townsend 1995:21); on one of these six subplots, the spiny chonta palm (*Astrocaryum murumuru* Mart. var. *murumuru*) also accounted for more than 50 percent of the total frequencies. Townsend indicated that Sirionó informants recognize three forest types (Ibibate, Kiarochu, and Ibéra). Because these forest types are not "detectable on aerial photographs" (Townsend 1995:20), field research is necessary to understand the variety and subtlety of human disturbance and molding of landscapes.

In terms of palms, the relative frequency (number of individuals of a species divided by all individuals multiplied by 100) is less in the Mound and Pampa inventories than in the plots sampled by Townsend. At the Mound Inventory, *A. phalerata* at 53 individuals is only 11.8 percent relative frequency (53 divided by 448 multiplied by 100) of the total number of trees in the size class greater than or equal to 10 centimeters DBH. This palm is much more frequent in the

Pampa Inventory, at 36 percent (153 divided by 425 times 100), though this is somewhat less than half the total—that is, slightly dissimilar from Townsend's results; her plots appear to have been located in a slightly less well-drained area than the Pampa Inventory. *A. phalerata* is the ecologically *most* important species in the Pampa Inventory and the second ecologically most important species in the Mound Inventory (table 7.4). In contrast, the spiny chonta palm with its highly desirable heart of palm, *A. murumuru* var. *murumuru,* accounts for exactly 25 percent of all individual trees in the Mound Inventory, and it is the ecologically most important species there. Although this palm is less frequent (at 3.5 percent) in the Pampa Inventory, it still has a relatively high importance value (rank eighth among 53 species).

The two forests are similar, yet different enough to be categorized by different names and concepts, as they are in Sirionó language and culture. The solitary palm, *A. phalerata,* which has high importance values on both inventories, being somewhat more important in the Pampa Inventory, is facultative in this sense (it can occur in more than one biome, defined in terms of environmental gradients and immediate, local features such as slope, light penetration, and water tolerance), tolerating some flooding and also appearing on the better-drained terra firme of the Mound Inventory. On sedimentary soils to the west of the Brazilian Shield, this species appears to replace the babaçu palm (*Attalea speciosa* Mart. ex Spreng.) in an allopatric manner (Henderson 1995:151); babaçu palms are associated with anthropogenic forests in eastern Amazonia and elsewhere south of the Amazon River. The palm has a multitude of uses, and its fruits traditionally were gathered as food by the Sirionó in every month of the year, fronds used as thatch, and cooked fruit employed as a laxative (Fernandez Distel 1987b:118; Holmberg [1950] 1985:48–50). *A. murumuru* var. *murumuru* is likewise facultative, being found "in periodically inundated areas or tidally inundated areas near the sea, along river margins, or occasionally in lowland rain forest on *terra firme.* Since the fruits are edible, it is likely that its range has been extended by humans" (Henderson 1995:246). Indeed, the heart of palm of juvenile *A. murumuru* var. *murumuru* is a delicacy and an important, nutritious food item of the Sirionó in every month of the year (Holmberg [1950] 1985:48–50; he calls it "palm cabbage"). The sumuqué palm, *Syagrus sancona* H. Karst, with an edible fruit, is common in the Pampa Inventory and somewhat less important on the Mound Inventory. According to Henderson, this palm is a rare and threatened species in Amazonia, and its distribution may be due to humans (1995:135).

In terms of dicotyledonous plants collected in the inventories, several indicate the possibility of anthropogenic patterns of distribution. Hog plum or yellow mombin (*Spondias mombin* L.), a tree with a juicy, tart, vitamin C–rich edible fruit that occurs on both inventories, is well known from both anthropogenic forests and riparian (i.e., floodable) forests elsewhere (Balée, 1994:132;

Clement, chapter 6, this volume). Sometimes it is cultivated in Amazonian dooryard gardens (Paz Rivera 2003:30; Smith et al. 1995:215) and is facultative, as are many species of old-forest fallows. Hog plum may have migrated from gardens, orchards, or fields on the mounds down the slope into the floodplain of the Pampa Inventory.

A number of wind-dispersed species are present on both sites. Anemochory is an efficient dispersal mechanism of pioneer species, some of which are long lived and grow to display significant height and high biomass. Although no single species of *Tabebuia* appears to be shared between the Pampa and Mound inventories, the genus itself has winged seeds, which facilitate dispersal into areas cleared or disturbed by humans (Gentry 1993:268). All *Tabebuia* species (tajibo or pau d'arco) are valuable timber. *Tabebuia serratifolia* (M. Vahl) Nicholson (known as yellow tabs in the horticultural landscaping market of southern California), which occurs on the Mound Inventory (appendix 7.1), is a well-known species found in secondary forests throughout Amazonia and much of South America (Balée 1994:277; Killeen, Garcia, and Beck 1993:149). Likewise, although no one species of *Cordia* (Boraginaceae) appears on both inventories, *Cordia alliodora* (Ruíz and Pavon) Oken, which occurs on the Mound Inventory, possesses dry and expanded corolla lobes that contribute to its wind dispersal (Gentry 1993:296), as with *Tabebuia*. The kapok tree (*Ceiba pentandra* Gaertn.), a pantropical species, is frequently associated with disturbed sites (as a pioneer species) or with riparian habitats (Whitmore 1990:104), possibly because its small seeds occur inside the kapok, accounting for dispersal by wind (cf. Gentry 1993:288; van Roosmalen 1985:56). The Guarayo, Tupí-Guaraní people to the south of the Sirionó, evidently planted *C. pentandra* and considered it to be a sacred tree, as did the Maya (Paz Rivera 2003:8). The moraceous *Cecropia concolor* Willd., which is shared on both inventories, is a common member of fallow forests in eastern Amazonia (Balée 1994:281). Another dicot with winged seeds is the combretaceous *Terminalia* sp. 1, which not only is shared between the two inventories (table 7.3), but is also the tenth ecologically most important species in the Mound Inventory (table 7.4). *Terminalia* sp. is tied for twelfth place with *Sapium glandulosum* (L.) Morong in the Pampa Inventory (appendix 7.2).

The euphorbiaceous *S. glandulosum*, an ancient species, is found in numerous habitats of eastern Bolivia, including secondary forest (Killeen, Garcia, and Beck 1993:312). The related and poisonous ochoó tree (*Hura crepitans* L.) is a taxon probably dating from the beginning of the formation of the Amazon floodplain forests millions of years ago (Bush 1994; Colinvaux et al. 1996); it is physiologically plastic, having coexisted with cold-indicator taxa in the Pleistocene period (Bush 1994:14; Bush et al.1990; Killeen, Garcia, and Beck 1993:303; Piperno, Bush, and Conlinvaux 1990). Ochoó (called *assacu* in Brazilian Portuguese) is commonly found in poorly drained Amazonian forests, but it is also present in secondary forest on terra firme (Killeen, Garcia, and Beck 1993:303).

TABLE 7.4 Ten Species from Mound and Pampa Inventories with Highest Importance Values (IVs) in Order of Importance

Rank	Mound Inventory	IV	Pampa Inventory	IV
1	*Astrocaryum murumuru* Mart. var. *murumuru*	46.21	*Attalea phalerata* Mart. ex Spreng.	83.27
2	*Attalea phalerata* Mart. ex Spreng	23.15	*Gallesia integrifolia* (Spreng.) Harms	41.70
3	*Hirtella triandra* Sw. ssp. *triandra*	25.47	*Ficus* sp. 1	21.66
4	*Gallesia integrifolia* (Spreng.) Harms	23.47	*Syagrus sancona* H. Karst	18.10
5	*Ampelocera ruizii* Klotzch	22.43	*Caesaria gossypiosperma* Briq.	15.78
6	*Hura crepitans* L.	15.38	*Hura crepitans* L.	12.10
7	*Dendropanax cuneatus* (DC.) Decne & Planch.	13.73	*Ampelocera ruizii* Klotzsch	10.88
8	*Sorocea guilleminiana* Gaudich.	12.25	*Astrocaryum murumuru* Mart. var. *murumuru*	9.21
9	*Ficus pertusa* L.f.	7.18	*Tabebuia* sp. 1	7.15
10	*Terminalia* sp. 1	6.93	*Jacaranda* sp. 1	7.01

Ochoó is mostly confined to the floodplain or riparian habitats in a natural state; thus, its appearance on the mound inventory makes up 5.12 percent of the total importance value there (i.e., 15.38/300 times 100), with a frequency of 29.09 percent (i.e., 15 individuals/55 total individuals times 100) suggests human interference (appendix 7.1; table 7.4). The sap of the ochoó tree is used as a fish poison by the Sirionó. The tree is also important on the Pampa Inventory (with 4.03 percent of the total importance value and a frequency of 9.43 percent; appendix 7.2, tables 7.3 and 7.4) and is the sixth ecologically most important species on both inventories (table 7.4).

The widespread *Genipa americana* L., bi or genipapo, which has an edible fruit that is also used in native body painting, is frequently cultivated and occurs in a wide variety of habitats in Amazonia (van Roosmalen 1985:379). The plant is present on both inventories. *Sapindus saponaria* L., shared on both inventories, has many indigenous uses in lowland South America and its range may have been increased by humans (Paz Rivera 2003:38–39, citing Brüchner 1989).

The shared species with different degrees of ecological importance suggest exchange of germplasm between the two forest types. In some cases, species with high ecological importance values on the terra firme of the Mound Inventory have more recently migrated downslope and thus have lower ecological importance in the Pampa Inventory (as with *A. murumuru* var. *murumuru*). In other cases, species with high ecological importance values in the Pampa Inventory may have migrated to the terra firme of the Mound Inventory and thus display much lower ecological importance values (as with *S. sancona*).

The human impact on the vegetation of the Mound Inventory becomes clear when noting species that occur only on terra firme. One of the most ecologically important of terra firme species is the moraceous *Sorocea guilleminiana* Gaudich, known as *turumbúri* in Sirionó. This culturally and ritually important tree produces a bright red fruit in bunches, from which the Sirionó make a highly appreciated fermented beverage September through December. Sirionó informants say it occurs only on mounds, and thus for the Sirionó it is the indicator species par excellence of anthropogenic mounds. Although it is not planted by humans per se (birds appear to be the zoochorous dispersal agents [Robert Langstroth, personal communication, 1999]), *Sorocea* sp. is present only because its typical habitat of permanent terra firme was created by humans in the remote past.

Both forests inventories contain many useful trees in a variety of categories, including food, fuel, construction material, medicine, stimulant, condiment, fish poison, and body adornment. The Mound Inventory has a somewhat higher proportion of useful species in all categories. Considering only the attribute of edibility (table 7.5), wild plant foods collected by the Sirionó occurring within the Mound Inventory include the fruits of turumbúri (*S. guilleminiana*), motacu (*A. phalerata*, fruits and heart of palm), sumuqué (*S. sancona,* fruits), chonta palm (*Astrocaryum murumuru* Mart. var. *murumuru*, heart of palm), hog plum (*S. mombin,* fruits), pacay (*Inga* spp., fruits), mururé (*Brosimum acutifolium* C.C. Berg, fruit and seeds), genipapo (*G. americana*, fruits), coquino (*Pouteria* sp. 1, which is quite possibly *Pouteria macrophylla*, a common denizen of old-forest fallows in eastern Amazonia, fruits), and aguaí (*Pouteria* sp. 2, fruits). The highly prized, large, sweet, pulpy fruits of the sapotaceous species are considered important food items in the traditional Sirionó diet in February, March, and April (Holmberg [1950] 1985:48–49). The frequency of fruit trees is 202 individual trees within 448 total trees, or 44.86 percent of the total. The introduced feral orange tree (*Citrus maxima* [Burman F.] Merr.), which does not tolerate flooding, but has thrived in the anthropogenic conditions of the mounds, may have been spread by the Jesuits.

The Pampa and Mound inventories share five trees having edible fruits or seeds: *S. mombin, A. murumuru, A. phalerata, S. sancona,* and. *G. americana* (table 7.5). The Pampa Inventory also has a few unidentified economic species as well as the well-known sterculiaceous (cacao family) *Guazuma ulmifolia* Lam., which has fruits considered to be edible throughout much of Amazonia. Fruit trees on the Pampa Inventory account for 41.86 percent of the ecological importance value of the inventory and a frequency of 50.58 percent. The figures

TABLE 7.5 Trees with Edible Fruits or Seeds in Sirionó Culture Measured on the Mound Inventory and the Pampa Inventory

Species at Mound	Relative Frequency	IV	Species at Pampa	Relative Frequency	IV
Spondias mombin L.	0.446%	3.93	Spondias mombin L.	0.705%	3.04
Astrocaryum murumuru Mart. var. murumuru	25.00%	46.21	Astrocaryum murumuru Mart. var. murumuru	3.52%	9.21
Attalea phalerata Mart ex. Spreng	11.83%	41.70	Attalea phalerata Mart ex. Spreng.	34.15%	83.27
Syagrus sancona H. Karst	1.56%	4.50	Syagrus sancona H. Karst	7.52%	18.10
Inga quaternata Poeppig. & Endl.	0.446%	1.29	Inga sp. 17	0.705%	1.78
Inga sp. 1	0.223%	0.67	Genipa americana L.	0.470%	1.45
Brosimum acutifolium C. C. Berg	0.223%	0.75	Pouteria sp. 16	0.235%	0.80
Sorocea guilleminiana	4.464%	12.25	Pouteria sp. 17	0.235%	3.22
Genipa americana L. Gaudich.	0.223%	0.94	Sapotaceae sp. 11	0.235%	1.65
Pouteria sp. 1 (coquino)	0.446%	1.32	Sapotaceae sp. 12	0.235%	0.72
Pouteria sp. 2 (aguaí)	0.223%	0.62	Guazuma ulmifolia Lam.	0.705%	2.42
Totals	45.08%	113.82		50.58%	125.66
(202 individual trees/448 = 44.86			(215/425 = 50.58)		

are therefore comparable with the Mound Inventory: both inventories are heavily rich in fruit trees. In summary, the forests of the Mound and Pampa inventories may not have been planted and cultivated by humans, but these orchards of fruit trees are genuine and present because of, not in spite of, human influence, intentionality, and management, specifically in the building of the mound and its associated structures and attributes.

Finally, the alpha diversity, especially the turumbúri tree (S. guilleminiana) in the Mound Inventory, can only be understood as caused by humans, given that species on the mounds normally occur only on terra firme (which, given the periodic geomorphological reworking of the floodplain, flooding, and high water table, are rare in this part of the Bolivian Amazon). The alpha diversity of the Pampa Inventory is high for the region as a whole, given that it is surrounded by savannas characterized by fewer than 20 species regardless of DBH (Townsend 1995:22). The beta diversity is also high in both inventories (when we consider the Mound and Pampa inventories to be different habitat types, defined principally by slope differentiation, itself an artifact of human disturbance), given that they share only 22.22 percent of their species. In other words, species diversity of both inventories is lower than the classic high-canopy forests in Amazonia, yet the adjacent inventories share only about 23 percent of their species. The analysis suggests that long-term human activities are important for understanding alpha and beta diversity at the Ibibate Mound Complex and vicinity.

Although both forests are anthropogenic, the formative processes that created the Mound and Pampa inventories differ in specifics. The forest of the Mound Inventory is a direct artifact of past human activities on the mounds (dwellings, garbage accumulation, gardens, fields, and orchards). When the mounds were occupied, there was less vegetation due to dense housing and activity areas. The historical (anthropogenic) processes that shaped the vegetation in the Pampa Inventory may have been different. When the Ibibate Mound Complex was occupied, the low-lying area around the mounds was maintained as open savanna and managed as gardens, fields, and orchards. Despite being prone to annual flooding, the domestic debris found in soil probes to the northwest of Mound 1 suggests that the area supported dispersed houses. Some of this debris may have been intentionally mixed into the soils to enhance fertility. The presence of forest in the Pampa Inventory is probably due to the cessation of annual burning and farming, which occurred with the removal of native peoples from the landscape. The forests on both inventories were managed in pre-Columbian times, as they are today.[1]

HISTORICAL ECOLOGY OF THE IBIBATE MOUND COMPLEX

The Ibibate Mound Complex is an excellent example of the complex temporal and biological dynamics of the landscape and the importance of humans with their associated cultural technologies in the shaping of that landscape. When the mounds were occupied, they and probably much of the surrounding landscape were relatively treeless. The presently occupied Ibiato Mound is a good analogy for what the Ibibate Mound Complex may have looked like as a low-density settlement (figure 7.7). Based on the density of artifact debris and information from archaeological excavations the Mendoza and Loma Alta de Casarabe mounds, we believe that the population density would have been higher in the pre-Columbian period. The number and density of inhabitants living on the Ibibate Mound Complex and in the immediate vicinity would have had a massive effect on the local ecology. Hunting, burning, gathering, fuel collecting, garbage and human waste disposal, and earthwork construction were potent anthropogenic processes. The removal of humans from the mound and the surrounding landscape during the Colonial period, with the exception of small surviving groups of trekking and foraging bands, would have had a dramatic effect on the development of the vegetation assemblage and ecological complexity. After the mounds were abandoned as a settlement (which means cessation of house building, hearth use, pottery firing, garbage and fill accumulation, weed clearance, and other domestic activities), long-lived, woody vegetation colonized the mound, encouraged by the artificially organically enriched and well-drained soils. Many of these species survived and thrived because they were originally established on the mounds in gardens and fields. The deep water-filled barrow pits around the mounds protected the tree vegetation from sporadic fires that trekking and hunting peoples set on the savannas. Over time, a dense gallery forest became established on the mounds and along the old river levee. What appears to be a pristine canopy forest is actually a product of anthropogenic formation over a considerable period of time. During the period of intensive occupation of the Ibibate Mound Complex and other mounds of the area, gallery forests and forest islands were probably greatly reduced by cultivated fields and gardens, settlements, paths, and roads. Tree coverage was probably limited to small patches of forest along the old river levees, orchards, and fields in the fallow cycle (figure 7.8). Any forest that did exist was probably anthropogenic within gardens, orchards, forest fallow, and secondary growth in old fields and settlements.

Aerial photographs and satellite imagery show that much of the presently forested areas were once savanna grassland. This is particularly the case for the low-lying areas of seasonally inundated forest immediately to the east of the Ibibate Mound Complex (figure 7.1). In addition, much of the San Pablo Forest was once open savanna. Strips of original gallery forest along old river-channel levees can be distinguished from areas of old savanna that are now covered with a lower canopy of trees (figure 7.1). This landscape is clearly the product of less-intense human management in recent times, in particular a reduction in annual burning of the savanna, since the massive demographic collapse caused after the arrival of Europeans.

In summary, ethnographers, botanists, archaeologists, and foresters have recently begun collaborative research to document the long-term history of human use of the Amazon basin and of human impact on the environment. The collaborative effort herein derives its sustenance and guiding theoretical organization ultimately from our own diverse backgrounds—in this case, archaeology and ethnography/ethnobotany. The Ibibate Mound Complex provides an excellent case of massive pre-Hispanic human disturbance of a local

FIGURE 7.8 Reconstruction of a typical mound and cultural landscape in the Bolivian Amazon. (Artwork by Daniel Brinkmeier)

context through the construction of a large mound complex, excavation of large moatlike barrow pits, and generations of settlement, gardening, and slash-and-burn farming on the site. The mound complex is still used by the Sirionó for collecting, hunting, and farming limited swidden fields (*chaco*). Balée created a detailed inventory of vegetation on and around the mound complex. Erickson's archaeological research provided context for these vegetation inventories. Specific results included mapping of the mounds and other earthworks; making surface collections of artifacts; and conducting subsurface testing to determine the age and use period of the mounds, the cultures associated, and the functions of the mounds. Limited archaeological survey was also conducted in the forest surrounding the mounds and in the savanna toward the Ibiato Mound Complex.

The massive effects of the long history of human transformation and creation of the environment can be seen in the landscape today. The sharp boundaries of forest-savanna and forest island–savanna are products of systematic anthropogenic burning. Many forest islands are artificial elevations created by human occupation and earthwork construction. Straight lines of trees on the savanna, some kilometers in length, grow on earthen causeways originally built for transportation, communication, and possibly water management. Many ponds (pozas) used for drinking water and fishing were created during construction of the large mounds for settlement and burial. The effect on transforming the soil horizons—improving soil drainage and fertility through earthwork construction and incorporation of pottery, burned clay, charcoal, ash, and other organic matter—was profound. In some cases, it can be demonstrated that the natural hydrology of the region was altered through the construction of canals and causeways. The presence and distribution of fauna—in particular, the game animals prized by the Sirionó hunters—was greatly determined by anthropogenic processes. The long history of human disturbance and transformation of the landscape expanded and enhanced the habitats for these animals. Through disturbance and management, humans produced the ecotones where game is most abundant, in particular the thick stands of palms within the forests and along the forest-savanna edge, artificial sources of drinking water, forest-fallow fields with stands of economically important species, and soils amenable to agriculture and settlement. We conclude, therefore, that the landscape of the Ibibate Mound Complex and the Sirionó Indigenous Territory in general is not a pristine forest-savanna environment, but rather a record of a long history of human manipulation embedded in the land that can be read as a palimpsest.

ACKNOWLEDGMENTS

Financial support of this research was generously provided by a National Science Foundation Research Grant to Clark L. Erickson, a Wenner-Gren Foundation Small Grant No. 5581 to William Balée, and a series of summer research grants from the Stone Center for Latin American Studies at Tulane University to William Balée. Grateful acknowledgment is made to these participants in the Mound Inventory of Ibibate in 1993: Chiro Cuellar, Segundo Quirindendu, Orione Álvares da Silva, and Richard Annas. We thank Silvia Bergeron, Chiro Cuellar, Rafael Eatosa, Nancy Cuellar, and Vanessa Mendoza for assistance in the Pampa Inventory of 1997. Determinations of taxa were generously made by Ron Liesner (various taxa), M. Nee (various taxa), R. E. Gereau (*Inga, Citrus*), C. Taylor (Rubiaceae), D. Daly *(Muelenbeckia)*, W. D. Stevens, D. Wasshausen, and A. S. Bradburn. Additional determinations were made by O. A. da Silva and S. Bergeron, to whom we are likewise indebted. We are also grateful to Sirionó ethnobiological consultants for sharing information regarding trees and the landscape of the Ibibate and Pampa inventories more generally, especially Chiro Cuellar, Nancy Cuellar, Dalia Sosa Eirubi, Hernán Eato, Susana Yicacere, María, Arturo Eanta, Carlos Eirubi, Chuchu Pepe, Daniel Mayachare, and Eloy Erachendu. Archaeologist Wilma Winkler was codirector of the Agro-Archaeological Project of the Beni. Winkler, John Walker, Kay Candler, and Marcello Canuto participated in the making of the detailed topographic map of the mound complex. Grateful acknowledgment is due to José Aguayo for assistance in writing the computer program for analysis of tree inventory data. Wendy Townsend and Zulema Lehms kindly made the initial introductions of the researchers to the Sirionó community at Ibiato in August 1993. We thank the Consejo Sirionó and the Sirionó community at Ibiato in general for authorization to do the research and for their kindness and hospitality during the research period. Oswaldo Rivera and Juan Albarracín of the Instituto Nacional de Arqueología helped obtain archaeological permits. The archaeological research was conducted under a multiyear agreement between the Instituto Nacional de Arqueología (Ministerio de Cultura) and the University of Pennsylvania. The director of the Corporation of the Department of the Beni (CORDEBENI), engineer Hans Schlink, enthusiastically provided valuable support to the project. During long discussions on the riverboat *Reina de Enin,* Heiko Prümers clarified many issues related to mound chronology, content, and formation processes. William Dickinson, Christina Tolis, and Samuel Voorhees—students in the freshmen seminar "Native Peoples and the Environment" (University of Pennsylvania, Spring 2002)—helped prepare the geographical information system and regional maps. We thank the Laboratorio del Centro de Investigación Agrícola Tropical, Santa Cruz, Bolivia, for the soil analysis and Johannes Lehmann for his help in interpreting them. Figures are based on topographic base maps of 1:100,000 and 1:50,000 scale. Aerial photographs were provided by the Instituto Geografico Militar, and LANDSAT TM and ETM imagery was provided courtesy of the Global Land Cover Facility of the University of Maryland and the Earth Explorer of the U.S. Geological Survey.

NOTES

1. See chapter 3, note 1, for a distinction between the use of *loma* in that chapter and the use in this chapter.
2. Alicia Duran Coirolo and Roberto Bracco Boksar (2000) provide an excellent survey of pre-Columbian mounds in the Americas.
3. Ibibate should not be confused with another name used frequently in this chapter, Ibiato (also spelled Iviato, Eviato), the location of the main Sirionó community and also a pre-Columbian mound with an elevation of 9.5 meters.
4. Mound 1 was mapped using a Topcon 302 © EDM Surveying Instrument. Mound 2 was mapped using Brunton© compass and tape. The maps and perspectives (figures 7.2, 7.3, and 7.7) were created using SURFER 3.2© and ArcView GIS 8.2© mapping and geographical information system software.
5. Detailed descriptions and illustrations of the artifacts will be presented in future reports.
6. Our surface collections also show general similarities to pottery styles in the Hernmarck, Velarde, and Masicito mounds excavated by Nordenskiöld (1910, 1913, 1924). These mounds are located 30–40 kilometers to the south and southwest of the Ibibate Mound Complex.
7. Although we did not study the faunal biodiversity at the Ibibate Mound Complex, the mounds provide a dry habitat for burrowing animals that thrive in the highest elevations of the mounds. The Sirionó state that the mounds are excellent hunting

APPENDIX 7.1 Analysis of Woody Plant Species ³ 10-Centimeters DBH on One Hectare of Forest in the Mound Inventory at the Ibibate Mound Complex, Bolivia

Family	Species	Number of Individuals	Number of Subplots of Occurrence	Basal Area (m²)	Relative Density	Relative Frequency	Relative Dominance	Importance Value (IV)
Achatocarpaceae	Achatocarpus praecox Griseb.	1	1	0.037	0.37	0.22	0.13	0.72
Anacardiaceae	Spondias mombin L.	2	2	0.758	0.75	0.45	2.73	3.93
Annonaceae	Annonaceae sp. 1	1	1	0.010	0.37	0.22	0.04	0.63
Annonaceae	Xylopia ligustrifolia Humb. & Bonp. ex Dunal	1	1	0.017	0.37	0.22	0.06	0.65
Apocynaceae	Aspidosperma sp. 1	2	2	0.029	0.75	0.45	0.10	1.3
Araceae	Indt. 04	1	1	0.014	0.37	0.22	0.05	0.64
Araliaceae	Dendropanax arboreus (L.) Decne. & Planchon	3	3	0.573	1.12	0.67	2.07	3.86
Araliaceae	Dendropanax cuneatus (DC.) Decne. & Planchon	21	16	0.850	5.97	4.69	3.07	13.73
Arecaceae	Astrocaryum murumuru Mart. var. murumuru (= Astrocaryum chonta)	112	32	2.571	11.94	25.00	9.27	46.21
Arecaceae	Attalea phalerata Mart. ex Spreng. (= Scheelea princeps)	53	31	5.074	11.57	11.83	18.30	41.70

APPENDIX 7.1 (continued)

Family	Species	Number of Individuals	Number of Subplots of Occurrence	Basal Area (m²)	Relative Density	Relative Frequency	Relative Dominance	Importance Value (IV)
Arecaceae	Syagrus sancona H. Karst.	7	6	0.195	2.24	1.56	0.70	4.50
Bignoniaceae	Bignoniaceae sp. 14	4	3	0.039	1.12	0.89	0.14	2.15
Bignoniaceae	Clytostoma ulegnum Kanzlin	5	4	0.052	1.49	1.12	0.19	2.80
Bignoniaceae	Tabebuia capitata (Bur. & K. Schum.) Sandw.	2	2	0.038	0.75	0.45	0.14	1.34
Bignoniaceae	Tabebuia serratifolia (M. Vahl) Nicholson	2	2	0.028	0.75	0.45	0.10	1.30
Bombacaceae	Ceiba pentandra Gaertn.	7	6	0.502	2.24	1.56	1.81	5.61
Boraginaceae	Cordia alliodora (Ruiz & Pavón) Oken	2	2	0.044	0.75	0.45	0.16	1.36
Cecropiaceae	Cecropia concolor Willd.	3	2	0.182	0.75	0.67	0.66	2.08
Celastraceae	Salacia impressifolia (Miers) A.C. Smith	2	2	0.023	0.75	0.45	0.08	1.28
Chrysobalanaceae	Hirtella triandra Sw. subsp. triandra	56	27	0.804	10.07	12.50	2.90	25.47
Clusiaceae	Calophyllum brasiliense Cambess.	5	5	0.939	1.87	1.12	3.39	6.38
Combretaceae	Terminalia sp. 1	7	7	0.766	2.61	1.56	2.76	6.93

Family	Species	Number of Individuals	Number of Subplots of Occurrence	Basal Area (m²)	Relative Density	Relative Frequency	Relative Dominance	Importance Value (IV)
Euphorbiaceae	*Hura crepitans* L.	16	11	2.139	4.10	3.57	7.71	15.38
Euphorbiaceae	*Sapium glandulosum* (L.) Morong	7	6	0.473	2.24	1.56	1.71	5.51
Fabaceae	Fabaceae sp. 1	1	1	0.050	0.37	0.22	0.18	0.77
Fabaceae	*Machaerium* sp. 1	1	1	0.021	0.37	0.22	0.08	0.67
Fabaceae	*Ormosia nobilis* Tul.	1	1	0.125	0.37	0.22	0.45	1.04
Fabaceae	*Swartzia jororii* Harms	4	3	0.118	1.12	0.89	0.43	2.44
Flacourtiaceae	*Casearia gossypiosperma* Briq.	3	3	0.030	1.12	0.67	0.11	1.90
INDT	Indt. 02	3	1	0.034	0.37	0.67	0.12	1.16
Lauraceae	Lauraceae sp. 1	5	5	0.147	1.87	1.12	0.53	3.52
Lauraceae	Lauraceae sp. 2	1	1	0.021	0.37	0.22	0.08	0.67
Lauraceae	Lauraceae sp. 3	1	1	0.048	0.37	0.22	0.17	0.76
Lauraceae	Lauraceae sp. 4	1	1	0.009	0.37	0.22	0.03	0.62
Mimosaceae	*Acacia* sp. 1	1	1	0.015	0.37	0.22	0.05	0.64
Mimosaceae	*Inga quaternata* Poeppig. & Endl.	2	2	0.026	0.75	0.45	0.09	1.29
Mimosaceae	*Inga* sp. 1	1	1	0.021	0.37	0.22	0.08	0.67
Moraceae	*Brosimum acutifolium* C. C.Berg	1	1	0.045	0.37	0.22	0.16	0.75

APPENDIX 7.1 (continued)

Family	Species	Number of Individuals	Number of Subplots of Occurrence	Basal Area (m²)	Relative Density	Relative Frequency	Relative Dominance	Importance Value (IV)
Moraceae	Ficus insipida Willd.	2	2	0.111	0.75	0.45	0.40	1.60
Moraceae	Ficus pertusa L.f.	6	5	1.102	1.87	1.34	3.97	7.18
Moraceae	Moraceae sp. 2	1	1	0.015	0.37	0.22	0.05	0.64
Moraceae	Sorocea guilleminiana Gaudich.	20	12	0.919	4.48	4.46	3.31	12.25
Olacaceae	Heisteria nitida Spruce ex Engler	1	1	0.042	0.37	0.22	0.15	0.74
Phytolaccaceae	Gallesia integrifolia (Spreng.) Harms	18	11	4.256	4.10	4.02	15.35	23.47
Polygonaceae	Coccoloba cujabensis Weddell	1	1	0.022	0.037	0.22	0.08	0.67
Polygonaceae	Triplaris americana Fisch. & Meyer ex C. Meyer	7	7	0.084	2.61	1.56	0.30	4.47
Rubiaceae	Calycophyllum sp. 1	1	1	0.056	0.37	0.22	0.20	0.79
Rubiaceae	Genipa americana L.	1	1	0.096	0.37	0.22	0.35	0.94
Sapindaceae	Sapindus saponaria L.	1	1	0.106	0.37	0.22	0.38	0.97
Sapindaceae	Talisia hexaphylla M. Vahl.	6	6	0.096	2.24	1.34	0.35	3.93
Sapotaceae	Pouteria sp. 1	2	2	0.032	0.75	0.45	0.12	1.32
Sapotaceae	Pouteria sp. 2	1	1	0.009	0.37	0.22	0.03	0.62
Sapotaceae	Sapotaceae sp. 12	2	2	0.110	0.75	0.45	0.40	1.60
Ulmaceae	Ampelocera ruizii Klotzsch	28	14	3.039	5.22	6.25	10.96	22.43
Verbenaceae	Vitex cymosa	2	2	0.768	0.75	0.45	2.77	3.97
TOTALS		448	268	27.730				300.0

APPENDIX 7.2 Analysis of Woody Plant Species ³ 10-Centimeters DBH on One Hectare of Forest in the Pampa Inventory at the Ibibate Mound Complex, Bolivia

Family	Species	Number of Individuals	Number of Subplots of Occurrence	Basal Area (m²)	Relative Density	Relative Frequency	Relative Dominance	Importance Value (IV)
Anacardiaceae	*Spondias mombin* L.	3	3	0.275	1.29	0.71	1.04	3.04
Apocynaceae	*Aspidosperma* sp. 1	5	5	0.142	2.16	1.18	0.54	3.88
Apocynaceae	*Aspidosperma* sp. 5	2	0.026	0.026	0.86	0.47	0.10	1.43
Araliaceae	*Dendropanax arboreus* Decne & Planchon	1	1	0.010	0.43	0.24	0.04	0.71
Araliaceae	*Schefflera morototoni* (Aubl.) M.S.F.	1	1	0.016	0.43	0.24	0.06	0.73
Arecaceae	*Astrocaryum murumuru* Mart. var. murumuru	15	11	0.249	4.74	3.53	0.94	9.21
Arecaceae	*Attalea phalerata* Mart. ex Spreng. (= *Scheelea princeps*)	153	34	8.619	14.66	36.00	32.61	83.27
Arecaceae	*Syagrus sancona* H. Karst	32	18	0.742	7.76	7.53	2.81	18.10
Bignoniaceae	*Clytostoma ulegnum* Kanzlin	1	1	0.013	0.43	0.24	0.05	0.72
Bignoniaceae	*Jacaranda* sp. 1	9	8	0.381	3.45	2.12	1.44	7.01
Bignoniaceae	*Tabebuia* sp. 1	9	9	0.303	3.88	2.12	1.15	7.15

APPENDIX 7.2 (continued)

Family	Species	Number of Individuals	Number of Subplots of Occurrence	Basal Area (m²)	Relative Density	Relative Frequency	Relative Dominance	Importance Value (IV)
Bignoniaceae	Bignoniaceae sp. 22	2	1	0.021	0.43	0.47	0.08	0.98
Bignoniaceae	Bignoniaceae sp. 38	1	1	0.009	0.43	0.2403	0.03	0.70
Bombacaceae	Ceiba pentandra Gaertn.	5	5	0.287	2.16	1.18	1.09	4.43
Boraginaceae	Cordia sp. 1	3	3	0.075	1.29	0.71	0.28	2.28
Cecropiaceae	Cecropia concolor Willd.	2	1	0.066	0.43	0.47	0.25	1.15
Clusiaceae	Calophyllum brasiliense Camb.	2	2	0.393	0.86	0.47	1.49	2.82
Combretaceae	Terminalia sp. 1	5	5	0.157	2.16	1.18	1.98	5.32
Euphorbiaceae	Hura crepitans L.	10	6	1.893	2.59	2.35	7.16	12.10
Euphorbiaceae	Sapium glandulosum (L.) Morong	5	5	0.524	2.16	1.18	1.98	5.32
Fabaceae	Swartzia jororii Harms	4	2	0.116	0.86	0.94	0.44	2.24
Flacourtiaceae	Casearia gossypiosperma Briq.	30	15	0.594	6.47	7.06	2.25	15.78
INDET 32	indt. 16	5	4	0.056	1.72	1.18	0.21	3.11
INDET 33	indt. 17	4	4	0.125	1.72	0.94	0.47	3.13
INDET 35	indt. 19	2	1	0.020	0.43	0.47	0.08	0.98

APPENDIX 7.2 (continued)

Family	Species	Number of Individuals	Number of Subplots of Occurrence	Basal Area (m²)	Relative Density	Relative Frequency	Relative Dominance	Importance Value (IV)
INDET 36	indt. 20	5	4	0.153	1.72	1.18	0.58	3.48
INDET 38	indt. 22	2	1	0.021	0.43	0.47	0.08	0.98
INDET 40	indt. 24	1	1	0.008	0.43	0.24	0.03	0.79
INDET 41	indt. 25	1	1	0.010	0.43	0.24	0.04	0.71
INDET 42	indt. 26	1	1	0.016	0.43	0.24	0.06	0.73
INDET 43	indt. 29	4	3	0.149	1.29	0.94	0.56	2.779
Lauraceae	Lauraceae sp. 1	1	1	0.012	0.43	0.24	0.05	
Lauraceae	Lauraceae sp. 26	3	2	0.054	0.86	0.71	0.20	1.77
Mimosaceae	Acacia sp. 1	1	1	0.008	0.43	0.24	0.03	0.70
Mimosaceae	Anadenanthera colubrina (Vell. Conc.) Benth.	2	2	0.407	0.86	0.47	1.54	2.87
Mimosaceae	Inga sp. 17	3	2	0.055	0.86	0.71	0.21	1.78
Mimosaceae	Mimosaceae sp. 25	1	1	0.029	0.43	0.24	0.11	0.78
Mimosaceae	Mimosaceae sp. 28	2	2	0.103	0.86	0.47	0.39	1.72
Moraceae	Ficus sp. 1	6	5	4.782	2.16	1.41	18.09	21.66
Phytolaccaceae	Gallesia integrifolia (Spreng.) Harms	31	19	2.026	8.19	7.29	7.67	23.15

Family	Species	Number of Individuals	Number of Subplots of Occurrence	Basal Area (m²)	Relative Density	Relative Frequency	Relative Dominance	Importance Value (IV)
Polygonaceae	Muehlenbeckia sp. 2	1	0.013	0.43	0.24	0.05	0.72	
Polygonaceae	Triplaris americana Fisch. & Meyer ex C. Meyer	6	6	0.090	2.59	1.41	0.34	4.34
Rubiaceae	Calycophyllum sp. 1	6	3	0.404	1.29	1.41	1.53	4.23
Rubiaceae	Genipa americana L.	2	1	0.145	0.43	0.47	0.55	1.45
Sapindaceae	Sapindus saponaria L.	1	1	0.012	0.43	0.24	0.05	0.72
Sapindaceae	Sapindaceae indt. 10	1	1	0.011	0.43	0.24	0.04	0.71
Sapindaceae	Talisia hexaphylla M. Vahl.	8	8	0.240	3.45	1.88	0.91	6.24
Sapotaceae	Pouteria sp. 16	1	1	0.035	0.43	0.24	0.13	0.80
Sapotaceae	Pouteria sp. 17	1	1	0.673	0.43	0.24	2.55	3.22
Sapotaceae	Sapotaceae sp. 11	1	1	0.259	0.43	0.24	0.98	1.65
Sapotaceae	Sapotaceae sp. 12	1	1	0.012	0.43	0.24	0.05	0.72
Sterculiaceae	Guazuma ulmifolia Lam.	3	3	0.110	1.29	0.71	0.42	2.42
Ulmaceae	Ampelocera ruizii Klotzsch	14	8	1.093	3.45	3.29	4.14	10.88
TOTAL		425	232	26.158				299.1

locations for the game they prize, probably attracted to the fruiting trees. The complex microtopography of the earthwork complex provides numerous artificially created ecotones between terrestrial and aquatic environments. The water-filled barrow pits or ponds provide stable habitats for fish and other aquatic species and year-round drinking water for terrestrial animals (also see Townsend 1995, 1996).

REFERENCES

Balée, W. 1988. Indigenous adaptation to Amazonian palm forests. *Principes* 32:47–54.

——. 1989. The culture of Amazonian forests. *Advances in Economic Botany* 7:1–21.

——. 1994. *Footprints of the Forest: Ka'apor Ethnobotany—the Historical Ecology of Plant Utilization by an Amazonian People.* New York: Columbia University Press.

Balée, W., and D. G. Campbell. 1990. Evidence for the successional status of liana forest (Xingu River basin, Amazonian Brazil). *Biotropica* 22 (1): 36–47.

Boom, B. M. 1986. A forest inventory in Amazonian Bolivia. *Biotropica* 18 (4): 287–294.

Brüchner, H. 1989. *Useful Plants of Neotropical Origin and Their Wild Relatives.* New York: Springer-Verlag.

Bush, M. B. 1994. Amazonian speciation: A necessarily complex model. *Journal of Biogeography* 21 (1): 5–17.

Bush, M. B., P. A. Colinvaux, M. C. Wiemann, D. R. Piperno, and K. B. Liu. 1990. Pleistocene temperature depression and vegetation change in Ecuadorian Amazonia. *Quaternary Research* 34:330–345.

Bustos, V. 1978a. *La arqueología en los llanos del Beni, Bolivia.* Documentos Internos del Instituto Nacional de Arqueología (INAR) no. 32/78. La Paz, Bolivia: INAR.

——. 1978b. Una hipótesis de relaciones culturales entre el altiplano y la vertiente oriental de los Andes. *Pumapunku* (La Paz) 12:115–126.

——. 1978c. *Investigaciones arqueológicas en Trinidad, Departamento del Beni.* Instituto Nacional de Arqueología (INAR) no. 22. La Paz, Bolivia: INAR.

——. 1978d. *Proyecto de excavaciones arqueológicas en el departamento del Beni, en lomas de la región de Cocharcas y Villa Banzar.* Documentos Internos del Instituto Nacional de Arqueología (INAR) no. 10/78. La Paz, Bolivia: INAR.

Campbell, D. G. 1989. Quantitative inventory of tropical forests. In D. G. Campbell and H. D. Hammond, eds., *Floristic Inventory of Tropical Countries,* 523–533. Bronx: New York Botanical Garden.

Campbell, D. G., D. C. Daly, G. T. Prance, and U. N. Maciel. 1986. Quantitative ecological inventory of *terra firme* and *várzea* tropical forest on the Rio Xingú, Brazilian Amazon. *Brittonia* 38 (4): 369–393.

Coimbra Sanz, G. 1980. *Mitologia sirionó.* Santa Cruz, Bolivia: Universidad Gabriel René Moreno.

Colinvaux, P. A., P. E. De Oliveira, J. E. Moreno, M. C. Miller, and M. B. Bush. 1996. A long pollen record from lowland Amazonia: Forest and cooling in glacial times. *Science* 274:85–88.

Denevan, W. M. 1966. *The Aboriginal Cultural Geography of the Llanos de Mojos of Bolivia.* Berkeley: University of California Press.

——. 2001. *Cultivated Landscapes of Native Amazonia and the Andes.* New York: Oxford University Press.

Dougherty, B., and H. Calandra. 1981. Nota preliminar sobre investigaciones arqueológicas en los Llanos de Moxos, Departamento del Beni, Republica de Bolivia. *Revista del Museo de la Plata* (La Plata), Sección Antropologica 53:87–106.

———. 1981–82. Excavaciones arqueológicas en la Loma Alta de Casarabe, Llanos de Moxos, Departamento del Beni, Bolivia. *Relaciones de la Sociedad Argentina de Antropología* (Buenos Aires) 14 (2): 9–48.

———. 1984. Prehispanic human settlement in the Llanos de Mojos, Bolivia. In J. Rabassa, ed., *Quaternary of South America and Antarctic Peninsula,* 2:163–199. Rotterdam: A .D. Balkema.

———. 1985. Ambiente y arqueología en el oriente Boliviano, la Provincia Itenez del Departamento del Beni. *Relaciones de la Sociedad Argentina de Antropología* (Buenos Aires) 16 (n.s.): 37–61.

Durán Coirolo, A., and R. Bracco Boksar, eds. 2000. *Arqueología de las tierras bajas.* Montevideo: Comisión Nacional de Arqueología, Ministerio de Educación y Cultura.

Erickson, C. L. 1980. Sistemas agrícolas prehispánicos en los Llanos de Mojos. *América Indígena* 40 (4): 731–755.

———. 1994. Raised fields as a sustainable agricultural system from Amazonia. Paper presented in the symposium "Recovery of Indigenous Technology and Resources in Bolivia" at the Eighteenth International Congress of the Latin American Studies Association, Atlanta, March 10–12.

———. 1995. Archaeological perspectives on ancient landscapes of the Llanos de Mojos in the Bolivian Amazon. In P. Stahl, ed., *Archaeology in the American Tropics: Current Analytical Methods and Applications,* 66–95. Cambridge, U.K.: Cambridge University Press.

———. 2000a. An artificial landscape-scale fishery in the Bolivian Amazon. *Nature* 408:190–193.

———. 2000b. Los caminos prehispánicos de la Amazonia boliviana. In L. Herrera and M. Cardale de Schrimpff, eds., *Caminos precolombinos: Las vías, los ingenieros y los viajeros,* 15–42. Bogota: Instituto Colombiano de Antropología y Historia.

———. 2000c. Lomas de ocupación en los Llanos de Moxos. In A. Durán Coirolo and R. Bracco Boksar, eds., *La arqueología de las tierras bajas,* 207–226. Montevideo, Uruguay: Comisión Nacional de Arqueología, Ministerio de Educación y Cultura.

———. 2001. Pre-Columbian roads of the Amazon. *Expedition* 43 (2): 21–30.

Erickson, C. L., K. L. Candler, J. Walker, W. Winkler, M. Michel, and D. Angelo. 1993. Informe sobre las investigaciones arqueológicas del Proyecto Agro-Arqueológico del Beni en el 1993. Unpublished report, University of Pennsylvania and the Instituto Nacional de Arqueología, Philadelphia and La Paz.

Erickson, C. L., K. L. Candler, W. Winkler, M. Michel, and J. Walker. 1993. Informe sobre las investigaciones arqueológicas del Proyecto Agro-Arqueológico del Beni en el 1992. Unpublished manuscript, University of Pennsylvania and the Instituto Nacional de Arqueología, Philadelphia and La Paz.

Erickson, C. L., J. Esteves, W. Winkler, and M. Michel. 1991. Estudio preliminar de los sistemas agrícolas precolombinos en el Departamento del Beni, Bolivia. Unpublished manuscript, University of Pennsylvania and the Instituto Nacional de Arqueología, Philadelphia and La Paz.

Erickson, C. L., W. Winkler, and K. Candler. 1997. Las investigaciones arqueológicas en la region de Baures en 1996. Unpublished report, University of Pennsylvania and the Instituto Nacional de Arqueología, Philadelphia and La Paz.

Faldín, J. 1984. La arqueología beniana y su panorama interpretivo. *Arqueología Boliviana* 1:83–90.

Fernández Distel, A. A. 1987a. Informe sobre hallazgos cerámicos superficiales en Eviato (Departamento del Beni, Republica de Bolivia). *Scripta Ethnologica* (Buenos Aires) 11:414–452.

Fernández Distel, A. A. 1987b. Vegetales silvestres útiles entre los Sirionó. *Scripta Etnologica* (Buenos Aires) 11:117–124.

Gentry, A. H. 1993. *A Field Guide to the Families and Genera of Woody Plants of Northwest South America (Colombia, Ecuador, Peru) with Supplementary Notes on Herbaceous Taxa.* Washington, D.C.: Conservation International.

Glaser, B., and W. Woods, eds. 2004. *Explorations in Amazonian Dark Earths in Time and Space.* Heidelberg: Springer-Verlag.

Hanagarth, W. 1993. *Acerca de la geoecología de las sabanas del Beni en el noreste de Bolivia.* La Paz, Bolivia: Editorial Instituto de Ecología.

Hanke, W. 1957. Einige Funde im Beni-Gebiet, Ostbolivien. *Archiv fur Volkerkunde* (Wien) 12:136–143.

Henderson, A. 1995. *Palms of the Amazon.* New York: Oxford University Press.

Holmberg, A. R. [1950] 1985. *Nomads of the Long Bow: The Sirionó of Eastern Bolivia.* Prospect Heights, Ill.: Waveland Press.

Killeen, T. J., E. Garcia, and S. G. Beck, eds. 1993. *Guía de árboles de Bolivia.* St. Louis and La Paz, Bolivia: Missouri Botanical Garden and Quipus S.R.L.

Langstroth, R. 1996. Forest islands in an Amazonian savanna of northeastern-Bolivia. Ph.D. diss., University of Wisconsin, Madison.

Lee, K. 1979. 7,000 años de historia del hombre de Mojos: Agricultura en pampas estériles: Informe preliminar. *Universidad Beni:*23–26.

Lehmann, J., D. C. Kern, B. Glaser, and W. I. Woods, eds. 2003 *Amazonian Dark Earths: Origin, Properties, Management.* Dordrecht, Netherlands: Kluwer Academic.

Lévi-Strauss, C. 1944. On dual organization in South America. *America Indígena* 4:37–47.

Monje Roca, R. 1981. *La nación de los Sirionós.* La Paz, Bolivia: Don Bosco.

Mori, S. A., B. M. Boom, A. M. de Carvalho, and T. S. dos Santos. 1983. Southern Bahian moist forests. *Botanical Review* 49:155–232.

Nordenskiöld, E. 1910. Archaologische Forschungen im Bolivianischen Flachland. *Zeitschrift fur Ethnologie* (Berlin) 42:806–822.

———. 1913. Urnengraber und Mounds im Bolivianischen Flachlande. *Baessler Archiv* (Berlin and Leipzig) 3:205–255.

———. 1916. Die Anpassung der Indianer an die Verhältnisse in den Uberschwemmungsgebieten in Südamerika. *Ymer* (Stockholm) 36:138–155.

———. 1924. *The Ethnography of South America as Seen from Mojos in Bolivia.* Comparative Ethnological Studies No. 3. Göteborg, Sweden: Pehrssons Förlag.

Paz Rivera, C. L. 2003. Forest-use history and the soils and vegetation of a lowland forest in Bolivia. Master's thesis, University of Florida, Gainesville.

Peters, C. M., M. J. Balick, F. Kahn, and A. B. Anderson. 1989. Oligarchic forests of economic plants in Amazonia: Utilization and conservation of an important tropical resource. *Conservation Biology* 3:341–349.

Pia, G. E. 1983. *Investigaciones arqueológicas en el oriente boliviano.* Trinidad: Proyecto de Investigación Oriente Boliviano.

Pinto Parada, Rodolfo. 1987. *Pueblo de leyenda.* Trinidad: Tiempo del Bolivia.

Piperno, D. R., M. B. Bush, and P. A. Colinvaux. 1990. Paleoenvironments and human settlement in late-glacial Panama. *Quaternary Research* 33:108–116.

Pires, J. M., and G. T. Prance. 1985. The vegetation types of the Brazilian Amazon. In G. T. Prance and T. E. Lovejoy, eds., *Key Environments: Amazonia,* 109–145. New York: Pergamon Press.

Prümers, H. 2000. *Informe de labores: Excavaciones arqueológicas en la loma Mendoza (Trinidad) (Proyecto "Lomas de Casarabe") 1ra Temporada, 1991.* Bonn: Comisión de Arqueología General y Comparada (KAVA), Instituto Alemán de Arqueología.

———. 2001. *Informe de labores: Excavaciones arqueológicas en la loma Mendoza (Trinidad) (Proyecto "Lomas de Casarabe") 2nd Temporada, 2000.* Bonn: Comisión de Arqueología General y Comparada (KAVA), Instituto Alemán de Arqueología.

———. 2002a. *Informe de labores: Excavaciones arqueológicas en la Loma Mendoza (Trinidad) (Proyecto "Lomas de Casarabe") 3ra Temporada, 2001.* Bonn: Comisión de Arqueología General y Comparada (KAVA), Instituto Alemán de Arqueología.

———. 2002b. Nota de Prensa. Unpublished document, Comisión de Arqueología Generaly Comparada (KAVA), Instituto Alemán de Arqueología, Bonn.

Riester, J. 1976. *En busca de la loma santa.* La Paz, Bolivia: Editorial Amigos del Libro.

Rydén, S. 1941. *A Study of the Sirionó Indians.* Göteborg, Sweden: Elanders Boktryckeri Aktiebolag.

———. 1964. Tripod ceramics and grater bowls from Mojos, Bolivia. In H. Becher, ed., *Beitrage zur Volkerkunde Süd Amerikas, Festgabe für Herbert Baldus,* Zum 65, 261–270. Hannover: Kommissionsuerlag, Münstermann-Druck.

Salomão, R. P., M. F. F. Silva, and N. A. Rosa. 1988. Inventário ecológico em floresta pluvial tropical de terra firme, Serra Norte, Carajás, Pará. *Boletim do Museu Paraense Emílio Goeldi, Sér. Bot.* 4 (1): 1–46.

Smith, N. J. H., E. A. S. Serrão, P. T. Alvim, and I. C. Falesi. 1995. *Amazonia: Resiliency and Dynamism of the Land and Its People.* New York: United Nations University Press.

Stearman, A. 1987. *No Longer Nomads: The Sirionó Revisited.* Lanham, Md.: Hamilton Press.

Townsend, W. R. 1995. Living on the edge: Sirionó hunting and fishing in lowland Bolivia. Ph.D. diss., University of Florida, Gainesville.

———. 1996. *Nyao Itó: Caza y pesca de los Sirionó.* La Paz, Bolivia: Instituto de Ecología, Universidad Mayor de San Andrés.

Van Roosmalen, M. G. M. 1985. *Fruits of the Guianan Flora.* Utrecht: Institute of Systematic Botany.

Vejarano Carranza, C. 1991. Informe viajes de campo con personas Proyecto Moxos. Unpublished manuscript, Universidad Técnica del Beni, Trinidad, Bolivia.

Walker, J. H. 2004. *Agricultural Change in the Bolivian Amazon.* Latin American Archaeology Reports. Pittsburgh: University of Pittsburgh.

Whitmore, T. C. 1990. *An Introduction to Tropical Rain Forests.* Oxford: Clarendon Press.

8

THE DOMESTICATED LANDSCAPES OF
THE BOLIVIAN AMAZON

CLARK L. ERICKSON

DOMESTICATION IS A comprehensive concept in anthropology referring to the cultural and genetic control of plants and animals and the processes of adopting farming and living in permanent settlements. Native Amazonians domesticated and cultivated a variety of crops (but few animal species) many millennia before the arrival of Europeans. In this chapter, I explore a simple hypothesis: that Amazonian peoples of the past invested more energy in domesticating entire landscapes than in domesticating individual plant and animal species. Through landscape engineering and the use of simple technology such as fire, the past inhabitants "domesticated" the forest, savanna, soil, and water of the Bolivian Amazon, which had profound implications for availability of game animals, economically useful plants, overall biomass, and regional biodiversity. Because the signatures of human activity and engineering are physically embedded in the landscape, archaeology can play a major role in studying these phenomena. The pre-Columbian peoples of the Bolivian Amazon built raised agricultural fields, practiced sophisticated water-management techniques, and lived in large, well-organized communities millennia before European contact. They rearranged soils, altered drainage, constructed earthworks, made marginal lands productive, and in some cases may have increased local biodiversity.

Two themes that are now recognized as myths—one of a pristine environment and the other of the ecologically noble savage—have long dominated the popular and scientific literature on Amazonia. The myth of the pristine environment is the belief that the Americas consisted largely of undisturbed nature before the arrival of Europeans, who subsequently destroyed the environment with their agriculture, mining, and city building. The myth of the ecologically noble savage is the idea that past and present indigenous peoples always existed in harmony with this undisturbed nature. In both myths, nature is imagined as being in a state of perpetual equilibrium with old, undisturbed forests as the ideal form.

Environments portrayed as being in this pristine state are often those with low populations of native peoples practicing "traditional" lifeways. Most or much of Amazonia was considered such a place. In the past few decades, however, research has shown that this state was recently created and not the product of timeless harmonies. Indeed, much of what has been viewed as "pristine wilderness" in the Amazon is the indirect result of massive depopulation after the arrival of Europeans. Within a century, Old World diseases, slavery, missionization, resettlement, and wars eliminated the great majority of the indigenous inhabitants from these landscapes. As a whole, Amazonia did not return to its sixteenth-century population level until the twentieth century. Historical ecologists have shown that before that depopulation occurred, native peoples directly determined much of Amazonia's environmental structure and content. Thus, present-day Amazonian landscapes were shaped by a complex history of past human activities and sudden demographic collapse.

In this chapter, I explore myths, the debunking of myths, and the re-creation of myths in light of contemporary thinking about the relationship between humans and nature in Amazonia. I critique the adaptationist and selectionist approaches that permeate most interpretations of human-environment interaction. Drawing on the insights of new ecology and historical ecology, I argue that the concept of domestication of landscapes provides a powerful alternative perspective. In cultural evolution, the domestication of plants and animals is an important criterion for ranking civilizations or "complex societies." For example, archaeologists have long held in esteem the early domestication of plants and animals and intensive agriculture by societies of the Near East and Asia. Scholars interested in the origins of agriculture rarely recognize the Amazon as a site of agricultural revolution or as a center of crop domestication. My goal in this chapter is to explore a simple hypothesis: that Amazonian peoples of the past invested more energy in domesticating landscapes as a whole than in domesticating individual species of plants and animals. I believe that this domestication of landscape was driven by social demands far beyond the subsistence level.

The new ecology, the archaeology of landscapes, and historical ecology are critical to any understanding of contemporary environments. These approaches highlight the long-term history of landscapes, humans' active role in determining the nature of contemporary environments, and viable models for management of resources and conservation of biodiversity based on indigenous knowledge systems. Daniel Janzen's (1998) metaphor of the gardenification of nature emphasizes that the so-called natural environments of the Americas are actually the historical product of human intentionality and ingenuity, creations that are imposed, built, managed, and maintained by the collective multigenerational knowledge and experience of Native Americans, a point made some time ago by William Denevan, William Balée, Darrell Posey, and others. I argue that understanding the environment as an indigenous *creation* is much more useful and accurate than the more common practice of describing humans as simply "adapting to," "impacting," "transforming," "altering," or "socializing" a static background.

THE CONCEPT OF HUMAN ADAPTATION

According to Emilio Moran, a prominent advocate of the adaptationist model, nongenetic human adaptation (or, more precisely, "human adaptability") "focuses on those functional and structural features of human populations that facilitate their coping with environmental change and stressful conditions" or their making adjustments in "response to constraints" (1982:4). In this perspective, "[a] human population in a given ecosystem will be characterized by strategic behaviors that reflect both present and past environmental pressures. In general the longer a population has been in a given environment, the greater its degree of adaptation to those environmental pressures" (Moran 1993:163). Thus, the diversity of cultures in the Amazon simply reflects a variety of adaptive strategies to a given, diverse set of environmental and historical conditions. In evolutionary ecology, certain "efficient" cultural practices confer a long-term Darwinian advantage on the members of the societies that choose them (Alvard 1994, 1995; Kuznar 2001; Piperno and Pearsall 1998; Rindos 1984; B. Smith 1995). Moran (1993) and his colleagues therefore rank societies as being well adapted or poorly adapted by various empirical criteria.

As Gould and Lewontin (1979) have argued, this approach risks turning evolutionary biology into a sequence of "just-so stories" in which all features of organisms are "explained" as adaptations to some presumed aspect of the environment. Similarly, this perspective can reduce rich and complex cultural systems to examples of adaptation. The adaptationist model has been justly criticized as "tautological, teleological, reductionist, progressive, and victim-blaming" (Goodman and Leatherman 1998:10). But it has long permeated interpretations of Amazonia's past and present and still does. In my view, this approach seriously limits both our understanding of past and present conditions as well as the resolution of the problems currently facing Amazonian peoples and the environment.

CONTEMPORARY PERSPECTIVES ON AMAZONIAN PEOPLES

In much of the literature, the foraging or hunter-gather-fisher folk of the past and present are considered to be "cold" societies—unchanging "people without history" (Wolf 1982) who have little or no impact on their surroundings. Farming and urban folk, by contrast, are granted the status of "hot" societies, capable of changing their environments and transforming the landscape. In this perspective, contemporary foraging societies in the Amazon are often depicted as representatives or remnants of earlier, simple human societies that occupy the bottom rung on the ladder of social evolution: stable adaptations to marginal environments.

Archaeology and historical ecology provide alternative explanations for the existence of the two contrasting lifeways. Almost forty years ago Donald Lathrap (1968) and Claude Lévi-Strauss (1963, 1968) argued that foragers in both the historical and contemporary ethnographic literature of the Amazon are in fact descendants of farmers who had left or been driven out of prime agricultural lands by more powerful indigenous peoples and adopted a foraging lifeway. In contrast, William Balée's (1994, 1995) "hypothesis of agricultural regression" posits that contemporary foraging societies are the legacy of warfare, disease, and colonization, which disrupted farming and village life. Indeed, Balée (1989) and Bailey and colleagues (1989) have shown that many contemporary hunter-gatherers do not so much exploit the fruits of undisturbed nature as harvest the products of abandoned swidden fields and gardens and implicitly or explicitly profit from the cooperation of their agricultural neighbors. Arguing that foragers could not survive without agriculturalists (the "foraging exclusion hypothesis" [Bailey et al. 1989]) can imply that the Amazon and other tropical regions were uninhabitable before agriculture—an idea that, although provocative, has been extensively critiqued (Colinvaux and Bush 1991; Piperno and Pearsall 1998). In this chapter, I argue for a variant of this idea: that Amazon foragers depend not so much directly on the past and present fruits of agriculture as on domesticated landscapes.

Popular approaches borrowed from evolutionary ecology, such as optimal foraging theory, consider Amazonian peoples to be rational, dynamic decision makers who consciously or unconsciously practice the most efficient short-term subsistence strategies for maintaining and reproducing themselves (Alvard 1994, 1995; Kuznar 2001). Subsistence behavior is thus explained in terms of adaptive fitness. Dietary shifts and rescheduling of activities under environmental constraints and stress eventually lead to the domestication of crops and adoption of agricultural economies (Piperno and Pearsall 1998). From this point of view, agriculture and sociopolitical complexity are more usefully regarded as the results of coevolutionary processes and mutualism than as the products of human intentionality and agency.

In the perspectives advanced by historical ecology and the archaeology of landscapes, the cultural ecological, adaptationist, and evolutionary models are turned on their heads. Foragers do not simply "map onto" resources from their slowly changing, naturally determined environments. Contemporary Amazonian foragers and farmers instead work with the products of a historical trajectory of human landscape creation. The availability, distribution, abundance, and productivity of the fauna, flora, soils, and other inorganic resources used by Amazonian peoples today were largely determined by previous, historically contingent, human activity—both the agriculture of the more immediate past and the domestication of landscape that occurred before it, long before domesticated crops and agriculture became recognizable in the archaeological record. The wild/domestic dichotomy so dear to ecologists, anthropologists, and conservationists (but often not to native peoples) is spurious and masks human agency in creating many neotropical landscapes.

Evidence for this early domestication of the neotropical environment is often subtle and indirect. Archaeologists emphasize the appearance of large sites, domesticated crops, pottery, agricultural hardware (such as stone axes, manioc grater blades, *manos,* and *metates*), and forest disturbance as evidence of full-blown agricultural economies and permanent settlement (Lathrap 1970, 1987; Lathrap, Gebhart-Sayer, and Mester 1985; Oliver 2001; Petersen, Neves, and Heckenberger 2001; Piperno and Pearsall 1998; Roosevelt 1980). These signatures were ubiquitous in Amazonia sites by late prehistory. The parallel and possibly earlier transformation of landscapes between sites is often overlooked.

Agriculture was simply a logical, intentional, historically contingent outcome of long-term intensive occupation, use, transformation, creation, and domestication of the Neotropics by humans. Lathrap's hypothesis (1977) that the roots of agriculture are to be found in the sedentary fishing societies on Late Pleistocene landscapes of the lowland Neotropics is slowly becoming accepted (Oliver 2001; Piperno and Pearsall 1998; Roosevelt, Douglas, and Brown 2002; Roosevelt et al. 1996). Lathrap argued that the transplanting and cultivating of economic species within the context of the house garden, a form of localized domestication of landscape, are central to understanding the Amazonian past. In this perspective, the first inhabitants of the region around 11,000–12,000 years ago had already begun the process of domesticating the landscape, specific economic plants, and society itself. Yen points out that most domesticated species were brought into already domesticated landscapes (1989:68–69). Farming and settled village life added new and often intensive strategies to a preexisting knowledge of landscape creation and management.

THE CONCEPT OF DOMESTICATION

The concept of domestication most often refers to the control of plants and animals by humans, a process that began many thousands of years ago in different parts of the world. Domestication and the practices of agriculture are considered major landmarks in the history of humankind—the basis for production of surplus, of transformation of the earth's surface, and of civilization.

Most scholars define *domestication* as the genetic alteration of plants and animals by humans to produce domesticated crop varieties, the basis of full-scale agriculture (Harlan 1992:63–64, 1995:30–31; Harris 1989, 1996; Rindos 1984). Agriculture also involves scheduled activities (e.g., plowing, planting, and harvesting calendars) and practices that harness energy for transforming the environment into productive land (e.g., irrigation and fertilization). The evidence for and explanations of the origins of plant and animal domestication have been discussed in detail (Harlan 1992, 1995; Harris 1996a; Harris and Hillman 1989; B. Smith 1995).

Scholars place a subtle premium on the number of fully domesticated species and the extent of land transformation through intensive agriculture as hallmarks of cultural development. Although possessing important crops such as maize and manioc that are suitable for intensive agriculture, farming in humid tropical regions such as Amazonia is characterized by numerous semidomesticates, or cultivation of both wild species and species propagated through cloning, swidden agriculture, and active management of standing forests (Balée 1994; Clement 1999 and chapter 6, this volume; Denevan 2001; Rival 1998, 2002; and others). In much of Amazonia, game, fish, and palm fruits rather than domesticated animals were sources of protein (Beckerman 1979). Human labor was emphasized over that of draft animals and labor-saving inventions. Based on these traits, many scholars imply that Amazonian societies did not quite "make it" in terms of agricultural advancement and achievement. Some blame this "failure" on the lack of draft animals and on the limitations of Stone Age technology (simple wooden digging sticks and stone axes). As I show later, however, the characterization of neotropical landscape creation as simple ignores the rich historical ecology of this region.

In the following discussion, I use the concept of domestication to refer to cultural activities that transform land or environment into landscape, a form of built environment,[1] thus redirecting the focus away from domesticates and toward landscape. The engineered built environments or domesticated landscapes of certain Amazonian peoples were as impressive as any Egyptian pyramid, Mesopotamian city, or Chinese terrace-irrigation system. In the Amazon, the transformation was driven by social demands: social group formation, domestic routines, territoriality, local environmental knowledge, gifting, and competitive feasting.

THE DOMESTICATION OF LANDSCAPE

The concepts of domesticated landscape, domesticated environment, humanized environment, or domiculture are more than simple metaphors. The terms first entered the mainstream anthropological literature in publications by R. A. Hynes and A. K. Chase (Chase 1989; Hynes and Chase 1982) and by Douglas Yen (1989).[2] Later scholars adopted the concept to explain the origins of agriculture and processes that lead to it (e.g., Clement 1999 and chapter 6, this volume; Harlan 1992, 1995; Harris 1989, 1996b; Kuznar 2001; Terrell et al. 2003; and others). Chase specifically defines *domiculture* as "hearth-based areas of exploitation (domuses), each carrying with it a package of resource locations, restrictions upon open-ended exploitation (religious prohibitions, strategic planning of delayed harvesting, etc.), and localized technologies to fit particular domuses" (1989:43). He clearly stresses the intentional, conscious knowledge systems and deliberate activities of humans that can, but do not necessarily, lead to genetic domestication of species (1989:44, 46–47). Thus, the domestication of landscape encompasses *all nongenetic, intentional, and unintentional practices and activities of humans that transform local and regional environments into productive, physically patterned, cultural landscapes for humans and other species.*[3]

The use of the ethnography and history of foragers and cultivators for modeling the evolutionary processes leading to genetic domestication of crops and the origins of agriculture has a long history (for Amazonia, see Clement 1999 and chapter 6, this volume; Hastorf 1998 and chapter 3, this volume; Lathrap 1977, 1987; Piperno and Pearsall 1998; Sauer 1952). In regard to plants, the processes leading to genetic domestication involve planting, transplanting, tending ("husbandry" or "mothering"), cultivation, weeding, transport outside natural habitats, and the use of fire as a management tool to enhance survival of economic species. Contemporary Amazonian peoples are constantly gardening the forest, weeding and pruning here and there, as they move across the landscape. One of the early major contributions by historical ecologists was pointing out that what looks like a "natural" environment rich in resources is often actually an environment constructed and managed as forest fallow in agroforestry regimes (Balée 1994; Denevan and Padoch 1988; Peters 2000; Posey and Balée 1989).

Many researchers have generally treated these practices that domesticate landscapes as distinct from and less sophisticated than the more "advanced" practices of intensive agriculture and genetic domestication. They view burning, collecting and foraging, keeping of wild pets, agroforestry, cultivation, horticulture, gardening, and settlement mobility as evolutionarily more "primitive" than full-blown agriculture. As such, ethnographic case studies are employed as analogy to explain intermediate stages in cultural evolution between foraging, on the one hand, and agriculture and the processes leading to farming, on the other (e.g., Harlan 1992, 1995; Harris 1989, 1996b; Ingold 1987, 1996; Rindos 1984; Shipek 1989; Wiersum 1997a, 1997b; and others). Some also stress the assumed behavioral contrasts between what foragers do to and think about the environment and what farmers do and think: foragers exert low or minimal impact on environments, whereas farmers exert high impact on environments (Ingold 1987, 1996). Drawing on Friedrich Engels, Ingold proposes a distinction between, on the one hand, agriculturalists and pastoralists who truly transform or "master" their environment through food production and, on the other, hunter-gatherers who merely "use it" while collecting and hunting their food (1987:71). "Like the architect's house, the farmer's field is artificial, *engineered* by human action," he states, whereas "the environment of the hunter-gatherer *is not constructed but co-opted, it is not artificial but 'natureficial'*" (72–73, emphasis in original).

In reading this literature, one notes the reluctance to grant agency and history to preagricultural and nonagricultural peoples and to appreciate the subtleties and importance of the domestication of landscape. Instead, the old nature/culture dichotomy is replicated in the form of a forager/farmer dichotomy (e.g., Harris 1989, 1996b: "evolutionary continuum of people-plant interactions"). But as Hynes, Chase, and Yen have stressed, landscape domestication is not only a precursor to plant and animal domestication, but also a sophisticated set of social strategies and practices of its own. As Yen stresses, "None of this [critique of the forager/farmer dichotomy] constitutes a denial of the derivation of agriculture from hunting-gathering roots; rather it questions that contemporary hunter-gatherers are the backwards relics of a single evolutionary line which most accounts of agricultural development seem to suggest" (1989:66). Burning, collecting, pruning, and other landscape-transforming practices are not performed by static, ahistorical societies as by-products of the road to agriculture. They are chosen strategies, goals, or ends in themselves (Chase 1989; Descola 1996; Hather 1996; Hynes and Chase 1982; Michon and De Floresta 1997; Rival 1998, 2002; Spriggs 1996; Terrell et al. 2003; Yen 1989; also see critiques raised by Lathrap [1968] and Lévi-Strauss [1963, 1968] regarding archaism and the foraging peoples of the Amazon).

Discussions about the domestication of landscape parallel discussions about the distinction between cultivation of plants (wild and domestic species) and agriculture (cultivation of domesticated species) raised by scholars interested in the origins of agriculture (Harris 1996a; Harris and Hillman 1989). Such research differs from historical ecology in that in the former the focus is on the cultivation of particular plants and animals rather than on the cultivation of landscapes. Yen highlights the contrast in stating that "the effect of the hunter-gatherer domestication of environment may be likened to a form of group selection, in which the plant targets are aggregated as interbreeding units, compared with the individual selection practiced by the agriculturalist, which establishes closer control over the plants' breeding systems and can result in the varietal differentiation of species into physiological types, e.g. wet and dry adaptations in rice and Colocasia taro" (1989:66). Terrell and colleagues (2003) implement these ideas methodologically in their "interactive matrix of species and harvesting tactics" (what they call a "provisions spreadsheet"), which includes numerous economic species affected by human activities in the process of domestication of landscape rather than the genetic domestication of selected plants and animals.

Some evolutionary ecologists and evolutionary anthropologists adopt the concept of domestication of landscape as Darwinian coevolution or mutualism of human-environment interaction. In this view, human culture is often reduced to a form of animal-environment coevolutionary interaction (a stance well critiqued in Balée 1989; Ingold 2000; and Terrell et al. 2003). In contrast to the original concept of domestication of landscape proposed by Hynes, Chase, and Yen and to the perspective of historical ecology presented here, evolutionary ecology includes little or no role for human intentionality and agency. Indeed, evolutionary ecologists regard much of what native peoples did or do to the environment over the long term as epiphenomenal, unintentional, and unconscious (see Alvard 1994, 1995; Kuznar 2001; Rindos 1984; and others). Most historical ecologists (and some ecologists), however, view humans as a "keystone species" rather than as simply another animal with mutualistic relationships (e.g., Kay and Simmons 2002; Mann 2002; O'Neil 2001; Terrell et al. 2003; and others).

I believe that the domestication of landscape occurred before, during, and after plants and animals were genetically domesticated and full-scale agriculture was developed. By placing a premium on the presence of specific domesticated crops, agricultural hardware, and long-term settlements, archaeologists and anthropologists have often missed the importance of landscape domestication. As an anthropological archaeologist and historical ecologist, I also argue that the intentional human behavior is more interesting than the unintentional. These ideas are not new. Certain archaeologists, ethnographers, geographers, and economic botanists who have studied traditional agriculture have long emphasized a human-centric landscape approach to understanding the long-term, historical relationship between humans and the environment (e.g., the numerous studies of land management by native peoples of the Americas summarized in Denevan 2001; Doolittle 2000; and Whitmore and Turner 2002).

In summary, the concept of domestication of landscape contributes to historical ecology in several ways. First, by redirecting attention away from the Neolithic Revolution, agriculture, and specific domesticated crops as being the most important transformation of environment, one can better appreciate the importance of human cultural activities that do not change the genetics of the specific species that lead to domestication, but that influence the presence, availability, and productivity of these species. Second, the focus on multiple species rather than on individual cultivated or cropped species redirects attention toward the landscape as a complex and historical context. Third, by unraveling the unproductive dichotomy between foragers who practice hunting, gathering, fishing, and cultivation and farmers who practice agriculture—a distinction assumed in the perspectives of cultural ecology, human ecology, cultural evolution, and evolutionary ecology—one can better understand more subtle, but important strategies of anthropogenic environmental change. Fourth, by rejecting the simple linear evolutionary continuum from foraging to agriculture, one can appreciate that the domestication of landscape can be an end in itself for the creation of productive landscapes. And fifth, moving beyond a critique of the myths of pristine environments and ecologically noble savages as well as beyond the often concomitant assumptions that all human activities affect the environment negatively, one can begin to appreciate human creativity and environmental transformation if one focuses on conscious activities, application of environmental knowledge, and engineering employed to domesticate landscapes for human use, all of which can often result in changes in biodiversity and in the spatial distribution and availability of economically useful species.

THE ANTHROPOGENIC ENVIRONMENT, BIODIVERSITY, AND HUMAN ACTIVITIES

Most historical ecologists now agree that the Amazonian rain forest is a form of cultural artifact, to at least some degree. We now know that Amazonian peoples practice sophisticated forms of indigenous cultivation, gardening, and agroforestry management. They move seeds and plants around the forest, create gap disturbance, encourage certain economic species, and remove others. The long rest or fallow period between cycles of burning and cultivation in tropical agriculture is actually part of long-term production and harvest strategies (Clement, chapter 6, this volume; Denevan and Padoch 1988; Posey and Balée 1989). The old equilibrium model of ecological succession in Amazonia—the cutting and clearing of a forest patch, cultivation of crops for 2–3 years, then the abandonment of the plot and (re)growth of the forest—is being replaced by more sophisticated and dynamic models that take into account historical contingency and human agency. Long-term human activities and disturbances, not laws of ecological succession, have determined the form and structure of the Amazonian tropical forest.

Simply defining *landscape* as the interaction between humans and environment—or setting up *anthropogenic* in binary opposition to *natural*—does not place enough emphasis on human intentionality. Coevolutionary models (Kuznar 2001; Piperno and Pearsall 1998; Rindos 1984; B. Smith 1995), "dialectical" approaches (Crumley 1994; Crumley and Marquardt 1987), self-organizing systems and resiliency theory (Redman and Kinzig 2003) often depict humans as being swept up in a long-term process that unconsciously modifies the environment.

These long-term anthropogenic processes are often seen as more important than short-term human activities in shaping the neotropical environment. For example, large areas of the Amazon are classified as "black earth" or dark earth, a rich human-produced anthrosol prized by farmers of the region. Amazonian Dark Earth sites are often discussed as an unconscious by-product of settlement by pre-Columbian peoples over thousands of years (Glaser and Woods 2004; Lehmann et al. 2003; Neves and Peterson, chapter 9, this volume; N. Smith 1980). In contrast, I have argued (Erickson 2003) that the phenomenon of Amazonian Dark Earths is actually an excellent example of the intentional domestication of soils—specifically, the creation of ideal conditions or habitat for certain microorganisms that improve and sustain fertility through the incorporation of organic matter and of carbon produced under low temperature into the soil, as recently documented by soil scientists.

Elizabeth Graham (1998) extends the concept of "built environment" to include all soils formed under repeatedly occupied and farmed landscapes of the Americas. In my view, the term applies to most of Amazonia. *Built environment* implies that created landscapes are planned in advance and passed down through generations in a process that might be called "landscape accumulation." Such built landscapes are produced by a conscious indigenous knowledge system operating in a historical context. Because they are historically contingent, they are often complex palimpsests of human activity (Erickson and Balée, chapter 7, this volume).

The relationship between humans and biodiversity in the Neotropics is hotly debated. Biologists, ecologists, geographers, and anthropologists are reaching a consensus that a certain level of disturbance or disequilibrium is critical for creating and maintaining biodiversity and environmental health (Blumler 1996, 1998; Botkin 1990; Connell 1978; O'Neil 2001; Stahl 1996, 2000, and chapter 4, this volume; Zimmerer and Young 1998). Although both natural and anthropogenic disturbances play a role in shaping nature, historical ecologists are concerned with patterned, varied, and sustained human activity.

Do anthropogenic processes exert positive or negative impacts on biodiversity? Whether human activities degrade or enhance biodiversity often depends on how *biodiversity* is defined and measured and at what temporal and geographical scale (Stahl 1996). Nonetheless, many scholars argue that there is a general answer: that indigenous peoples' activities tend to reduce species richness and ecological health (e.g., Redman 1999). In their critique of the myth of the ecologically noble savage, numerous researchers now claim that native peoples hunt game animals to extinction, degrade the environment, and waste precious resources (e.g., Kay and Simmons 2002; Krech 1999; Redford 1991; Stearman 1994). As seen through the lens of evolutionary ecology (including optimal foraging theory), native peoples stress short-term, selfish benefits over long-term goals and thus do not practice resource conservation (Alvard 1994,

1995). Other researchers argue the opposite: that natives enhance biodiversity as resource managers (Balée 1989, 1994, 1995, 1998a, 1998b; Brookfield 2001; Denevan and Padoch 1988; Maffi 2001; Posey and Balée 1989).

Both sides are beginning to agree, however, that there is no "natural" baseline or benchmark of pristine wilderness that should be used as a standard for comparisons (Bennett 1962; Denevan 1992; Hunter 1996; Stahl 1996). The question of whether human activities are positive or negative becomes complicated if humans played a major role in creating the very landscapes where biodiversity and nature are said to exist.

The environment of the Amazon region of tropical South America was long considered to be of limited potential. It was commonly believed that in the past, as in the present, the social and political organization of indigenous peoples was simple, that populations were nomadic or widely dispersed over the landscape, and that subsistence was based on hunting, gathering, and small-scale agriculture. The extent to which contemporary tropical forest foragers and small-scale farmers are appropriate models for understanding the Amazonian has been questioned, however (see summary in Heckenberger, chapter 10, this volume; Stahl 2002). Because population densities are low today, many scholars have assumed that early human inhabitants of the Neotropics were "cold," ahistorical societies of foragers who did not significantly alter the environments in which they lived. In the 1950s and 1960s, it was said that the scale of agriculture and settlements was constrained by poor soils (Meggers 1954), and in the 1970s researchers considered the lack of animal protein to be the constraint (Beckerman 1979; Gross 1975).

In the 1960s, the discovery of massive raised-field systems, causeways, canals, occupation mounds and other earthworks, and urban settlements in many parts of Amazonia challenged this perspective (Denevan 1966, 2001). By now, it is generally recognized that most of the 5 to 6 million inhabitants of Greater Amazonia in the centuries before European conquest were agriculturalists living in large permanent settlements. The transformation of the Amazon basin from lightly settled "wilderness" to densely populated agriculture is usually attributed to farmers of late prehistory. I argue instead that the changeover took much longer than has been commonly supposed and that its major focus was not the development of intensive agriculture, but rather landscape transformation, specifically the domestication of landscape. From their interpretation of lake cores, Piperno and Pearsall (1998) believe that landscape transformation began as early as 11,000 years ago. Roosevelt and colleagues (1996) show that the inhabitants of the Monte Alegre rock shelter on the central Amazon were altering that environment more than 11,000 years ago. The effects of hunting and gathering on the Amazonian environment have been greatly underestimated. Anthropologists have recently shown that small mobile groups of foragers such as the Nukak (Politis 1996) and the Hoti (Zent and Zent 2004) have had profound, massive, and permanent impact on forests through their frequent shifts of residence and discard of edible palm fruits. Thus, agriculture is only one of many anthropogenic processes that shaped the neotropical environment in prehistory (Balée 1994; Descola 1996; Rival 1998; and others).

The debate about the relationship between human activity and biodiversity over time is complex. Ideally, one would be able to compare anthropogenic landscapes with natural landscapes. Most natural scientists assume some natural, pristine, or prehuman baseline or benchmark upon which comparisons can be made: a past or present wilderness devoid of humans and their activities. However, is there a single documented region or locale in the Amazon that has not been disturbed, shaped, transformed, or produced by humans to some degree since the end of the Pleistocene (11,000–10,000 years ago)? I argue that there is no pristine or natural environment for comparison. One might respond that the prehuman environment can be reconstructed on the basis of paleoenvironmental research in Late Pleistocene and Early Holocene period contexts, but the relevancy of such environmental reconstructions would be limited. I doubt that Late Pleistocene ice age environments can be chosen as a natural, pristine, or prehuman benchmark to compare with anthropogenic landscapes, to address contemporary debates about human activities and biodiversity, or to guide contemporary conservation efforts in the Amazon.

THE ARCHAEOLOGY OF DOMESTICATED LANDSCAPES IN THE BOLIVIAN AMAZON

In the Bolivian Amazon, known locally as the Llanos de Mojos (grassland plains of Mojos or Moxos), the domestication of landscape included interrelated and overlapping human activities that over time created a complex, highly structured, engineered cultural landscape (figure 8.1). This domestication of landscape included burning, transplanting, constructing roads, farming, establishing mound and forest island settlements, and creating artificial wetlands that permanently altered topography, soil structure and fertility, hydrology, faunal and floral community structure, local climate, and biodiversity (figure 8.2).

Pre-Columbian farmers heavily modified the savanna and forest landscape of the Bolivian Amazon, creating over time a complex, highly structured, engineered cultural landscape. Erland Nordenskiöld (1910, 1913, 2003) at the turn of the twentieth century and William Denevan (1963, 1966, 2001) and George Plafker (1963) in the early 1960s discovered that large expanses of the savannas of the Llanos de Mojos were covered with massive earthworks, including raised fields, canals, causeways, reservoirs, dikes, and mound settlements constructed by the pre-Hispanic inhabitants of the zone.[4] Denevan and his colleagues have found similar raised fields throughout the Americas in the flooded savannas of

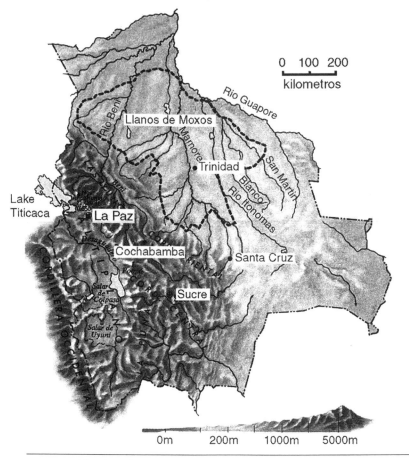

FIGURE 8.1 The Llanos de Mojos, Bolivia.

Colombia, Ecuador, Venezuela, and Surinam, as well as in the Andean region of Peru, Bolivia, Colombia, and Ecuador (Denevan 2001). More recently, scholars have documented other cases of massive landscape transformation (Glaser and Woods 2004; Heckenberger et al. 2003; Lehmann et al. 2003; Raffles and WinklerPrins 2003).

The Llanos de Mojos (or Moxos) is located in the southwestern headwaters of the Amazonian drainage basin. The region corresponds roughly to the modern political boundaries of the Department of the Beni of Bolivia. The area is a relatively flat landscape of forest along rivers and higher ground *(bosque, galeria,* and *islas de monte)* (20,000 square kilometers) and savanna grasslands (pampa), scrub and palm forest, and wetlands (90,000 square kilometers) (Denevan 2001). Much of the low-lying lands are covered with shallow floodwaters during the rainy season. During the rest of the year, dry conditions prevail, and water

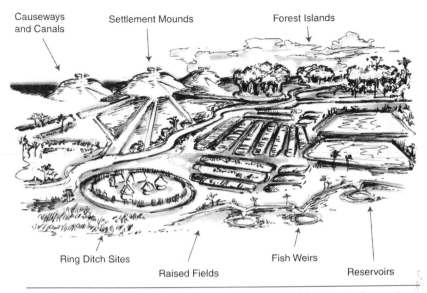

Causeways and Canals

Settlement Mounds

Forest Islands

Ring Ditch Sites

Raised Fields

Fish Weirs

Reservoirs

FIGURE 8.2 Major anthropogenic features of the domesticated landscapes of the Bolivian Amazon. (Artwork by Daniel Brinkmeier)

becomes a scarce commodity. Poor soils, the lack of high ground, and the alternation of seasonal floods with dry conditions make farming difficult.

By burning, transplanting, moving vast amounts of earth, establishing mound and forest island settlements, and creating artificial wetlands, the pre-Columbian inhabitants of the Llanos de Mojos permanently altered the region's topography, hydrology, soil structure and fertility, faunal and floral community structure, local climate, and biodiversity, creating a productive anthropogenic landscape. A major part of this transformation of the environment was the construction of raised fields, causeways, canals, reservoirs, mounds, forest islands, ring ditch sites, fish weirs, ponds, and other structures. We now know that this domesticated landscape sustained large populations organized in large villages and towns dispersed within the savannas and forests (Denevan 2001; Erickson 1995; Erickson and Balée, chapter 7, this volume; Walker 2004).

The Bolivian Amazon provides an interesting case study in historical ecology. Tropical savannas and forests are often conceptualized as binary categories. As Fairhead and Leach (1996) point out, much of the contemporary literature in the natural sciences and development theory treats tropical savannas as marginal environments or even, in more extreme cases, as degraded tropical forests. Many scholars still view savannas as the end result of deforestation and overexploitation of the tropical environment and thus as worthless and of little value to study. Yet these "degraded" landscapes supported some of the densest populations and the most elaborate sociopolitical institutions in the Amazon during late prehistory. The native peoples (Mojo, Baure, and others) of the Bolivian Amazon and their ancestors are prime examples of dense, well-organized, indigenous societies living in a constructed, engineered landscape.

BURNING

Burning is now recognized as a major factor in the creation and maintenance of savannas in addition to its use in general forest management (Pyne 1998). Historical records attest that in the past native peoples systematically burned the savannas of the Bolivian Amazon (Denevan 1966; Hanagarth 1993; Langstroth 1996), as do the ranchers, farmers, and hunters who live there now. Large fire fronts, often extending for kilometers, sweep across the savanna during the dry season (figure 8.3). Those who burn today stress that fire removes dead grasses, encourages new grasses for livestock and game animals, keeps the forest at bay, and "cleans" the landscape. Most insist that the 90,000 square kilometers of savanna exist only because of regular anthropogenic burning, although the annual flooding and soils may play some role (Denevan 1966; Langstroth 1996).[5]

I long assumed that raised fields, causeways, canals, and other nonmound earthworks were most appropriate for the present-day savanna and wetlands and that slash-and-burn agriculture, gardening, and agroforestry (which leave little archaeological footprint) were practiced in the forests in a way similar to what

FIGURE 8.3 A 2-kilometer-long fire front during the annual burning of the savannas by ranchers and farmers of the Bolivian Amazon. Anthropogenic fire has been used as an environmental management tool for thousands of years.

farmers do today. Thus, based on Denevan's pioneering research, I had expected to find pre-Columbian earthworks in the open savannas, where their construction would be most efficient due to lack of forest and annual flooding. Over the past 13 years, though, colleagues and I have identified many raised fields, causeways, and canals under tall, mature, continuous canopy forest. Large areas of fish weirs and causeway-canal networks are now completely covered with *Mauritia flexuosa* palm forest in Baures in the northeastern Bolivian Amazon.

This complex data set suggests that these locations were either former savanna used for human activities and later colonized by trees or former forest that had been intentionally cleared by humans for farming and other activities, but later (re)grew after agricultural abandonment. Massive depopulation of the region in the Colonial period due to epidemics, mission failure, and exploitive labor demand may account for the regeneration of forest on what was previously open savanna. The phenomenon also suggests that if forest grows there today, either it grew there "naturally" in the distant prehuman past or humans created favorable conditions for its establishment. In the Bolivian Amazon, colleagues and I have found a strong association between areas that are open savanna today and areas where burning has been continuous since the arrival of Europeans. By contrast, areas that were previously savanna with extensive pre-Columbian earthworks are now heavily forested. In some cases, certain anthropogenic characteristics of the landscape (discussed later; see also Erickson and Balée, chapter 7, this volume) facilitated the expansion of the forest, which since the arrival of Europeans has invaded large areas of the savanna.

RAISED FIELDS

Raised fields are large earthen planting surfaces elevated above the seasonally flooded savannas and wetlands (figure 8.4). Experiments and ethnographic analogy demonstrate that raised fields serve a wide variety of functions, including localized drainage, improved soil conditions (by aerating the soil, mixing soil horizons, and doubling the organic topsoil), water management (for drainage and irrigation), nutrient production, capture, and recycling in the adjacent canals (through sediment sinks, organic muck production, and management of economic faunal and floral resources).

Until recently, the pre-Columbian raised fields of Mojos remained unstudied (Denevan 1963, 1966; Plafker 1963). Denevan recorded 35,000 raised fields with 6,000 hectares of platform surfaces, based on his interpretation of aerial photographs of savanna (2001:246).[6] Analysis of recent aerial photographic coverage has identified the faint traces of raised fields over a much larger area of savanna, and ground survey has located many additional raised fields under dense tree canopy. Raised-field form and size are variable, ranging from 1.5–6 meters wide, 6–300 meters long, and 0.3–1.0 meter in height along the Apere River to 5–20 meters

FIGURE 8.4 Pre-Columbian raised fields north of Santa Ana de Yacuma, Bolivia. Field platforms (lighter areas) measuring 15–20 meters wide, 50–150 meters long, and 0.5–1 meter tall are separated by canals of similar dimensions.

wide, 300 meters long, and 0.5–1.0 meter in height along the Iruyañez River (Denevan 2001:241–246; Dougherty and Calandra 1984; Erickson 1995; Michel 1993; Walker 2004). Stratigraphic profiles in excavation trenches through raised fields and canals document a considerable volume of earth moved during the construction and maintenance of raised fields. Field patterning is highly structured in some areas; in others, the fields are more informally organized. Discrete groupings of fields, bounded or unbounded by causeways and canals, may reflect pre-Hispanic land-tenure systems and the social organization of farmers who constructed and maintained these fields (Erickson 1995; Walker 2004). Regional distinctions in the types of earthworks that are present suggest cultural and technological diversity (Denevan 1966, 2001; Erickson 1995).

Whereas previous projects studied settlement and burial mounds, more recent archaeological survey, mapping, and excavations focus on the agricultural earthworks and the associated hydraulic infrastructure. Trenches excavated in pre-Hispanic raised fields and causeways have provided valuable information on the internal structure of the earthworks, construction techniques, rebuilding phases, sedimentation rates, original functions, crops cultivated, soil fertility, and the chronology of field construction, use, and abandonment. Preliminary analysis of pollen from field canals has identified the presence of cocoyam (gualusa, *Xanthosoma* sp.), guayusa (*Ilex* sp., mate, used for a caffeine-rich drink), and urucu (*Bixa orellana* L., used for red body paint), all of which almost certainly were cultivated (Erickson 1995). Based on historical documents and raised-field experiments, manioc, sweet potatoes, peanuts, beans, squash, and possibly maize were probably the major crops.

Earthwork stratigraphy shows a complex succession of construction, use, maintenance, and expansion episodes. Through radiocarbon dating of raised fields and associated settlements, colleagues and I (Erickson 1995; Erickson et al. 1991; Erickson and Walker n.d.; Walker 2004) have documented human occupation and use of the flooded savannas by 900 BC and the establishment and expansion of raised-field agriculture from 400 BC until the arrival of Europeans, when the system was abandoned.

Experimental archaeology has also contributed to our understanding of the function, labor investment in construction and maintenance, crop yields, and sustainability of raised-field agriculture (figure 8.5). A total of 900 person-days were needed to construct a single hectare of fields and canals using metal bladed shovels and picks tools. Our raised-field experiments at the Biological Station of the Beni produced bountiful harvests of manioc (Erickson 1994, 1995). Experiments by Bolivian agronomists recorded yields of 12–24 metric tons per hectare of manioc, 14 of squash, 12 of sweet potatoes, 0.5 of beans, and 0.2 of maize (Arce 1993; C. Pérez 1995; T. Pérez 1996; Stab and Arce 2000:320, table 16.1). The yields surpassed local production in slash-and-burn fields (with the exception of maize, which was probably cultivated on better soils). This finding is surprising because soil samples analyzed from this experimental site were found to be poor for agriculture (Jacob n.d.; Stab and Arce 2000). Although some of the experimental fields have provided substantial production for a number of years, the sustainability of the system without organic inputs is unknown. Additional raised-field experiments demonstrate that yields can be increased by improving soil conditions through incorporating canal sediments, dung, and a mulch of water hyacinth *(Eichhornia azurea)*, an aquatic plant that thrives in raised-field canals (C. Pérez 1995; T. Pérez 1996; Saavedra forthcoming; Stab and Arce 2000).[7] The Bolivian agronomists and I believe that with sufficient labor input and under proper management, raised-field agriculture is probably productive and sustainable; thus, the technology could have supported large dense populations during its 2,000 years of use.

The volume of earth moved and restructured in the construction of raised fields is impressive (Erickson 1995). The construction of raised fields and canals altered and restructured soil horizons to a depth of 0.5–1.0 meter, and the original vegetation was removed, burned, or buried. Our mapping shows that for any given block of raised fields, the area occupied by canals and platforms is roughly equal. Thus, the construction of a hectare of raised fields involved moving 2,500—5,000 cubic meters of earth (or 250,000–500,000 cubic meters of earth per square kilometer of raised fields). The alternating topography of platforms and canals replaced a relatively flat landscape and substantially changed

FIGURE 8.5 Experimental raised fields with manioc at the Biological Station of the Beni. Manioc is growing on the platforms. Note the continuous terrestrial-aquatic ecotone surrounding each field platform. (Photograph by Robert Langstroth)

local drainage patterns. The microrelief of platforms and canals dramatically increased the culturally usable area of this highly productive terrestrial and aquatic ecotone or interface (also long recognized to be high in biodiversity). For example, construction of 1.0 hectare of raised fields (0.5 hectares of platforms and canals each) with platforms measuring 5 by 20 meters creates 2.5 linear kilometers of terrestrial-aquatic ecotone (one square kilometer of raised fields creates 250 linear kilometers of terrestrial-aquatic ecotone).

The ridge and swale topography created by raised-field construction also plays a role in the presence and structure of forests today. In many cases, the annual burns, which keep the savannas open, are restricted by the moisture in the old canals between platforms, allowing trees to become established on platforms. Often the resultant anthropogenic forests are highly patterned, with the trees growing in straight, orchardlike rows spaced by the alternation of platform and canal. Our ground survey has shown that most sharp linear boundaries between forest and savanna are due to pre-Columbian canals protecting forests from burns (Erickson 2001:21, figure 6).

CAUSEWAYS AND CANALS

The raised fields, mounds, and forest islands are often associated with complex networks of large, long causeways and canals (Denevan 1966, 1990; Erickson

2000b, 2001). Causeways are constructed of earth removed from canals on one or both sides. On the ground, these causeways are low structures of 0.25–1.0 meter in elevation, 4–6 meters wide, and often 2–5 kilometers long. Most are badly eroded, and many are covered with trees and bushes, a sharp contrast to the surrounding grass-covered savanna (figure 8.6). These pre-Columbian earthworks served for transportation and communication between villages and towns located on mounds, forest islands, and gallery forests (figure 8.7).

Baures in northeastern Bolivia has dense networks of long causeways and canals that cross the savannas, wetlands, and forested islands (Eder [1772] 1985; Erickson 2000b, 2001; Lee 1995). Some segments of old causeways between local settlements and ranches are still used today for communication and transportation during the rainy season. The Baures Hydraulic Complex, located between the San Joaquin and San Martin rivers, has the densest concentration of these features. There are thousands of linear kilometers of causeways and canals in this zone, most of which are remarkably straight and several kilometers long. Many cross one another and some connect to other causeways. In a number of cases, two to four causeways run parallel (figure 8.6). Foot traffic would have used the raised roadways for communication and transportation between settlements and between settlements, rivers, and agricultural fields, and throughout much of the year canoe traffic would have been possible in the adjacent canals (Denevan 1966, 1990). In addition, the builders' obsession with straightness over long distances, the "overengineering" of the designs and construction, and the sheer number of these features suggest that they may have had ritual and political functions, possibly associated with astronomy, calendrics, or specific ceremonies.

In terms of domestication of landscape, I and others suggest that these earthworks were also part of an integrated system of water management both locally and regionally (Erickson 1980, 2000b; Erickson, Winkler, and Candler 1997; Lee 1979, 1995, n.d.). Geographic information system (GIS) analysis of detailed topographic mapping in several agricultural raised-field sites has begun to address this hypothesis. Causeways appear to have been used to block the flow of water, impounding large bodies of shallow water within raised-field blocks. Opening and closing sections of causeways could have maintained optimal water levels for field cultivation. For example, a single 1-meter-tall causeway that connected the levees of two rivers 2 kilometers apart would create a shallow 10-square-kilometer lake that retained 5 million cubic meters of water (based on an average slope of 20 centimeters per kilometer).[8]

Nordenskiöld (1916) and Denevan (1966) reported cases of the use of artificial canals constructed across the necks of long river meanders and between rivers to shorten canoe travel time. Nordenskiöld proposed that these activities may have eventually changed river courses and created new oxbow lakes. In Baures, colleagues and I have documented a number of pre-Columbian, historical, and

FIGURE 8.6 Two parallel causeways and associated canals (the dark linear features) covered with palms crossing the savanna between forest islands west of Baures, Bolivia. Because of irregular burning in this unpopulated area, *Mauritia flexuosa* have encroached on the savanna.

FIGURE 8.7 Causeway *(background)* and canal *(foreground)* in use for transportation and communication between fields and settlement mounds *(on horizon)*. (Artwork by Daniel Brinkmeier)

ethnographic canals as river meander cutoffs and connections between rivers (Erickson 2000b, 2001; Erickson, Winkler, and Candler 1997). The rivers of the central Llanos de Mojos tend to flow south to north, making east–west travel difficult by canoe. The pre-Columbian inhabitants solved this problem by carving channels between rivers and other natural water bodies for their canoes (Denevan 1966, 1990; Nordenskiöld 1916). Although most of these channels are less than a kilometer long, Pinto Parada (1987) mapped a continuous aquatic transportation network of 120 linear kilometers of artificial and natural water bodies between San Ignacio and Casarabe.

MOUNDS

Raised fields and associated earthworks sustained large populations organized as hamlets, villages, towns, and possibly urban centers dispersed across the savannas and forests (Erickson 2000c; Walker 2004). Pre-Columbian inhabitants of the Bolivian Amazon constructed large artificial mounds *(lomas)* of earth and domestic rubbish (Denevan 1966; Erickson 2000c; Erickson and Balée, chapter 7, this volume; Pinto Parada 1987). These mounds range in size from the huge Ibibate Mound Complex (18 meters tall and covering 9 hectares) (Erickson and Balée, chapter 7, this volume) to isolated small mounds and related forest islands that cover the savannas and gallery forests of the Bolivian Amazon (Erickson 2000c; Langstroth 1996; Walker 2004).[9] Based on a survey of ranchers, Lee (1995) estimates a total of 10,000 settlement mounds (including forest islands) in the Bolivian Amazon. Many of the larger mounds were occupied continuously for hundreds, possibly thousands, of years, only to be abandoned during the population collapse during the Colonial period. Today they are still valued as dry and fertile locations for settlement, corrals, fields, orchards, and gardens. Archaeological excavations and surface collections have recovered a rich inventory of pottery, animal bones, shell, and stone axes from mounds and forest islands, thus demonstrating that they were important locations for settlement (Bustos 1978a, 1978b, 1978c, 1978d; Denevan 1966; Dougherty and Calandra 1981, 1981–82, 1983, 1984; Erickson 2000c; Langstroth 1996; Nordenskiöld 1910, 1913; Paolillo 1987; Pinto Parada 1987; Prümers 2000, 2001, 2002a, 2002b). Many mounds were also used as cemeteries for burying and commemorating the dead, as indicated by large burial urns and human bones. The larger mounds, especially those surrounded by smaller settlements and nonmound earthworks, may have also had ceremonial and political functions within regions.

The construction of mounds had a major impact on the local environment. Their soils contain so much organic and inorganic domestic debris that Langstroth (1996) has referred to them as "sherd soils." These domesticated anthropogenic soils are deep, well drained, and rich in organic matter, so they support considerable biodiversity and are prized for farming (Erickson and Balée, chapter 7, this volume). Rich soils often extend beyond the actual mound, indicating soil improvement through long-term anthropogenic activities. The excavation of earth used in the fill of mounds created large water-filled barrow pits, which supported a thriving aquatic community that provided many economic resources for farmers and collectors.

In summary, these mounds and forest islands (discussed in the next section) functioned as areas for settlement, burial, fortification, and ritual; included gardens fields, orchards, and hunting locations; and served as political boundaries and territorial markers; or a combination thereof (Erickson 2000c).

FOREST ISLANDS

Thousands of forest islands (islas de monte) are found in the savannas of the Bolivian Amazon (Erickson 2000c; Langstroth 1996; Walker 2004). Most forest islands are slightly raised above the flooded savanna (0.5–1.0 meters) and range in area from a small cluster of trees to many hectares (figures 8.2, 8.7, and 8.8). Langstroth (1996) has shown that many forest islands were originally formed on the slight elevation of abandoned river levees and were protected from annual burns by what remains of the old meandering channels. I believe that most forest islands in the Bolivian Amazon are largely anthropogenic—locations of small farming communities, orchards, and house gardens that are often surrounded by raised fields, fish weirs, and causeways and canal networks. Colleagues and I have tested and excavated many intact forest islands and investigated others that were disturbed by road construction. All contain refuse debris from long-term pre-Columbian settlement. Many contain dense stands of economically valuable species. Because of their drainage and enhanced anthropogenic soils, forest islands are sought out by local farmers and ranchers as prime locations for settlements, gardens, orchards, corrals, and slash-and-burn agriculture.

RING DITCH SITES

Various forms of small (often one to two per forest island) and large (some surrounding entire forest islands) ring ditches occur in Baures (figure 8.8). In 1995 and 1996, our brief surveys located and mapped nine separate ring-ditched sites on forested islands near Baures (Erickson 2002; Erickson, Winkler, and Candler 1997). The enclosed area of each is estimated to range from 1 hectare to 5 hectares, and can include up to three village sites on a single forest island. The ditches are impressive earthworks of up to 4 meters deep and 10 meters wide, sometimes with steep sidewalls, and have diameters of between 150 and 350 meters. A number of sites also have multiple concentric moat rings. The Jasiaquiri and Bella Vista sites have a series of encircling linear canals that enclose areas of several square kilometers.

FIGURE 8.8 A ring ditch site at Jasiaquiri, a large forest island in Baures, Bolivia. The ditch (the darker oval of trees), measuring 5–10 meters wide and up to 4 meters deep, encloses approximately 3 hectares.

The ditches can be round, oval, square, rectangular, D shaped, or irregular in plan. Based on the presence of pottery fragments, it is surmised that some of the ditches enclosed settlements; but others are more enigmatic in function (Erickson 2002; Erickson, Winkler, and Candler 1997). Although unmarked by earthworks, circular villages are a common organizational plan for Amazonian settlements in the historical and ethnographic record (Erickson 2003; Wust and Barreto 1999). The moatlike encircling ditches suggest a function of defending the settlement or, minimally, restricting access to it. The deep ditches around some, but not all, of the pre-Columbian sites would have been excellent barriers against enemies. Eder ([1772] 1985), a missionary living in the Baures region in the early 1700s, reported moat and palisade villages used for defense against raiding by other groups. Other possible functions include use of the sites as elite residences, cemeteries, ritual spaces, and gardens. Many of the larger ring ditch sites are the nodes of radiating networks of straight causeway and canal, suggesting functions as population, political, and ritual centers. Alceu Ranzi (2003) considers the ring ditch villages to be geoglyphs, comparable to the better-known Nazca Lines of Peru, because of their geometry and monumentality.

Many of the forest islands with ring ditch sites are farmed today with short fallow cycles with no apparent decline in crop production. To date, no raised fields have been located in the region of Baures. Kenneth Lee (1995) estimates the total extent of these ring ditch sites, causeways, canals, fish weirs, and other earthworks in Baures to be 12,000 square kilometers, although the area of continuous distribution earthworks is probably smaller. Similar ring ditch sites have been reported for Riberalta in Bolivia and for Acre and the Upper Xingu region in Brazil (Arnold and Prettol 1988; Heckenberger, chapter 10, this volume; Heckenberger, Petersen, and Neves 1999; Heckenberger et al. 2003; Pärssinen et al. 2003; Peterson, Neves, and Heckenberger 2001; Ranzi 2003; Saunaluoma et al. 2002).

FISH WEIRS AND RELATED STRUCTURES

Unlike the Mojo peoples, who based their agriculture on raised fields, the Baure people of northeastern Bolivia intensively farmed the forest islands and gallery forests. The imprint of Baure farming is still recognized and exploited by the local folk who call these forests "chocolate islands" (*chocolatales*) because of the orchardlike concentrations of chocolate (domesticated or feral *Theobroma* sp.) found there. The region with the densest remains of earthworks is presently unoccupied.

During a survey of Baures in 1995 and 1996, colleagues and I studied a particular form of narrow linear earthwork that we refer to as "zigzag" structure because of its characteristic footprint (figure 8.9). The zigzag earthworks, clearly distinguished from the grassy savanna because of scrubby vegetation growing on them, are 1 meter wide, 20–30 centimeters tall, and up to 3 kilometers long. Dense networks of zigzag structures cover the savanna between the larger linear causeways and canals that divide the savanna into roughly rectilinear blocks. We found funnel-like narrow openings marked by parallel earthen walls of 1–3 meters long where these structures change direction.

Based on the form, location, and associations of these structures, my colleagues and I are convinced that they functioned as fish weirs during the rainy season (Erickson 2000a; Erickson, Winkler, and Candler 1997). The weirs are similar to those reported for indigenous groups throughout Amazonia. The weirs, combined with the larger causeways, would have impounded a thin sheet of water over a large area. The openings would have allowed excess water to flow across the savanna. The Amazon is home to many fish species that thrive in shallow floodwaters of tropical savannas. Basket or woven textile nets placed at the mouths of the narrow openings in the weirs could have been used to harvest fish (figure 8.10). These openings, possibly lined with logs, were used to pass heavy dugout canoes over the weirs without damaging the weirs. Artificial circular ponds with a diameter of 10–30 meters and a depth of 1–2 meters are associated with most weir openings. These ponds are still teeming with small fish in the dry season. Recent studies have shown that savanna fisheries of the Bolivian Amazon are extremely productive where standing

FIGURE 8.9 Fish weirs (the two irregular dark lines in the center) and causeway-canal *(upper right)* on the savanna between forest islands west of Baures, Bolivia.

FIGURE 8.10 A landscape of fish weirs (zigzag earthworks), artificial ponds *(center)* ringed by *Mauritia flexuosa* palms, and forest islands with settlements *(in background)*. Basket traps are placed in the openings in the fish weirs *(insert lower left)*. (Artwork by Daniel Brinkmeier)

water can be maintained (Hanagarth 1993). We also identified another aquatic species that may have been "cultivated" in the weir structures. Tens of thousands of shells of an edible gastropod have been found and are associated with the weir structures. *Pomacea*, the apple snail that plagues aquarium owners worldwide, was an important food source of the Baure during the Colonial period.

The weirs also aid in the harvest of game animals. Animal trails run alongside the weirs and channel the peccaries, tapirs, deer, and agoutis that are attracted to the palms *(Mauritia flexuosa)* growing on forest islands, causeways, and weirs. My hunter-guides hunt these animals from their canoes during the wet season. The technique is simply to paddle up to groups of deer and peccary swimming across the savanna and kill them with a blow to the head. These so-called wild game animals are the harvested products of a landscape domesticated more than 500 years ago.[10]

The earthworks provided a sophisticated means of regulating water levels within the savannas to enhance and manage seasonal aquatic resources. Fish weirs cover approximately 550 square kilometers. Using this simple but elegant technology, the Baure converted much of the savanna landscape into a huge aquatic farm that produced abundant, storable, and sustainable yields of animal protein. Hence, there was no need to improve protein production through genetic domestication, the common path taken by other societies.

The permanent impact of the artificial creation of this landscape is highlighted by what happened after the dense indigenous populations were removed from the landscape in the seventeenth century. The pre-Columbian earthworks still structure the abandoned landscape in highly complex patterns of vegetation, fauna and flora, soils, and hydrology—often misinterpreted as a "pristine environment" by nongovernmental organizations, ecologists, conservationists, tour guides, and government officials. Absent the annual burns, much of the savanna has been colonized by dense stands of forest. These landscapes are dominated by the economically important burití palm, *Mauritia flexuosa*—a sea of hundreds of square kilometers of starch and protein.[11]

THE ISSUE OF ORIGINS

Historical ecology and the archaeology of landscapes assume that human activity takes place in landscapes shaped by previous inhabitants. This assumption raises the issue of origins of the human and environment relationship. The first inhabitants of the Neotropics encountered a "natural" environment in the Early Holocene. These early foragers sought out the economic resources provided by wetlands and naturally disturbed environments. The faunal and floral remains from the Monte Alegre site on the central Amazon dating to 11,000 years ago and from the Peña Roja site along the Caquetá River in Colombia dating to

9,300–8,700 years ago provide evidence of this strategy, in addition to strongly suggesting early anthropogenic disturbance of the forest and floodplain around these sites (Mora 2003; Roosevelt et al. 1996).

Early inhabitants of the Amazon basin had at their disposal a powerful pre-industrial technology for modifying, transforming, managing, and creating desired landscapes—fire. Neotropical scholars have largely ignored human use of fire as a tool for creating anthropogenic landscapes except when combined with slash-and-burn agriculture. Anyone doubting the power of fire to transform landscapes should read Stephen Pyne's excellent fire histories of Australia, California, and the Great Plains (also Blackburn and Anderson 1993; Kay and Simmons 2002). In Pyne's view, humans "use their fire power to reshape the planet, to render it more suitable to their needs. In effect, humans began to cook the earth. They reworked landscapes in their ecological forges" (1998:64). Evidence of fire histories for parts of the Amazon basin and Central America begin by 11,000 BP (Piperno and Pearsall 1998). Natural-caused and human-caused fires can be distinguished. As Pyne (1998) points out, the signature, timing, scale, and function of anthropogenic fire are unique. Besides the role of fire in later slash-and-burn agriculture, recent studies of Amazonian Dark Earths demonstrate that pre-Columbian peoples discovered the importance of soil carbon and used low-temperature incomplete burning to produce charcoal that they incorporated into soils, thus creating black earths *(terra preta)* (Glaser and Woods 2004; Lehmann et al. 2003).

In the Bolivian Amazon, I initially assumed that my guides were pathological pyromaniacs by their keen interest in setting fire to everything in our path. To the locals, especially ranchers, a good burn is one that sweeps across a broad expanse of the savanna and clears off the old grass for new grass and opens up the forest understory for weedy species and fruit trees. Regular burns mean more grass for livestock and desired game, as well as shifts in biodiversity. The entire Bolivian Amazon is covered with thick smoke in the dry season as ranchers, farmers, and hunters burn the savannas—something that has been going on for thousands of years. Ironically, the very technology that created and maintained the landscape might soon end because of conservationists' and Green politicians' enthusiastic efforts in recently passing laws controlling and banning much of the annual burns of the savanna (Superintendente Agrario 2000).

GREEN INDIANS, CONSERVATION BIOLOGY, AND LANDSCAPES WITH HISTORY

To those spearheading efforts to preserve nature and protect biodiversity and natural resources, descriptions of nature as a product of human activities and as a cultural invention are considered ecologically damaging (see Soulé and Lease 1995) or at the least overemphasized (see Vale 2002a, 2002b, 2002c, 2002d). Scholars who critique the concept of nature and wilderness (e.g., Cronon 1996) have been accused of reducing nature and wilderness to cultural and linguistic categories and meanings (Soulé and Lease 1995). Critics argue that postmodernism and critical theory open the floodgates of hyperrelativism, ignore scientific knowledge, and undermine any positive advances made by conservationists to protect the environment.

Native peoples and their relationship to the environment are often at the center of these debates. The idea that humans have been disturbing, burning, clearing, hunting, and domesticating nature for tens of thousands of years undermines the core political philosophy of groups such as the Nature Conservancy, Conservation International, and the World Wildlife Fund. Many believe that archaeologists and anthropologists' recent critique of biologists' long-held assumptions is more insidious—that it is used or can be used to justify the rape of nature by developers, industry, and agrobusiness or even to reject native peoples' land claims (Meggers 2001, 2003; Soulé and Lease 1995; Vale 2002a).

I believe that it is far more damaging to deny the environment and native peoples their histories.[12] The denial of agency to foragers and small-scale farmers to transform and create landscapes is based on the myth of the ecologically noble savage that still dominates archaeological thinking. Amazonian peoples have been building and managing the environment for a very long time. Many of their intentional and unintentional activities could conceivably be described as deforestation, massive soil erosion, extinction of species, reduction of biodiversity, and environmental degradation (Denevan 1992; Kay and Simmons 2002; Krech 1999; Redman 1999). But characterizations of past and present native peoples as either agents of environmental degradation or as ecologically noble savages are—despite their apparent opposition—based on contemporary Western values, aesthetics, and assumptions. Native peoples did not tiptoe through the forest, nor did they live in harmonious equilibrium with nature. Somehow, they were able to sustain huge populations for long periods of time on landscapes that natural scientists classify variously as "marginal," "fragile," and "pristine." They are also responsible for what we now call nature in the Neotropics. Natural history is best understood in reference to human history.

TOWARD A HUMAN-CENTRIC UNDERSTANDING OF NATURE

The Amazon basin was not a pristine environment in 1492 and probably has not been since the first humans arrived there around 12,000 years ago or earlier. Most of us would agree that humans were and are a factor in shaping the present Amazonian landscape. We probably disagree on the degree of human causality, however. I have pointed out the arbitrariness of any comparison between human causation and some imagined "natural" or "pristine" baseline or benchmark. In the perspective of the new ecology, environmental perturbations, climate change, and catastrophe should be considered normal and *necessary* for the overall health of ecosystems (Blumler 1996, 1998; Botkin 1990; Stahl 1996). Archaeologists and historical ecologists point out that early humans in Amazonia simply added a more sustained and profound level of perturbation for at least 11,000 years (Denevan 1992; Kay and Simmons 2002; Stahl 1996).

The nature/culture dichotomy, ahistorical models of the natural sciences that stress equilibrium and order, and the anthropological concept of human adaptation have limited our understanding of the Amazonian environment. Any attempt to understand how nature came to be and what it will be in the future must consider human action in its long-term historical trajectory. A historicized, politicized, and humanized ecology provides a solid foundation for proactive change (Botkin 1990; Brosius 1999; Escobar 1999; Janzen 1998; Kay and Simmons 2002; Zimmerer and Bassett 2003). Archaeology of landscapes and historical ecology provide a powerful multiscalar, historical, people-centric perspective by which to understand the long-term dialectical relationship between humans and the environments they created. If we accept the idea that human agents have played and continue to play a primary role in creating landscape, there is hope that active human intervention informed by this perspective can confront contemporary issues such as global warming, loss of biodiversity, and unsustainable development.

DISCUSSION AND CONCLUSION

The Bolivian Amazon is an example of a totally domesticated landscape—a humanized landscape. The inhabitants participated in what ecologist Dan Janzen (1998) has referred to as the "gardenification of nature" or what my colleagues and I call "domestication of landscape." Farmers decided which trees would be grown on the landscape and where they would be grown; even today, the results are more like a garden or orchard than natural vegetation communities.

The permanent long-term effects on biodiversity created by the native peoples of the Bolivian Amazon were substantial and long lasting. Wetlands were expanded through intentional water management at a regional scale. The alternating ridge and swale topography locally created by raising field platforms, fish weirs, mounds, and forest islands and by cutting canals, barrow pits, ponds and reservoirs produced heterogeneous microenvironments for terrestrial and aquatic life—millions of linear kilometers of rich terrestrial-aquatic ecotones or edges in what was previously a relatively homogeneous, flat environment. In terms of biological productivity, expanding productive ecotones and introducing a patchwork of artificial landforms increased the presence and available biomass of selected economic species, both wild and domesticated. However, the systematic replacement of natural vegetation by economic and domestic species presumably changed vegetation composition substantially. We will never know because there is no pristine environment for comparison.

Conservation is often defined as intentional practices of short-term constraints on behavior for the long-term benefit of maintaining or improving biodiversity (e.g., Alvard 1994, 1995). The degree of planning, design, labor organization, and technology inherent in the complex, highly organized, engineered landscapes of the Bolivian Amazon show clear intentionality and forethought. For landscape archaeologists and historical ecologists working with the coarse and fragmentary archaeological and environmental record, recovering evidence of short-term decisions is often difficult. That the earthworks were used for hundreds and in some cases thousands of years suggests that the knowledge of how to manage the environment was passed down through generations of farmers who both benefited from past inputs and contributed to the landscape-domestication process.

From a long-term perspective, the rich biodiversity recognized in the forests and wetlands of the Bolivian Amazon today is because of, not in spite of, the pre-Columbian farmers who replaced nature with a cultural agroscape or anthropogenic landscape. Were these activities sustainable? Sustainability usually implies harvesting the interest without reducing the principle while maintaining a certain degree of quality of life. Archaeologically, sustainability can be measured in terms of the time depth of continuous intensive agriculture and high human carrying capacity on a given landscape. Again, the 2,000-year record of pre-Columbian intensive agriculture, earthwork construction, and urbanized populations in the Bolivian Amazon strongly suggests sustainability.

The use of fish weirs, raised fields, and other production strategies ended with the arrival of Europeans and their diseases, to which the locals had no resistance, as well as ensuing missionization, enslavement, imposition of new crops and livestock, and civil wars, rather than with the onset of overpopulation, environmental degradation, or unsustainable practices. For the past 300 years, this landscape has been relatively unpopulated. Forests expanded over vast areas of the anthropogenic savanna where annual burning was discontinued. This is not an example of a landscape reverting "back to nature"; the present vegetation is the historical legacy of past human activities.

The international community of conservationists, natural scientists, and most of my anthropologist colleagues consider what the pre-Columbian inhabitants of the Bolivian Amazon did as destructive and degrading of the natural environment. In contrast, although I am cautious about interpreting the landscape aesthetics and values of native peoples before 1492, I am convinced that the ideal landscape for the inhabitants of the Bolivian Amazon was a nicely gridded landscape of earthworks, roads, and settlements on a relatively treeless plain. This built environment was as productive and sustainable and probably equally species rich as the forests that exist there today.

In this case, historical ecology demonstrates that Amazonian peoples did not "adapt to" and were not "constrained by" or "limited by" the natural environment in Amazon, but rather created those very environments in which they lived and thrived. This domestication of the landscape was an intentional act, at least as it pertains to the engineering and knowledge used to transform the landscape in the pre-Columbian period. Through the perspective of archaeology of landscapes and historical ecology, we are beginning to understand how this impressive environment came to be—its human history—and to propose viable models of land use for sustainable development and conservation of biodiversity. Historical ecologists and now some biologists recognize that biodiversity is increasingly now found in the "countryside," or what historical ecologists would call the anthropogenic landscape. Past peoples constructed and maintained these landscapes; thus, solutions must include active management by present and future peoples based on this complex historical ecology.

ACKNOWLEDGMENTS

The author thanks Wilma Winkler (coinvestigator of the Agro-Archaeological Project of the Beni), Kenneth Lee, Ricardo Bottega, Anita Bruckner, Edwin Bruckner, Conrad Bruckner, Georghina Brochetti, Hans Schlink, Oswaldo Rivera, Juan Albarracín, and local government authorities of Trinidad, San Ignacio, Santa Ana, Baures, and the Sirionó Indigenous Territory who made this research possible. Wilma Winkler, Alexei Vranich, John Walker, Marcello Canuto, Kay Candler, Dante Angelo, and Marcos Michel made up the archaeological team during various field seasons. I also learned from the discussion of historical ecological topics with colleagues William Denevan, Robert Langstroth, William Woods, Johannes Lehmann, William Balée, Peter Stahl, John Walker, Jeffrey Quilter, Charles Mann, Frances Hayashida, and Jason Yaeger. Grants from the National Science Foundation, the Heinz Charitable Trust Foundation, the American Philosophical Society, Research Funds of the University of Pennsylvania Museum, and CORDEBENI supported fieldwork and analysis. I especially thank organizer Jeffrey Quilter and the participants of the Cultural Production of Nature in the Tropics Roundtable at Dumbarton Oaks (1999) for comments and discussion on my first presentation of these ideas. Early versions of this chapter were presented at various professional conferences and seminars, including the Penn Humanities Forum (2000), the Symposium on Neotropical Historical Ecology (2002), the Watson Armour III Spring Symposium (2003), and the annual meeting of the

Society for American Archaeology (2004). I especially thank Peter Stahl, William Balée, and Charles Mann for their detailed comments, insights, and debate on the topic of this chapter.

NOTES

1. Several scholars have expanded the concept of domestication beyond plants and animals. For example, Peter Wilson (1988) and Ian Hodder (1990) argue that major social transformation in the Neolithic was not the domestication of specific plants and animals and the adoption of agriculture, but rather the domestication of human society. It was through the permanent, built environment (Wilson focused on village architecture, Hodder on the house) that social roles, relationships, and meaning were inscribed in people's lives. The concept helps us widen the definition of *domestication* to consider nongenetic issues such as settling in permanent communities; dealing with neighbors; marking fields, village, and territory; and thinking about the environment. The built environment, in this case limited to architecture, becomes both a "model of" and "model for" society as well as a dynamic container of human action. James Snead and Robert Preucel's discussion of the "domestication of nature" and the "naturalization of the social" as an expression of the "dialectic between society and nature" in cultural landscapes (1999:171–174) overlaps with some of the ideas expressed in this chapter.

2. The terms *domestication of landscape* (Terrell et al. 2003), *landscape domestication* (Cunningham 1997), *domestication of environment* (Yen 1989), *domesticated environment* (Blackburn and Anderson 1993), and *humanized landscape* (Butzer 1979:148, 1990:48; Denevan 1992a; Vale 2002a, 2002b, 2002c; Zelinsky 1973:16) refer to roughly the same concept. I thank Bill Denevan for pointing out the early references to me. Some scholars trace the origins of the concept further back in time (Harlan 1992, 1995; Ingold 1996; Terrell et al. 2003:349, n. 13; and others). Douglas Yen (1989:61) and others attribute the concept of domestication of landscape to Edgar Anderson's (1952) "dungheap hypothesis" for the origins of agriculture, adopted from Darwin. Anderson's concept focused on localized anthropogenic conditions where weedy species could thrive and become the focus of genetic domestication. David Rindos (1984) frames some aspects of the concept of domestication of landscape within "incidental" and "specialized" domestication categories of his coevolutionary model, which stresses the unconscious.

3. Whereas my definition stresses the conscious patterning and structure imposed on the landscape by humans, other scholars stress ecology and population in their definitions (Clement 1999; Terrell et al. 2003). Charles Clement defines *landscape domestication* as "a conscious process by which human manipulation of the landscape results in changes in landscape ecology and in the demographics of its plant and animal populations, resulting in a landscape more productive and congenial for humans" (1999:190; see also Clement, chapter 6, this volume, and McKey et al. 1993:22–23). He subdivides domestication of landscape into a continuum of intensity of manipulation: pristine, promoted, managed, cultivated (swidden/fall, monoculture) (1999a:191–192). Clement gives credit to Hynes and Chase (1982) and Chase (1989) for the concept.

4. Erland Nordenskiöld (2003) first reported the raised fields of the Llanos de Mojos in 1924 based on observations from fieldwork in 1908–1909. William Denevan (1963, 1966, 2001) and George Plafker (1963) brought the fields' importance in Amazonian archaeology to scholars' notice. In addition to colleagues' and my project research,

archaeological research on raised fields includes Bernardo Dougherty and Horacio Calandra (1984), Marcos Michel (1993, 2000), and John Walker (2000, 2001, 2004). Mounds are important symbols in indigenous myths (e.g., Riester 1976) and local public imagination (e.g. Pinto Parada 1987). Mounds have been the traditional focus of archaeologists since Nordenskiöld's (1910, 1913) first excavations at the turn of the twentieth century. More recently, numerous national and international projects have explored mounds through survey, mapping, and excavation, including research conducted by Ricardo Bottega, Victor Bustos (1978a, 1978b, 1978c, 1978d), William Denevan (1966), Bernardo Dougherty and Horacio Calandra (1981, 1981–82, 1983, 1984), Juan Faldín (1984), Alicia Fernández Distel (1987), Wanda Hanke (1957), Kenneth Lee (1979, 1995, n.d.), Rodolfo Pinto Parada (1987), Stig Rydén (1941), and Mario Vilca. More recently, archaeologist Heiko Prümers and his German-Bolivian team have mapped and meticulously excavated the Mendoza Mound (Prümers et al. 2000, 2001, 2002a, 2002b) near Casarabe. Robert Langstroth's (1996) study of the vegetation, soils, and formation of mounds and forest islands is a landmark study.

5. The stability and change in the forest-savanna boundary for the period of prehuman and human occupation in the Bolivian Amazon (Hanagarth 1993; Langstroth 1996; Mayle, Burnbridge, and Killeen 2000) and in Amazonia in general (Piperno and Pearsall 1998) are still under debate.

6. Denevan estimates a total of 100,000 raised fields spread unevenly throughout 180,000 square kilometers of the Llanos de Mojos (2001:246).

7. Kenneth Lee (1979, 1995, n.d.) first suggested the role of water hyacinth as a green manure in raised field agriculture.

8. In an earlier publication, I calculated a figure of 40 square kilometers of surface area and retention of 20 million cubic meters of water based on a slope of 5 centimeters per kilometer (Erickson 2000b:24). Although this is the case in some northern areas of the Bolivian Amazon, a slope of 18–20 centimeters per kilometer is average (Denevan 1966).

9. Large mounds are found along the Mamoré, Ibare, Caimanes, Tijamuchí, Apere, Matos, Isiboro, Blanco, and Secure rivers and their tributaries.

10. Charles Bennett (1962) and Olga Linares (1976) have also discussed how Native Americans enhanced the number of game animals through anthropogenic activities (also see Stahl, chapter 4, this volume).

11. For the importance of *Mauritia flexuosa* for Amazonian peoples, see Gragson 1992; Hiraoka 1999; and N. Smith 1999.

12. A point also made by Charles Kay and Randy Simmons, who go as far as to suggest that this denial of history, agency, and anthropogenic environment is racist (2002:xi).

REFERENCES

Alvard, M. S. 1994. Conservation by native peoples: Prey choice in a depleted habitat. *Human Nature* 5 (2): 127–154.

——. 1995. Intraspecific prey choice by Amazonian hunters. *Current Anthropology* 36: 789–818.

Anderson, E. 1952. *Plants, Man, and Life*. Boston: Little Brown.

Arce Z., J. 1993. Evaluación y comparacion de rendimientos de cuatro cultivos en tres anchuras de camellones (campos elevados) en la Estación Biológica del Beni (Prov. Ballivián, Dpto. Beni). Unpublished thesis, Trinidad, Universidad Técnica del Beni.

Arnold, D. E., and K. A. Prettol. 1988. Aboriginal earthworks near the mouth of the Beni, Bolivia. *Journal of Field Archaeology* 15:457–465.

Bailey, R., G. Head, M. Jenike, B. Owen, R. Richtman, and E. Zechenter. 1989. Hunting and gathering in tropical rain forests: Is it possible? *American Anthropologist* 91:59–83.

Balée, W. 1989. The culture of Amazonian forests. *Advances in Economic Botany* 7:1–21.

———. 1994. *Footprints of the Forest: Ka'apor Ethnobotany—the Historical Ecology of Plant Utilization by an Amazonian People.* New York: Columbia University Press.

———. 1995. Historical ecology of Amazonia. In L. Sponsel, ed., *Indigenous Peoples and the Future of Amazonia,* 97–110. Tucson: University of Arizona Press.

———, ed. 1998a. *Advances in Historical Ecology.* New York: Columbia University Press.

———. 1998b. Historical ecology: Premises and postulates. In W. L. Balée, ed., *Advances in Historical Ecology,* 13–29. New York: Columbia University Press.

Beckerman, S. 1979. The abundance of protein in Amazonia: A reply to Gross. *American Anthropologist* 81:533–560.

Bennett, C. F., Jr. 1962. The Bayano Cuna Indians: An ecological study of livelihood and diet. *Annals of the Association of American Geographers* 52:32–50.

Blackburn, T. C., and K. Anderson. 1993. Managing the domesticated environment. In T. Blackburn and K. Anderson, eds., *Before the Wilderness: Environmental Management by Native Californians,* 15–25. Menlo Park, Calif.: Ballena Press.

Blumler, M. A. 1996. Ecology, evolutionary theory, and agricultural origins. In D. R. Harris, ed., *The Origins and Spread of Agriculture and Pastoralism in Eurasia,* 25–50. Washington, D.C.: Smithsonian Institution Press.

———. 1998. Biogeography of land-use impacts in the Near East. In K. Zimmerer and K. Young, eds., *Nature's Geography: New Lessons for Conservation in Developing Countries,* 215–236. Madison: University of Wisconsin Press.

Botkin, D. 1990. *Discordant Harmonies: A New Ecology for the Twenty-First Century.* New York: Oxford University Press.

Brookfield, H. 2001. *Exploring Agrodiversity.* New York: Columbia University Press.

Brosius, J. P. 1999. Analyses and interventions: Anthropological engagements with environmentalism. *Current Anthropology* 40 (3): 277–309.

Bustos, V. 1978a. *La arqueología en los llanos del Beni, Bolivia.* Documentos Internos del Instituto Nacional de Arqueología (INAR) no. 32/78. La Paz, Bolivia: INAR.

———. 1978b. Una hipótesis de relaciones culturales entre el altiplano y la vertiente oriental de los Andes. *Pumapunku* (La Paz) 12:115–126.

———. 1978c. *Investigaciones arqueológicas en Trinidad, Departamento del Beni.* Instituto Nacional de Arqueología (INAR) no. 22. La Paz, Bolivia: INAR.

———. 1978d. *Proyecto de excavaciones arqueológicas en el departamento del Beni, en lomas de la región de Cocharcas y Villa Banzar.* Documentos Internos del Instituto Nacional de Arqueología (INAR) no. 10/78. La Paz, Bolivia: INAR.

Butzer, K. W. 1979. This is Indian country. *Geographical Magazine* 52:140–148.

———. 1990. The Indian legacy in the American landscape. In M. P. Conzen, ed., *The Making of the American Landscape,* 27–50. Boston: Unwin Hyman.

Chase, A. K. 1989. Domestication and domiculture in northern Australia: A social perspective. In D. R. Harris and G. C. Hillman, eds., *Farming and Foraging: The Evolution of Plant Exploitation,* 42–54. Boston: Unwin Hyman.

Clement, C. R. 1999. 1492 and the loss of Amazonian crop genetic resources. I. The relation between domestication and human population decline. *Economic Botany* 53 (2): 188–202.

Colinvaux, P. A., and M. B. Bush. 1991. The rain forest ecosystem as a resource for hunting and gathering. *American Anthropologist* 93:153–163.

Connell, J. H. 1978. Diversity in tropical forests and coral reefs. *Science* 199:1302–1310.

Cronon, W., ed. 1996. *Uncommon Ground: Rethinking the Human Place in Nature.* New York: Norton.

Crumley, C. L., ed. 1994. *Historical Ecology: Cultural Knowledge and Changing Landscapes.* Santa Fe: School of American Research.

Crumley, C. L., and William H. Marquardt, eds. 1987. *Regional Dynamics: Burgundian Landscapes in Historical Perspective.* San Diego: Academic Press.

Cunningham, A. B. 1997. Landscape domestication and cultural change: Human ecology of the Cuvelai-Etosha region. *Madoqua* 20 (1): 37–48.

Denevan, W. M. 1963. Additional comments on the earthworks of Mojos in northeastern Bolivia. *American Antiquity* 28:540–545.

——. 1966. *The Aboriginal Cultural Geography of the Llanos de Mojos of Bolivia.* Berkeley: University of California Press.

——. 1990. Prehistoric roads and causeways in lowland tropical America. In C. Trombold, ed., *Ancient Road Networks and Settlement Hierarchies in the New World,* 230–242. Cambridge, U.K.: Cambridge University Press.

——. 1992. The pristine myth: The landscape of the Americas in 1492. *Annals of the Association of American Geographers* 82:369–385.

——. 2001. *Cultivated Landscapes of Native Amazonia and the Andes.* Oxford: Oxford University Press.

Denevan, W. M., and Christine Padoch, eds. 1988. *Swidden-Fallow Agroforestry in the Peruvian Amazon.* Advances in Economic Botany no. 5. Bronx: New York BotanicalGarden.

Descola, P. 1996. *In the Society of Nature: A Native Ecology in Amazonia.* Cambridge, U.K.: University of Cambridge Press.

Doolittle, W. 2000. *Cultivated Landscapes of Native North America.* Oxford: Oxford University Press.

Dougherty, B., and H. Calandra. 1981. Nota preliminar sobre investigaciones arqueológicas en los Llanos de Moxos, Departamento del Beni, Republica de Bolivia. *Revista del Museo de la Plata, Seccion Arqueológica* 53:87–106.

——. 1981–82. Excavaciones arqueológicas en la Loma Alta de Casarabe, Llanos de Moxos, Departamento del Beni, Bolivia. *Relaciones de la Sociedad Argentina de Antropología* (Buenos Aires) 14 (2): 9–48.

——. 1983. Archaeological research in northeastern Beni, Bolivia. *National Geographic Society Research Reports* 21:129–136.

——. 1984. Prehispanic human settlement in the Llanos de Moxos, Bolivia. In Jorge Rabassa, ed., *Quaternary of South America and Antartica,* 2:163–199. Rotterdam: A. A. Balkema.

Eder, F. J. [1772] 1985. *Breve descripción de las reducciones de Mojos.* Cochabamba: Historia Boliviana.

Erickson, C. L. 1980. Sistemas agrícolas prehispánicos en los Llanos de Mojos. *America Indígena* (Mexico City) 40 (4): 731-755.

——. 1994. Raised fields as a sustainable agricultural system from Amazonia. Paper presented in the symposium "Recovery of Indigenous Technology and Resources in Bolivia" at the Eighteenth International Congress of the Latin American Studies Association, Atlanta, March 10–12.

——. 1995. Archaeological perspectives on ancient landscapes of the Llanos de Mojos in the Bolivian Amazon. In P. Stahl, ed., *Archaeology in the American Tropics: Current Analytical Methods and Applications,* 66–95. Cambridge, U.K.: Cambridge University Press.

——. 2000a. An artificial landscape-scale fishery in the Bolivian Amazon. *Nature* 408:190–193.

Erickson, C. L. 2000b. Los caminos prehispánicos de la Amazonia boliviana. In L. Herrera and M. Cardale de Schrimpff, eds., *Caminos precolombinos: Las vías, los ingenieros y los viajeros,* 15–42. Bogota: Instituto Colombiano de Antropología y Historia.

———. 2000c. Lomas de ocupación en los Llanos de Moxos. In A. Durán Coirolo and R. Bracco Boksar, eds., *La arqueología de las tierras bajas,* 207–226. Montevideo, Uruguay: Comisión Nacional de Arqueología, Ministerio de Educación y Cultura.

———. 2001. Pre-columbian roads of the Amazon. *Expedition* 43 (2): 21–30.

———. 2002. Large moated settlements: A late Precolumbian phenomenon in the Amazon. Paper presented at the Second Annual Meeting of the Society for the Anthropology of Lowland South America (SALSA), St. Johns College, Annapolis, Maryland, June 6–7.

———. 2003. Historical ecology and future explorations. In J. Lehmann, D. C. Kern, B. Glaser, and W. I. Woods, eds., *Amazonian Dark Earths: Origin, Properties, Management,* 455–500. Dordrecht, Netherlands: Kluwer Academic.

Erickson, C. L., J. Esteves, W. Winkler, and M. Michel. 1991. Estudio preliminar de los sistemas agrícolas precolombinos en el Departamento del Beni, Bolivia. Unpublished manuscript, University of Pennsylvania and the Instituto Nacional de Arqueología, Philadelphia and La Paz.

Erickson, C. L., and J. Walker. n.d. The archaeology of landscapes in the Bolivian Amazon. Manuscript in preparation.

Erickson, C. L., W. Winkler, and K. Candler. 1997. Las investigaciones arqueológicas en la region de Baures en 1996. Unpublished report, University of Pennsylvania and the Instituto Nacional de Arqueología, Philadelphia and La Paz.

Escobar, A. 1999. After nature: Steps to an antiessentialist political ecology. *Current Anthropology* 40 (1): 1–30.

Fairhead, J., and M. Leach. 1996. *Misreading the African Landscape: Society and Ecology in the Forest-Savanna Mosaic.* Cambridge, U.K.: Cambridge University Press.

Faldín, J. 1984. La arqueología beniana y su panorama interpretivo. *Arqueología Boliviana* 1:83–90.

Fernández Distel, A. A. 1987. Informe sobre hallazgos cerámicos superficiales en Eviato (Departamento del Beni, Republica de Bolivia). *Scripta Ethnologica* 11:414–152.

Glaser, B., and W. Woods, eds., 2004. *Explorations in Amazonian Dark Earths in Time and Space.* Heidelberg: Springer-Verlag.

Goodman, A. H., and T. L. Leatherman. 1998. Traversing the chasm between biology and culture: An introduction. In A. H. Goodman and T. L. Leatherman, eds., *Building a Biocultural Synthesis: Political-Economic Perspectives on Human Biology,* 3–41. Ann Arbor: University of Michigan Press.

Gould, S. J., and R. Lewontin. 1979. The spandrels of San Marcos and the Panglossian paradigm: A critique of the adaptationist programme. *Proceedings of the Royal Society of London,* Series B, 205:581–598.

Gragson, Ted L. 1992. The use of palms by the Pume Indians of southwestern Venezuela. *Principes* 36:133–142.

Graham, E. 1998. Metaphor and metamorphism: Some thoughts on environmental meta-history. In W. Balée, ed., *Advances in Historical Ecology,* 119–137. New York: Columbia University Press.

Gross, D. 1975. Protein capture and cultural development in the Amazon basin. *American Anthropologist* 77 (3): 526–549.

Hanagarth, W. 1993. *Acerca de la geoecología de las sabanas del Beni en el noreste de Bolivia.* La Paz, Bolivia: Editorial Instituto de Ecología.

Hanke, W. 1957. Einige Funde im Beni-Gebiet, Ostbolivien. *Archiv fur Volkerkunde* (Wien) 12:136–143.

Harlan, J. R. 1992. *Crops and Man.* 2d ed. Madison, Wisc.: American Society of Agronomy, Crop Science Society of America.

——. 1995. *The Living Fields: Our Agricultural Heritage.* Cambridge, U.K.: Cambridge University Press.

Harris, D. R. 1989. An evolutionary continuum of people-plant interaction. In D. R. Harris and G. C. Hillman, eds., *Farming and Foraging: The Evolution of Plant Exploitation,* 11–26. Boston: Unwin Hyman.

——. 1996a. Introduction: Themes and concepts in the study of early agriculture. In David Harris, ed., *The Origins and Spread of Agriculture and Pastoralism in Eurasia,* 1–24. Washington, D.C.: Smithsonian Institution Press.

——, ed. 1996b. *The Origins and Spread of Agriculture and Pastoralism in Eurasia.* Washington, D.C.: Smithsonian Institution Press.

Harris, D. R., and G. C. Hillman, eds. 1989. *Farming and Foraging: The Evolution of Plant Exploitation.* Boston: Unwin Hyman.

Hastorf, C. 1998. The cultural life of early domestic plant use. *Antiquity* 72:773–782.

Hather, J. G. 1996 The origins of tropical vegeculture: Zingiberaceae, Araceae, and Dioscoreaceae in Southeast Asia. In D. Harris, ed., *The Origins and Spread of Agriculture and Pastoralism in Eurasia,* 538–550. Washington, D.C.: Smithsonian Institution Press.

Heckenberger, M. J., A. Kuikuro, U. Tabata Kuikuro, J. C. Russell, M. Schmidt, C. Fausto, and B. Franchetto. 2003. Amazonia 1492: Pristine forest or cultural parkland? *Science* 301:1710–1714.

Heckenberger, M. J., J. Petersen, and E. G. Neves. 1999. Village size and permanence in Amazonia: Two archaeological examples from Brazil. *Latin American Antiquity* 10 (4): 353–376.

Hiraoka, M. 1999. Miriti *(Mauritia flexuosa)* palms and their uses and management among the ribeirinhos of the Amazon estuary. In C. Padoch, J. M. Ayres, M. Pinedo-Vasquez, and A. Henderson, eds., *Varzea: Diversity, Development, and Conservation of Amazonia's Whitewater Floodplains,* 169–186. Advances in Economic Botany no. 13. Bronx: New York Botanical Garden.

Hodder, I. 1990. *The Domestication of Europe: Structure and Contingency in Neolithic Societies.* Oxford: Blackwell.

Hunter, M. 1996. Benchmarks for managing ecosystems: Are human activities natural? *Conservation Biology* 10 (3): 695–697.

Hynes, R. A., and A. K. Chase. 1982. Plants, sites, and domiculture: Aboriginal influence upon plant communities in Cape York Peninsula. *Archaeology in Oceania* 17:38–90.

Ingold, T. 1987. *The Appropriation of Nature: Essays on Human Ecology and Social Relations.* Iowa City: University of Iowa Press.

——. 1996. Growing plants and raising animals: An archaeological perspective on domestication. In David Harris, ed., *The Origins and Spread of Agriculture and Pastoralism in Eurasia,* 12–24. Washington, D.C.: Smithsonian Institution Press.

——. 2000. *The Perception of the Environment: Essays in Livelihood, Dwelling, and Skill.* London: Routledge.

Jacob, J. n.d. Report on soils from raised fields in the Llanos de Mojos. Unpublished manuscript, University of Pennsylvania, Philadelphia.

Janzen, D. 1998. Gardenification of wildland nature and the human footprint. *Science* 279:1312.

Kay, C. E., and R. T. Simmons, eds., 2002. *Wilderness and Political Ecology: Aboriginal Influences and the Original State of Nature.* Salt Lake City: University of Utah Press.

Krech, S. 1999. *The Ecological Indian: Myth and History.* New York: W. W. Norton.

Kuznar, L. 2001. Ecological mutualism in Navajo corrals: Implications for Navajo environmental perceptions and human/plant coevolution. *Journal of Anthropological Research* 57:17–39.

Langstroth, R. 1996. Forest islands in an Amazonian savanna of northeastern-Bolivia. Ph.D. diss., University of Wisconsin, Madison.

Lathrap, D. W. 1968. The "hunting" economies of the tropical forest zone of South America: An attempt at historical perspective. In R. B. Lee and I. DeVore, eds., *Man the Hunter,* 23–29. Chicago: Aldine.

———. 1970. *The Upper Amazon.* New York: Praeger.

———. 1977. Our father the caiman, our mother the gourd: Spinden revisited, or a unitary model for the emergence of agriculture in the New World. In C. Reed, ed., *Origins of Agriculture,* 713–751. The Hague: Mouton.

———. 1987. The introduction of maize in prehistoric eastern North America: The view from Amazonia and the Santa Elena Peninsula. In W. F. Keegan, ed., *Emergent Horticultural Economies of the Eastern Woodlands,* 345–371. Occasional Paper no. 7. Carbondale: Southern Illinois University, Center for Archaeological Investigations.

Lathrap, D. W., A. Gebhart-Sayer, and A. Mester. 1985. The roots of the Shipibo art style: Three waves on Imiriacocha, or there were Incas before the Incas. *Journal of Latin American Lore* 11 (1): 31–119.

Lee, K. 1979. 7,000 años de historia del hombre de Mojos: Agricultura en pampas esteriles: Informe preliminar. *Universidad Beni*:23–26.

———. 1995. Apuntes sobre las obras hidráulicas prehispánicas de las llanuras de Moxos: Una opción ecológica inédita. Unpublished manuscript, Trinidad, Bolivia.

———. n.d. El baúl del gringo. Unpublished manuscript, Trinidad, Bolivia.

Lehmann, J., D. C. Kern, B. Glaser, and W. I. Woods, eds. 2003 *Amazonian Dark Earths: Origin, Properties, Management.* Dordrecht, Netherlands: Kluwer Academic.

Lévi-Strauss, C. 1963. The concept of archaism in anthropology. In C. Lévi-Strauss, ed., *Structural Anthropology,* 1:101–119. Harmondsworth, U.K.: Penguin.

———. 1968. The concept of "primitiveness." In R. B. Lee and I. DeVore, eds., *Man the Hunter,* 349–352. Chicago: Aldine.

Linares, O. 1976. "Garden hunting" in the American tropics. *Human Ecology* 4:331–349.

Maffi, L., ed., 2001. *On Biocultural Diversity: Linking Language, Knowledge, and the Environment.* Washington, D.C.: Smithsonian Institution Press.

Mann, C. 2002. 1491: Before it became the New World, the Western Hemisphere was vastly more populous and sophisticated than has been thought. *Atlantic Monthly* (March): 41–53.

Mayle, F., R. Burnbridge, and T. Killeen. 2000. Millenial-scale dynamics of southern Amazonian rain forests. *Science* 290:2291–2294.

McKey, D., O. F. Linares, C. R. Clement, and C. M. Hladik. 1993. Evolution and history of tropical forests in relation to food availability—background. In C. M. Hladik, A. Hladik, O. F. Linares, H. Pagezy, A. Semple, and M. Hadley, eds., *Tropical Forests, People, and Food: Biocultural Interactions and Application to Development,* 17–24. Man and Biosphere Series no. 13. New York: UNESCO and Parthenon.

Meggers, B. J. 1954. Environmental limitations on the development of culture. *American Anthropologist* 56:801–824.

———. 2001. The continuing quest for El Dorado: Round two. *Latin American Antiquity* 12:304–325.

———. 2003. Natural vs. anthropogenic sources of Amazonian biodiversity: The continuing quest for El Dorado. In G. A. Bradshaw and P. A. Marquet, eds., *How Landscapes Change,* 89–105. Berlin: Springer-Verlag.

Michel, M. 1993. Prospección arqueológica de San Ignacio de Moxos: Provincia Moxos, Departamento del Beni, Bolivia. Thesis, Carrera de Arqueología, Universidad Mayor de San Andrés, La Paz, Bolivia.

———. 2000. Desarollo temprano de la agriculura de campos elevados en los Llanos de Moxos, Depto. de Beni, Bolivia. In P. Ledergerber-Crespo, ed., *Formativo sudamericano: Homenaje a Alberto Rex Gonzales y Betty J. Meggers,* 271–281. Quito, Ecuador: Editorial Abya-Yala.

Michon, G., and H. De Foresta. 1997. Agroforests: Pre-domestication of forest trees or true domestication of forest ecosystems? *Netherlands Journal of Agricultural Science* 45 (4): 451–462.

Mora, S. 2003. *Early Inhabitants of the Amazonian Tropical Rain Forest: A Study of Humans and Environmental Dynamics.* Latin American Archaeology Reports no. 3. Pittsburgh: University of Pittsburgh.

Moran, E. 1982. *Human Adaptability.* Boulder, Colo.: Westview Press.

———. 1993. *Through Amazonian Eyes: The Human Ecology of Amazonian Populations.* Iowa City: University of Iowa Press.

Nordenskiöld, E. 1910. Archaologische Forschungen im Bolivianischen Flachland. *Zeitschrift fur Ethnologie* (Berlin) 42:806–822.

———. 1913. Urnengraber und Mounds im Bolivianischen Flachlande. *Baessler Archiv* (Berlin and Leipzig) 3:205–255.

———. 1916. Die Anpassung der Indianer an die Verhältnisse in den Uberschwemmungsgebieten in Südamerika. *Ymer* (Stockholm) 36:138–155.

———. 2003. *Indios y blancos en el nordeste de Bolivia.* La Paz: IFEA.

Oliver, J. 2001. The archaeology of forest foraging and agricultural production in Amazonia. In C. McEwan, C. Barreto, and E. Neves, eds., *Unknown Amazon: Culture in Nature in Ancient Brazil,* 50–85. London: British Museum Press.

O'Neil, R.V. 2001. Is it time to bury the ecosystem concept? (with full military honors of course!). *Ecology* 82 (12): 3275–3284.

Paolillo, A. 1987. La cultura delle lomas. In G.M. Ronzoni, ed., *Bolivia, lungo e sentieri incantati,* 146–171. Rome: Erizzo Editrice.

Pärssinen, M., A. Ranzi, S. Saunaluoma, and A. Siiriäinen. 2003. Geometrically patterned ancient earthworks in the Rio Blanco region of Acre, Brazil: New evidence of ancient chiefdom formations in Amazonian interfluvial *terra firme* environments. In M. Pärssinen and A. Korpisaari, eds., *Western Amazonia—Amazônia Ocidental: Multidisciplinary Studies on Ancient Expansionistic Movements, Fortifications, and Sedentary Life,* 135–172. Renvall Institute Publications no. 14. Helsinki: Renvall Institute for Area and Cultural Studies, University of Helsinki.

Pérez, C.A. 1995. Niveles de fertilizantes orgánica con tarope (*Eichhornia azurea* [Sw.] Kunth) en el cultivo de maíz en campos elevados en la Estación Biológica del Beni. Thesis, Universidad Técnica del Beni, Trinidad, Bolivia.

Pérez, T.A. 1996. Evaluación de los fertilizantes orgánicos en yuca (*Manihot esculenta* Cr.) en campos elevados de la Estación Biológica del Beni. Thesis, Universidad Técnica del Beni, Trinidad, Bolivia.

Peters, C. 2000. Precolumbian silviculture and indigenous management of neotropical forests. In D. Lentz, ed., *Imperfect Balance: Landscape Transformations in the Precolumbian Americas,* 203–223. New York: Columbia University Press.

Petersen, J.B., E. Neves, and M.J. Heckenberger. 2001. Gift from the past: *Terra preta* and prehistoric Amerindian occupation in Amazonia. In C. McEwan, C. Barreto, and E. Neves, eds., *Unknown Amazon: Culture in Nature in Ancient Brazil,* 86–105. London: British Museum Press.

Pinto Parada, R. 1987. *Pueblo de leyenda.* Trinidad: Tiempo del Bolivia.

Piperno, D., and D. Pearsall. 1998. *The Origins of Agriculture in the Lowland Neotropics.* San Diego: Academic Press.

Plafker, G. 1963. Observations on archaeological remains in northeastern Bolivia. *American Antiquity* 28:372–378.

Politis, G. G. 1996. Moving to produce: Nukak mobility and settlement patterns in Amazonia. *World Archaeology* 27 (3): 492–511.

Posey, D. A., and W. Balée, eds. 1989. *Resource Management in Amazonia: Indigenous and Folk Strategies.* Advances in Economic Botany no. 7. Bronx: New York Botanical Garden.

Prümers, H. 2000. *Informe de labores: Excavaciones arqueológicas en la Loma Mendoza (Trinidad) (Proyecto "Lomas de Casarabe") 1ra Temporada, 1991.* Bonn: Comisión de Arqueología General y Comparada (KAVA), Instituto Alemán de Arqueología.

——. 2001. *Informe de labores: Excavaciones arqueológicas en la Loma Mendoza (Trinidad) (Proyecto "Lomas de Casarabe") 2nd Temporada, 2000.* Bonn: Comisión de Arqueología General y Comparada (KAVA), Instituto Alemán de Arqueología.

——. 2002a. *Informe de labores: Excavaciones arqueologicas en la Loma Mendoza (Trinidad) (Proyecto "Lomas de Casarabe") 3ra Temporada, 2001.* Bonn: Comisión de Arqueología General y Comparada (KAVA), Instituto Alemán de Arqueología.

——. 2002b. Nota de Prensa. Unpublished document. Comisión de Arqueología General y Comparada (KAVA), Instituto Alemán de Arqueología, Bonn.

Pyne, S. J. 1998. Forged in fire: history, land, and anthropogenic fire. In William Balée, ed., *Advances in Historical Ecology,* 62–103. New York: Columbia University Press.

Raffles, H., and A. WinklerPrins. 2003. Further reflections on Amazonian environmental history: Transformations of rivers and streams. *Latin American Research Review* 38 (3): 165–187.

Ranzi, A. 2003. Geoglifos patrimônio cultural do Acre. In M. Pärssinen and A. Korpisaari, eds., *Western Amazonia—Amazônia Ocidental: Multidisciplinary Studies on Ancient Expansionistic Movements, Fortifications, and Sedentary Life,* 135–172. Renvall Institute Publications no. 14. Helsinki: Renvall Institute for Area and Cultural Studies, University of Helsinki.

Redford, K. H. 1991. The ecologically noble savage. *Cultural Survival Quarterly* 15 (1): 46–48.

Redman, C. 1999. *Human Impact on Ancient Environments.* Tucson: University of Arizona Press.

Redman, C., and A. Kinzig. 2003. Resilience of past landscapes: Resilience theory, society, and *longue durée. Conservation Ecology* 7 (1): 14.

Riester, J. 1976. *En busca de la loma santa.* La Paz: Editorial Amigos del Libro.

Rindos, D. 1984. *Origins of Agriculture: An Evolutionary Perspective.* New York: Academic Press.

Rival, L. 1998. Domestication as a historical and symbolic process: Wild gardens and cultivated forests in the Ecuadorian Amazon. In William Balée, ed., *Advances in Historical Ecology,* 232–250. New York: Columbia University Press.

——. 2002. *Trekking Through History: The Huarani of Amazonian Ecuador.* New York: Columbia University Press.

Roosevelt, A. C. 1980. *Parmana: Prehistoric Maize and Manioc Subsistence along the Amazon and Orinoco.* New York: Academic Press.

Roosevelt, A. C., J. Douglas, and L. Brown. 2002. The migrations and adaptations of the first Americans: Clovis and pre-Clovis viewed from South America. In N. Jablonski, ed., *The First Americans: The Pleistocene Colonization of the New World,* 159–235.

Memoirs of the California Academy of Science no. 27. San Francisco: California Academy of Science.

Roosevelt, A. C., M. Lima da Costa, C. Lopes Machado, M. Michab, N. Mercier, H. Valladas, J. Feathers, W. Barnett, M. Imazio da Silveira, A. Henderson, J. Sliva, B. Chernoff, D. S. Reese, J. A. Holman, N. Toth, and K. Schick. 1996. Paleoindian cave dwellers in the Amazon: The peopling of the Americas. *Science* 272:373–384.

Rydén, S. 1941. *A Study of the Sirionó Indians.* Göteborg, Sweden: Humanistiska Fonden, Elanders Boktryckeri Aktiebolag.

Saavedra, O. Forthcoming. Comportamiento poblacional de la *Eichornia carssipes* en el sistema de camellones. In F. Valdez and J.-F. Bouchard, eds., *Sistemas agrícolas andinos basados en el drenaje y / o la elevación del nivel de la superficie cultivada.* Quito, Ecuador: Editorial Abya-aylla.

Sauer, C. 1952. *Agricultural Origins and Dispersals.* New York: American Geographical Society.

Saunaluoma, S., J. Faldín, A. Korpisaari, and A. Siiriäinen. 2002. Informe preliminar de las investigaciones arqueológicas en la región de Riberalta, Bolivia. In A. Siiriäinen and A. Korpisaari, eds., *Reports of the Finnish-Bolivian Archaeological Project in the Bolivian Amazon—Noticias del proyecto arqueológico finlandés-boliviano en la Amazonia boliviana,* 1:31–52. Helsinki, Finland: Department of Archaeology, University of Helsinki.

Shipek, F. C. 1989. An example of intensive plant husbandry: The Kuneyaay of southern California. In D. R. Harris and G. C. Hillman, eds., *Farming and Foraging: The Evolution of Plant Exploitation,* 159–170. Boston: Unwin Hyman.

Smith, B. D. 1995. *The Emergence of Agriculture.* New York: Scientific American Library.

Smith, N. J. H. 1980. Anthrosols and human carrying capacity in Amazonia. *Annals of the Association of American Geographers* 70 (4): 553–566.

——. 1999. *The Amazon River Forest: A Natural History of Plants, Animals, and People.* New York: Oxford University Press.

Snead, J., and R. Preucel. 1999. The ideology of settlement: Ancestral Keres landscapes in the northern Rio Grande. In W. Ashmore and A. B. Knapp, eds., *Archaeologies of Landscape: Contemporary Perspectives,* 169–200. Oxford: Blackwell.

Soulé, M., and G. Lease, eds., 1995. *Reinventing Nature? Responses to Postmodern Deconstruction.* Washington, D.C.: Island Press.

Spriggs, M. 1996. Early agriculture and what went before in island Melanesia: Continuity or intrusion? In D. Harris, ed., *The Origins and Spread of Agriculture and Pastoralism in Eurasia,* 524–537. Washington, D.C.: Smithsonian Institution Press.

Stab, S., and J. Arce. 2000. Pre-Hispanic raised-field cultivation as an alternative to slash-and burn agriculture in the Bolivian Amazon: Agroecological evaluation of field experiments. In O. Herrera-MacBryde, F. Dallmeier, B. MacBryde, J. A. Comiskey, and C. Miranda, eds., *Biodiversidad, conservación y manejo en la región de la Reserva de la Biosfera Estación Biológica del Beni, Bolivia / Biodiversity, Conservation, and Management in the Region of the Beni Biological Station Biosphere Reserve, Bolivia,* 317–327. Washington, D.C.: Smithsonian Institution, SI/MAB Biodiversity Program.

Stahl, P. W. 1996. Holocene biodiversity: An archaeological perspective from the Americas. *Annual Review of Anthropology* 25:105–126.

——. 2000. Archaeofaunal accumulation, fragmented forests, and anthropogenic landscape mosaics in the tropical lowlands of prehispanic Ecuador. *Latin American Antiquity* 11 (3): 241–257.

——. 2002. Paradigms in paradise: Revising standard Amazonian prehistory. *Review of Archaeology* 23:39–51.

Stearman, A. M. 1994. "Only slaves climb trees": Revisiting the myth of the ecologically noble savage in Amazonia. *Human Nature* 5 (4): 339–357.

Superintendente Agrario. 2000. Procedimiento para la autorización de quema controladas de pastizales. Available at: www2.entelnet.bo/si-a/pdf/anexoquemas.pdf.

Terrell, J. E., J. P. Hart, S. Barut, N. Cellinese, A. Curet, T. Denham, C. M. Kusimba, K. Latinis, R. Oka, J. Palka, M. E. D. Pohl, K. O. Pope, P. R. Williams, H. Haines, and J. E. Staller. 2003. Domesticated landscapes: The subsistence ecology of plant and animal domestication. *Journal of Archaeological Method and Theory* 10 (4): 323–368.

Vale, T., ed. 2002a. *Fire, Native Peoples, and the Natural Landscape*. Washington, D.C.: Island Press.

——. 2002b. The pre-European landscape of the United States: Pristine or humanized? In T. Vale, ed., *Fire, Native Peoples, and the Natural Landscape*, 1–40. Washington, D.C.: Island Press.

——. 2002c. Preface. In T. Vale, ed., *Fire, Native Peoples, and the Natural Landscape*, xiii–xv. Washington, D.C.: Island Press.

——. 2002d. Reflections. In T. Vale, ed., *Fire, Native Peoples, and the Natural Landscape*, 295–301. Washington, D.C.: Island Press.

Walker, J. H. 2000. Raised field abandonment in the upper Amazon. *Culture and Agriculture* 22 (2): 27–31.

——. 2001. Work parties and raised field groups in the Bolivian Amazon. *Expedition* 43 (3): 9–18.

——. 2004. *Agricultural Change in the Bolivian Amazon*. Pittsburgh: Latin American Archaeology Reports, University of Pittsburgh.

Whitmore, T. M., and B. L. Turner II. 2002. *Cultivated Landscapes of Middle America on the Eve of Conquest*. Oxford: Oxford University Press.

Wiersum, K. F. 1997a. From natural forest to tree crops: Co-domestication of forest and tree species, an overview. *Netherlands Journal of Agricultural Science* 45 (4): 425–438.

——. 1997b. Indigenous exploitation and management of tropical forest resources: An evolutionary continuum in forest-people interactions. *Agriculture, Ecosystems, and Environment* 63:1–16.

Wilson, P. J. 1988. *The Domestication of the Human Species*. New Haven, Conn.: Yale University Press.

Wolf, E. R. 1982. *Europe and the People Without History*. Berkeley: University of California Press.

Wust, I., and C. Barreto. 1999. The ring villages of central Brazil: A challenge for Amazonian archaeology. *Latin American Antiquity* 10 (1): 3–23.

Yen, D. E. 1989. The domestication of environment. In D. R. Harris and G. C. Hillman, eds., *Farming and Foraging: The Evolution of Plant Exploitation*, 55–78. Boston: Unwin Hyman.

Zelinsky, W. 1973. *The Cultural Geography of the United States*. Englewood Cliffs, N.J.: Prentice-Hall.

Zent, E., and S. Zent. 2004. Amazonian Indians as ecological disturbance agents: The Hoti of the Sierra de Maigualida, Venezuelan Guayana. In T. J. S. Carlson and L. Maffi, eds., *Ethnobotany and Conservation of Biocultural Diversity*, 79–112. Bronx: New York Botanical Garden.

Zimmerer, K. S., and T. J. Bassett, eds. 2003. *Political Ecology: An Integrative Approach to Geography and Environment-Development Studies*. New York: Guilford Press.

Zimmerer, K. S., and K. Young. 1998. Introduction: The geographical nature of landscape change. In K. Zimmerer and K. Young, eds., *Nature's Geography: New Lessons for Conservation in Developing Countries*, 3–34. Madison: University of Wisconsin Press.

We have a general number one, that is, you can't consider
yourself as number one, without considering me too as the
same, and all the other young people.

—CHARLES DICKENS, OLIVER TWIST

9

POLITICAL ECONOMY AND PRE-COLUMBIAN LANDSCAPE TRANSFORMATIONS IN CENTRAL AMAZONIA

EDUARDO G. NEVES AND JAMES B. PETERSEN

ONE OF THE major recent advances in Amazonian anthropology is the recognition that past indigenous populations consistently transformed the regional landscape through the management of natural resources (Cleary 2001; Denevan 1992a, 2001; Posey 1985; Stahl 2002). Accordingly, *landscape manage-ment* can be defined here as "the human manipulation of inorganic and organic components of the environment that brings about a net environmental diversity greater than that of so-called pristine conditions, with no human presence" (Balée 1994:116).

This awareness of landscape management should be understood in light of what is now a prolonged debate about the role of environmental forces in the shaping of past social dynamics in Amazonia, and vice versa. A consequence of this change in perspective is a pendulum swing, a conceptual shift from extreme forms of environmental determinism to perspectives that focus on the landscape as a primarily historical construct, the visible imprint of past human agency. Such a conceptual shift has freed Amazonist archaeologists, ethnographers, and ethnohistorians to investigate environmental factors in the explanation of past social practices and developments. It also brings a shift in scale, where the focus of analysis becomes not so much the whole Amazon basin—with all related generalities about cultural development in the humid tropics—but rather the particular historical trajectories that are identifiable in specific contexts, including archaeological ones (Erickson, chapter 8, Erickson and Balée, chapter 7, and Heckenberger, chapter 10, this volume).

This chapter pursues these ideas in more detail. We attempt to show that human history in Amazonia has not been uniform in its unfolding over time, but rather has been punctuated by episodes of apparently long-term stability interrupted by what seem to be—from a present standpoint—rather sudden and dramatic events of sociocultural change. Human landscape management in Amazonia has been exerted within this long-term historical context. Therefore, it should likewise be recognized that landscape transformation has occurred in separate pulses over time, in intimate correlation with differing structures, functions, and changes in a series of past social formations.

Our argument develops as follows. First, we show how chronological data firmly place the most enduring and visible signs of pre-Columbian Amazonian landscape transformation as fairly recent phenomena. Second, a case study is presented to review the timing and pace of the creation of anthropic dark soils known as Amazonian Dark Earth (ADE), or *terra preta* (black earth), in the central Amazon. Third, we show how ADE formation in the central Amazon is best understood as the outcome of past sociopolitical processes that formerly prevailed there (Denevan, chapter 5, and Graham, chapter 2, this volume). Finally, we discuss how some forms of landscape transformation—such as ADE and anthropogenic forests—can be seen as the results of a dialectical process between people and nature. These exchanges entailed, on one hand, the formation of human-modified landscapes and, on the other, the transformation of the people themselves by the environments and landscapes they modified (Raffles 2002:47). In other words, we advocate a particularistic, historically oriented perspective based on the premise that landscapes need to be "temporalized" (Ingold 2000:208)—that is, landscape changes need to be understood in light of the social forces and symbolic values within which they were produced.

AMAZONIAN LANDSCAPES FROM A HISTORICAL PERSPECTIVE

Discussions about human transformations of landscapes in Amazonia should be based on a clear understanding of the structure and dynamics of the different ecosystems established by the end of the Pleistocene, when human occupation of the region began. In other words, one must reconstruct the basic environmental conditions ultimately transformed by human actions to better understand and evaluate the impact of these modifications.

Considerable debate exists regarding the natural history of Amazonian environments, including at least two major and often conflicting approaches. On one side, some paleoecologists propose that the tropical rain forest and all its diversity are ancient and stable in Amazonia, going through relatively few structural modifications over many millennia during the Late Pleistocene and Holocene epochs (Colinvaux 2001:30; Colinvaux, de Oliveira, and Bush 2000). On the other side, some scholars recognize a highly dynamic history of forest transformation from place to place over time, including the formation of forest refugia during drier periods of the Pleistocene. Such major changes are correspondingly said to have been key determinants for the great ecological diversity that is characteristic of contemporary Amazonia (Van der Hammen and Hooghiemstra 2000). This general debate is probably far from being settled, however, due to the lack of past environmental data. For example, few sites suitable for the collection of ancient pollen have been identified so far and nearly all of them lie along the periphery of the Amazon basin (Colinvaux 2001:20; Roosevelt 2000:468–471).

In both scenarios, the first human inhabitants of Amazonia during late glacial times found regionally diverse landscapes approaching the ecological diversity demonstrable today. The late glacial period (13,000–10,000 BP) witnessed increased rainfall, water levels, and sedimentation in various river valleys (Van der Hammen and Hooghiemstra 2000:738–739). At the time, late glacial species composition probably was not fundamentally different from present conditions, but different relative proportions were likely characteristic across the region. From the beginning of the Holocene epoch (after around 10,000 BP), rising sea levels and corresponding changes in the Amazon River and its tributaries inundated large areas of floodplains and created deep lakes *(rias)* found, for instance, in the lower Xingu, lower Tapajós, and lower Coari rivers. Early archaeological sites representing these ancient riverine settlements are currently under water.

Available data indicate that the ecological context for the early human settlement of Amazonia was a tropical rain forest, in broad terms, in the nonriverine settings known as interfluvial uplands, interfluves, or *terra firme*. Forests there are characterized by high floral diversity, a high percentage of low-density scattered species, and low floristic similarity among neighboring patches (Nelson and de Oliveira 2001:140). Thus, interfluvial upland ecosystems characteristically exhibit a great degree of biological diversity and a low density of individual species. These ecological conditions set the stage for the long-term history of human resource management within Amazonia in particular and within the humid tropics in general, as was noted by scholars such as Carl Sauer (1969:24) and Donald Lathrap (1977) several decades ago.

The earliest generally accepted dates for human occupation in Amazonia come from the site of Pedra Pintada Cave, situated in the middle-lower Amazon, and dating to around 11,000 radiocarbon years BP (Roosevelt et al. 1996; Roosevelt, Douglas, and Brown 2002:199). Such early sites are extremely rare, as currently known, however. The longest more or less continuous archaeological sequence in Amazonia comes from the tidal estuary and the lower course of the Amazon River (Prous 1992:471). Following the early dates from Pedra Pintada, Early to Middle Holocene epoch dates are known from the Taperinha shell mound near the mouth of the Tapajós River, at around 7000 BP (Roosevelt 1995, 2000; Roosevelt et al. 1991). Elsewhere, the shell mounds of the Mina phase along the lower Xingu River, Marajó Island, and the Atlantic shore are dated to around 5500 to 4000 BP (Roosevelt 1995; Simões 1981).

The longest more or less continuous ceramic-age sequence is known from Marajó Island at the mouth of the Amazon, dating from around 3500 BP to 500 BP or so (Meggers and Danon 1988; Roosevelt 1987, 1991). This long sequence culminated with a cultural explosion near the mouth of the Amazon River. Several distinct ceramic complexes were spread over Marajó Island and the continental north shore of the Amazon River from about 1500 BP to 500 BP, when Europeans arrived and began the devastation of the indigenous peoples (Guapindaia 2001:171–173).

Elsewhere in Amazonia, the available cultural sequences are not so long as at Marajó Island, or they are not as well delineated. For instance, at the Peña Roja and San Jacinto sites in the middle Caquetá River basin in Colombia, other Early Holocene epoch occupations are known, dating to around 9200–8000 BP (Cavelier et al. 1995; Mora et al. 1991; Oliver 2001:57). The same is true on the upper Orinoco River of Venezuela, with dates before 8000 BP (Barse 1995, 2003). Other open-air Preceramic occupations of the Early Holocene epoch are found in widespread locations across Amazonia such as the Carajás Hills in the southeastern corner (Magalhães 1994) and the upper Madeira River basin in the southwestern portion. Open-air sites on the upper Madeira River date also back to around 8000 BP (Miller et al. 1992). Recent work in central Amazonia has demonstrated occupations associated with bifacial lithic technology directly dated as old as 7700 BP (Petersen et al. 2004). Still other evidence of presumed Early to Middle Holocene epoch occupation in Amazonia has been recently summarized (Meggers and Miller 2003), but none of these finds is well understood.

These data demonstrate that the initial occupation of Amazonia was not exclusively restricted to areas adjacent to the large rivers, but also occurred in diverse settings early on, including riverine and nonriverine upland areas. Though limited, the evidence now available challenges the often-cited idea that hunter-gatherers could not survive in tropical forest settings without the development of farming (see discussion in Mercader 2003:2–3; Roosevelt 1999:89; Roosevelt, Douglas, and Brown 2002). Moreover, it also helps falsify Lathrap's (1968) classic culture-historical model for the regressive development of some hunter-gatherer economies within Amazonia as the result of population pressure built up in riverine floodplain areas, at least within early contexts. Lathrap's model assumes that farmers had to develop hunting and gathering (foraging) economies in Amazonia when they relocated from prime farmlands to the uplands as a result of ecological limitations that prevented stable farming in such infertile areas. The lack of evidence for population pressure and resource competition in riverine and near-riverine Amazonia during the Early Holocene epoch indicates that other factors must be sought to explain the presence of early hunter-gatherers in the interfluvial uplands (Denevan, chapter 5, this volume).

Admittedly, the specific economic activities of Preceramic peoples in Amazonia are still poorly documented. The current evidence nonetheless indicates that early settlers in the area were not specialized hunters of large game, as has been posited elsewhere. Instead, they were generalized, broad-spectrum foragers, as shown by recovery of numerous seeds from palms and other economic plants, as well as of plentiful fish and small mammal bones in the few well studied sites (Cavelier et al. 1995; Oliver 2001; Roosevelt, Douglas, and Brown 2002). At the Pedra Pintada site, archaeologists recovered evidence of "palm nuts, legume seeds, Brazil nuts, small and large fish bones, shellfish, turtles, tortoise, lizards, medium-sized rodents, and birds" (Roosevelt, Douglas, and Brown 2002:202; Roosevelt et al. 1996). Such a generalist pattern is found among various contemporary Amazonian hunter-gatherers such as the Nukak, Maku, and other groups in the northwest Amazonia, as well as among hunter-gatherers elsewhere (Politis 2001; Silverwood-Cope 1990). Thus, evidence of forest and broad landscape management among contemporary Amazonian hunter-gatherers reveals some of the basic parameters related to early Amazonian economies during the Late Pleistocene and Early-Middle Holocene epochs.

Rather than mechanically project ethnographic contexts into the past, we suggest that at least some historically known foragers represent long-term historical developments and continuation of ancient lifestyles. These lifestyles are not necessarily regressions into archaic, preagricultural economies, adopted by former farmers pushed into marginal ecological settings, but rather ongoing phenomena carried over from pre-Columbian times (Politis 2001). They may well represent continuation, albeit not frozen in space and time, of ancient, well-adapted lifestyles that first emerged among the pioneer populations in the region during the Late Pleistocene and Early-Middle Holocene.

Notably, hunter-gatherer societies in contemporary Amazonia can be divided into at least two broad groups. One group is composed of societies located mostly in upland settings in the northwest Amazon, the upper Orinoco and upper Amazon basins, such as the Maku, Nukak, and Waorani. Until recently these people were generalist hunter-gatherers in interfluvial upland areas. They have exhibited high settlement mobility and a specialization in the exchange of forest products for farm crops and other goods from sedentary riverine groups (Politis 2001; Ramos, Silverwood-Cope, and Oliveira 1980). Their mobility is likely very ancient, whereas the trade in crops is a more recent adjustment to the presence of farmers in adjacent contexts. The second broad group is composed of Tupí-Guaraní-speaking hunter-gatherers, who live largely in an arc along the southern, western, and eastern fringes of Amazonia (Balée 1994:209). They are specialized hunter-gatherers, who include, among others, the Sirionó, the Yuquí, the Aché (who actually live beyond Amazonia proper in the Paraná basin), the Parakanã, and the Guajá (see Cormier, chapter 11, this volume). Among the first broad group, evidence suggests that their foraging patterns are quite old (Neves 2001; Oliver 2001). In contrast, individual societies within the broad Tupí-Guaraní group have oscillated between agriculture and foraging over time, depending on internal and external factors (Fausto 2001:174).

In the northwest Amazon, the Nukak ultimately transform their landscape through the long-term process of cyclical use and abandonment of their camps. This subtle transformation is based on the consequent sprouting of seeds abandoned around Nukak huts and the transplantation of seedlings during the Nukaks' overland treks—both of which have produced culturally modified settings, containing human-enriched concentrations of useful plants, including crops and noncrops (Politis 1996:156–157, 2001). Comparable forms of plant management are described among some contemporary farmers in Amazonia, such as the Ka'apor and Kayapó (Balée 1994; Posey 1985, 1986).

The major consequence of this combination of intentional and unintentional actions is the enrichment of natural concentrations of economic plant species in given spots, most obviously palm species (Balée 1988). Morcote-Ríos and Bernal have recently shown that there is a general increase in the number and diversity of palm species found in archaeological sites in Amazonia (and elsewhere) that are dated as early as between 9000 and 5000 BP (2001:311). For example, at the Peña Roja site on the Caquetá River, archaeological strata dated between 9300 and 9000 BP have produced palm remains from the genera *Astrocaryum, Oenocarpus, Mauritia,* and *Attalea* (Cavelier et al. 1995). Species within these genera are widely consumed by humans in contemporary Amazonia, including *Astrocaryum aculeatum* (tucumã or murumuru), *Oenocarpus* sp. (bacaba), *Mauritia flexuosa* (burití or moriche), and *Attalea maripa* (inajá) (Clement, chapter 6, and Erickson and Balée, chapter 7, this volume).

Palm management is therefore one of the most ancient and visible forms of natural-resource management among past and present hunter-gatherers and other indigenous peoples in Amazonia, although palm seeds may be disproportionately preserved archaeologically because of their relative durability. One way or another, the generalist lifestyle of early hunter-gatherers provided the basis for the development of later agricultural systems within Amazonia (Oliver 2001; Piperno and Pearsall 1998). Available archaeological data show that these early developments, first hunter-gatherers and later farmers, were stable for many millennia, until around 2,000 to 3,000 years ago. Some of the most widely consumed palms in contemporary Amazonia were apparently introduced around this time, along with a gamut of clearly domesticated crops.

Not accidentally, at least one of these species is also probably the only fully domesticated palm used by native peoples in Amazonia: *Bactris gasipaes* (peach palm, chonta, pejibaye, or pupunha) (Clement, chapter 6, this volume). *Euterpe oleracea* (açaí), another palm extensively consumed today, appears rather late in the archaeological record (see Brondizio, chapter 12, this volume, on its present economic importance). Although the sample size is small and should be treated cautiously, the earliest available dates for these palms in Amazonia are quite recent: 1565 ± 35 BP in the Araracuara area on the Caquetá River in Columbia (Morcote-Ríos and Bernal 2001:319). On Marajó Island at the mouth of the Amazon, a direct radiocarbon date of 1,000 ± 90 BP for an açaí seed was obtained at the Teso dos Bichos site, related to the Marajoara culture (Roosevelt 1991:314). At the Lago Grande site, near the Solimões River in central Amazonia, layers containing abundant carbonized palm seeds, probably açaí, were associated with ceramic griddles of the Paredão phase that cross-date to the late first millennium AD (Donatti 2003).

The relatively recent dates for peach palm and açaí are not surprising. They match a larger data set that points to a radical shift in economic and social patterns in Amazonia and in eastern and central lowland South America more generally starting sometime around 2,500 to 3,000 years ago (Petersen, Neves, and Heckenberger 2001). The most visible forms of landscape transformation in the archaeological record of Amazonia are attributable to this period. These forms of landscape transformation include, among others, large areas of anthropogenic ADE soils, raised fields, artificial mounds, defensive ditches, roads, causeways, irrigation ditches, and other artificial water channels (Denevan 2001; Erickson 2000a, 2000b, and chapter 8, this volume; Erickson and Balée, chapter 7, this volume; Heckenberger, chapter 10, this volume; Heckenberger, Petersen, and Neves 1999; Heckenberger et al. 2003; Woods and McCann 1999).

Based on the archaeological and ethnographic evidence briefly summarized earlier, landscape management and transformation within Amazonia can be divided into two nonmutually exclusive and broad categories: low-intensity and high-intensity landscape management. Low-intensity landscape management, as suggested here, is quite ancient, probably as early as the establishment of human occupation in Amazonia, and it is still practiced by many indigenous groups. Low-intensity landscape management cumulatively may well have made a substantial imprint on landscapes in Amazonia, but its visibility and clear association with particular societies is difficult to assess archaeologically. What one sees in the present is the cumulative outcome of individual, small-scale interventions over many decades and even centuries, and in many cases individual outcomes are difficult to isolate from natural transformations. For instance, at Igarapé Guariba, near the mouth of the Amazon River, large channels were opened by local groups through small-scale, short-term mobilization of labor, which took advantage of the natural erosive action of the daily tidal and annual water-level variation in the area (Raffles 2002).

High-intensity landscape management is seemingly a much more recent phenomena in Amazonia, beginning after 3,000 ago and more visible after the beginning of the first millennium AD. In many cases, it represents the archaeological correlate of the development of stable, sedentary, socially complex lifeways across Amazonia. These changes may be related to the development of full agricultural economies in the area, even if the processes of plant domestication and small-scale farming may be much older (Oliver 2001; Piperno and Pearsall 1998).

High-intensity landscape management also represents the constitution of cultural places and territories loaded with material and symbolic meanings. The emergence of cultural places is a historical process and thus should be understood in relationship to the history of the societies that transformed these places and that were transformed in turn. The following section presents data from archaeological sites that we and our colleagues have studied over the past decade in the central Amazon. This case study shows how human management created a mosaic landscape and how the history of this landscape can be understood only within the broader perspective of the pulses of political centralization and decentralization among Amerindian societies regionally.

SOCIAL FORMATIONS AND LANDSCAPE TRANSFORMATION IN CENTRAL AMAZONIA AFTER 2500–2000 BP

Since 1995, we have directed an intensive archaeological survey and excavation program called the Central Amazon Project (CAP). Roughly 1,000 square kilometers in size, the CAP research area is adjacent to the junction of the Negro and Solimões rivers, forming the Amazon River proper in Brazilian perspective (although Americans and others often label the Solimões River as the Amazon River). The CAP area is situated near the city of Manaus in the heart of the Amazon River basin (Heckenberger, Petersen, and Neves 1999; Neves 2000; Neves et al. 2003, 2004; Petersen, Neves, and Heckenberger 2001; Petersen et al. 2004) (figure 9.1).

The CAP area is composed of both white-water and black-water river settings, along with extensive interfluvial uplands between them. The alluvial landscape adjacent to the Solimões River represents classic Amazonian floodplain (*várzea*), comprising a wide range of microhabitats such as seasonally variable lakes, grasslands, flooded forests (*igapó*), muddy beaches, and levees for farming (Denevan 1996:657). The width of the Solimões River (the Brazilian name for the upper Amazon River) floodplain in the CAP area varies from a few hundred meters to several kilometers directly along the river channel. At the south margin of the Solimões River, near the modern city of Manaquiri, the floodplain is even wider, reaching 20–30 kilometers beyond the river channel. The Negro River, in contrast, has no true várzea along it. Within the CAP

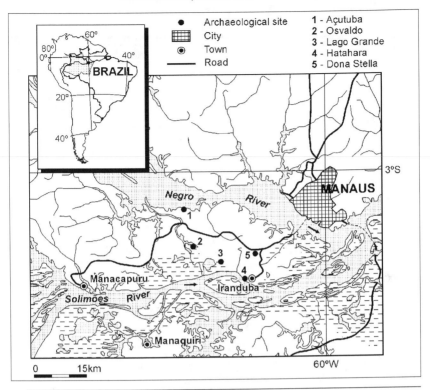

FIGURE 9.1 Map of the Central Amazon Project (CAP) study area in Amazonas State, Brazil.

research area, the alluvial plain of the Negro is narrower than the Solimões, covered either by seasonally flooded igapó forests or by white sandy beaches during times of low water. Overall, the local width of the maximum Negro River channel ranges from 1 to 6 kilometers from margin to margin.

High bluffs adjacent to the floodplains of both the Solimões River and the Negro River are exposed to erosion. Archaeological sites are normally located on these bluffs within the CAP study area, as well as in other locations away from the major rivers. According to Denevan (1996), in the central Amazon basin generally sites are actually not located on the floodplain as previously believed, but rather are found adjacent to it on the bluffs well above the highest water level, as seen elsewhere (Sternberg 1998). Therefore, water-level fluctuation is not a limiting factor in the establishment of human settlements along the main rivers (Meggers 1996).

The interfluvial uplands of the CAP study area are composed of rolling hills periodically dissected by small tributary creeks (*igarapés*). Fluvial erosion has also shaped these hills over a long time, forming high bluffs adjacent to the creeks and their narrow floodplains. Most soils are naturally acidic yellow

Latosols or Oxisols. Extensive patches of sandy, unfertile, well-drained Podzols also occur, where a characteristic somewhat stunted vegetation type (*campinaranas,* or *campinas*) is found (Daly and Mitchell 2000; Klinge 1965; Neves 2004).

CAP fieldwork in still being actively pursued, but our research has identified 67 archaeological sites so far. Five of these sites (Açutuba, Dona Stella, Hatahara, Lago Grande, and Osvaldo) have been extensively and intensively sampled through archaeological excavations and digital mapping, but others have been only lightly tested and surface collected. Sites are distributed both along major rivers and in the interfluvial uplands between the rivers. A recent survey in a subunit of the overall CAP study area, roughly 10 by 10 kilometers in size, documented diverse archaeological sites in the interfluvial uplands. However, interfluvial site sizes are significantly smaller than those oriented to the major floodplains of the Negro and Solimões rivers (Lima 2003).

One of the important results of the CAP research in the Negro-Solimões confluence area has been the establishment of a provisional chronology for the pre-Columbian occupation (Heckenberger, Neves, and Petersen.1998; Heckenberger, Petersen, and Neves 1999; Neves 2000; Neves et al. 2004; Petersen, Heckenberger, and Neves 2003; Petersen et al. 2004). The CAP chronology is supported by more than 70 radiocarbon dates and by cross dating of archaeological strata, based on recognizably diagnostic artifacts. It refines the basic chronology for the central Amazon as proposed more than 35 years ago by Peter Hilbert (1968). Most important, the CAP chronology includes Preceramic evidence not previously recognized. Unfortunately, the final Amerindian inhabitants of the area during the early Colonial period are poorly known due to early and widespread European disease, enslavement, and other disruption. Some evidence suggests that the Mura and perhaps the Arawakan-speaking Manao may have been living in or near the CAP area before Europeans arrived locally (Nimuendajú 1981; Porro 1994; Whitehead 1994:38–39; Wright 1999:347–370).

The CAP chronology starts with recently identified Preceramic occupations found in the interfluvial uplands. Dated as early as around 7700 BP with several dates at the Dona Stella site, the locally known occupations are sometimes buried under more than a meter of sandy Podzols and are visible only in eroded areas and road cuts. More than 10 such sites are currently known (Costa 2002; Neves 2003; Petersen et al. 2004).

Subsequent ceramic occupations within the CAP study area, around 2500 to 500 BP in general, are better dated and more fully understood. They are characterized by sites of variable size, some of them along the major rivers and as large as 90 hectares. Each of these large riverine sites normally has more than one recognizable time period represented, with three to four ceramic subdivisions recognized thus far in the CAP area. The earliest ceramic sites are related to the previously defined Manacapuru phase (Hilbert 1968) (or newly defined Açutuba and Iranduba phases), which is now dated between the fourth century BC and the ninth century AD (Heckenberger, Petersen, and Neves 1999; Hilbert 1968; Neves 2003; Petersen et al. 2004). Although the Manacapuru phase can be recognizably subdivided into several sequential phases based on pottery, the ceramics in this phase in general were not strictly a local development because they had widespread external connections. They share stylistic and technological traits with other complexes from the lower Amazon and perhaps as far as the lower Orinoco basin and even the Caribbean (Petersen, Heckenberger, and Neves 2003; Petersen et al. 2004).

Due to these similarities, the local Manacapuru ceramics are related more broadly either to the Amazon Incised tradition or horizon (Hilbert 1968; Meggers and Evans 1983) or to a more general Amazon Barrancoid tradition (Boomert 2000; Brochado and Lathrap 1982; Heckenberger, Petersen, and Neves 1999; Heckenberger et al. 1998; Petersen, Heckenberger, and Neves 2003; Petersen et al. 2004). This question, important for the cultural history of northern South America, remains to be resolved.

Manacapuru ceramics are chronologically followed by ceramics from the Paredão phase (Hilbert 1968). The Paredão phase is apparently a local complex and dated between the eighth and tenth or eleventh centuries AD, but it may have had a briefer duration (Donatti 2003; Hilbert 1968; Neves 2000; Nordenskiöld 1930; Petersen et al. 2004). Paredão phase sites are becoming better known due to the rapid expansion of the city of Manaus, which has led to the recent discovery and sometimes the destruction of some of these sites (Silva 2003).

Finally, the latest ceramics in the CAP area are assigned to the Guarita phase or subtradition, in either case part of the Amazon Polychrome tradition (Hilbert 1968). The Guarita phase is dated locally between the ninth and sixteenth centuries AD (Heckenberger, Petersen, and Neves 1999; Neves 2000; Petersen, Heckenberger, and Neves 2003; Petersen et al. 2004; Simões 1974; Simões and Kalkmann 1987). It also may be ultimately subdivided. A summary of the chronology of ceramic occupation in the CAP area is presented in table 9.1.

TABLE 9.1 Synthetic Ceramic Chronology of the CAP Study Area

Ceramic Complex	Temporal Length in the Area	General Affiliation
Manacapuru phase	400 BC–AD 900	Amazonian Barrancoid tradition
Paredão phase	AD 700–1000	local development?
Guarita phase	AD 1000–1600	Amazon Polychrome tradition

Source: Heckenberger, Petersen, and Neves 1999; Hilbert 1968; Neves 2000; Petersen et al. 2004.

Ceramic occupations in central and lower Amazonia are intimately associated with permanent and still visible transformations of the landscape. As noted
earlier, the most visible and most extensive of these transformations by far are
diverse anthropogenic ADE soils. As elsewhere in Amazonia, ADE represents
highly fertile soil in a region of typically infertile soils. It is highly sought after
by nonindigenous farmers today for the cultivation of cash crops targeted for
the rapidly expanding Manaus market (Glaser and Woods 2004; Lehmann et al.
2003; Neves et al. 2003, 2004; Petersen, Neves, and Heckenberger 2001; Woods
and McCann 1999). Other, more subtle forms of landscape transformation are
likewise evident in the CAP area. They include the construction of funerary
mounds and other kinds of artificial mounds, as well as of ditches at some sites,
probably with a defensive function. These forms of landscape transformation
are both the intentional and unintentional impact of pre-Columbian societies
in the area.

Most of the ceramic sites identified so far are associated with ADE, with
rather dramatic sediment alteration in some cases (Heckenberger, Petersen, and
Neves 1999:table 2; Woods and McCann 1999). Although researchers generally agree that these soils are anthropogenic—in the sense that they result from
human action in the past—there is no consensus on the specific cultural practices that generated them (e.g., see Glaser and Woods 2004; Lehmann et al.
2003). Some degree of unintentional *and* intentional formation pertained.

Chronology places the general emergence of ADE in Amazonia some time
around 2000 BP, if not earlier. Preceramic ADE layers in sites from the Massangana phase on the Jamari River in southwestern Amazonia, with dates
from 4780 ± 90 to 2640 ± 60 BP, are the earliest cases of ADE (Miller et al.
1992:37−38, 53−54). If the dating and ADE characterization are correct,
these ADEs are the earliest in Amazonia. The chronology of ADE formation documented by the CAP otherwise matches those of most other areas,
with dates back to the late first millennium BC and, more surely, to the first
millenium AD.

In any case, the relatively sudden appearance of ADE in the archaeological
record has been linked to general demographic growth and to the development of chiefdomlike social formations within Amazonia (Petersen, Neves,
and Heckenberger 2001). However, ADE was not formed continuously over
time. Its formation was dependent on type, density, and duration of various
intensive human activities. In many cases, periods of ADE formation were
interrupted at given sites due to differential settlement usage. There have been
few truly comprehensive and archaeologically based discussions about the processes responsible for pre-Columbian settlement abandonment and thus for
consequent interruption in the formation of ADE. The few available discussions have typically borrowed from modern ethnographic analogies based on
only partially representative small sociopolitical groups who exhibit a relatively

high degree of mobility and often frequent long-distance relocation. These groups almost surely represent one reduced end of the full spectrum of past native settlement practices, with many others lost due to historic disruptions. As with any form of human landscape transformation, ADEs have complex life histories deeply intertwined with the histories of the people who created them (Erickson 2003:476–482).

Data from the CAP may help us to understand these variable histories because many ADE sites seem to have been abandoned well before European colonization, with a few notable exceptions. The CAP data on site formation and abandonment suggest that the inferred chiefdoms in the central Amazon basin were cyclical or centrifugal social formations, characterized by alternate processes of political centralization and decentralization. Political centralization and social aggregation in the archaeological record of central Amazonia are represented by the simultaneous occupation of large sites, some of them several to many hectares in size, as well as of other smaller contemporaneous sites (Heckenberger, Petersen, and Neves 1999; Petersen, Neves, and Heckenberger 2001). Decentralization and political desegregation is demonstrable, however, by the sudden abandonment of some of these sites, both large and small (Neves et al. 2003, 2004).

To illustrate this point, we present data from two of the sites recently mapped and excavated by the CAP: Lago Grande and Hatahara. Both sites are located on high bluffs adjacent to the floodplain of the Solimões River, not far upstream from the city of Manaus and the Negro-Solimões confluence (figure 9.1). In the case of the Lago Grande site, the adjacent várzea floodplain is large and under water almost half the year. The flooded area forms a large lake, known as Lago Grande (Petersen, Neves, and Heckenberger 2001:figures 3.8 and 3.9). The southern and western sides of the lake are levees that are flooded during the wet season, thus connecting Lago Grande to the Solimões River and interfluvial waterways. In contrast, steep bluffs of varying height bound the northern and eastern sides of the lake, well above the water at all times. Viewed from the air, the Lago Grande bluffs have an irregular contour, forming bays and peninsulas, the latter being where several archaeological sites are found, including the Lago Grande site (Donatti 2003).

The Lago Grande site is the largest archaeological settlement located on the northern bank of the lake, although it is smaller than some other ceramic sites (figure 9.2). Topographic mapping and a series of auger tests show that the Lago Grande site is roughly 4 hectares in size, situated on a peninsula extending from adjacent high ground into the lake. It is a small to medium-size site within the overall CAP site inventory (figure 9.3).

The Lago Grande site is currently covered by thick secondary forest, so that most mapping and archaeological testing there is based on transects cleared through the brush. Topographic mapping shows that this site was

FIGURE 9.2 Aerial view of the Lago Grande site in the CAP study area.

a horseshoe-shaped, circular village. The radiocarbon dates for the Lago Grande site (table 9.2) demonstrate that this settlement was minimally occupied for about 200 years during the ninth and tenth centuries AD (Neves et al. 2004). These dates were obtained from the excavation of several 1-by-1-meter test units spread across the site (Donatti 2003).

Polychrome Guarita phase and Barrancoid Manacapuru phase ceramics are represented both on the surface and in the excavations (suggesting possible trade connections and both earlier and later occupations). However, Lago Grande is primarily a Paredão phase site between the earlier and later occupations, as indicated by the analysis of the ceramics recovered from one of the excavated test units. Therefore, it represents an example of a relatively long-term occupation in contrast to the short-term occupations ethnographically documented in Amazonia (e.g., Meggers 1996; Neves 1995). The stratigraphy of the Lago Grande site shows deeply buried ADE to more than 160 centimeters below the surface (figure 9.4).

The Lago Grande chronology indicates that the ADE was formed at different rates from place to place on the site. A darker stratum is buried under almost 1 meter of soil; it is approximately 30 centimeters thick and seemingly was formed rather quickly, within a few decades or so. The upper strata are lighter in color and include lower densities of ceramics.

At the Lago Grande site, occupation peaked around the beginning of the ninth century AD. From that time onward, the settlement was apparently continuously

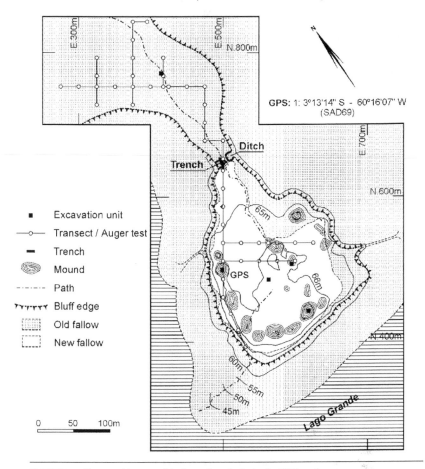

GPS: 1: 3°13'14" S - 60°16'07" W
(SAD69)

- Excavation unit
- Transect / Auger test
- Trench
- Mound
- Path
- Bluff edge
- Old fallow
- New fallow

0   50   100m

FIGURE 9.3 Topographic map of the Lago Grande site in the CAP study area.

occupied, although less intensively, until it was abandoned around the end of the tenth century AD. Given the abundance of carbonized plant remains from the excavations and the site's close proximity to the resource-rich floodplain of the Solimões River, it is unlikely that this site was abandoned for subsistence reasons. The presence of an artificial ditch constructed at a strategic location around the ninth century AD, blocking access to the village from the adjacent uplands, suggests that increasing conflict may ultimately have accounted for site abandonment, although some later reuse during the Guarita phase is evident (Neves et al. 2004).

Similar to the case of Lago Grande, the Hatahara site is also located on the top of a steep bluff, well above maximum high water and adjacent to the floodplain of the Solimões River. In this case, however, the immediately adjacent floodplain is much narrower, being only a few hundred meters wide

TABLE 9.2 Lago Grande Site: Noncalibrated Radiocarbon Dates (1 Sigma Standard Deviation), Units 1, 3, and 4

Sample Number–Unit Number	Depth (cm) Below Surface	Date (Years BP)	Lab Number
319–1	36	1050 ± 40	Beta 143600
563–3	37	960 ± 40	Beta 178921
111–4	47	1110 ± 40	Beta 178922
514–3	50–60	950 ± 60	Beta 178919
1340–4	56	1050 ± 40	Beta 178925
324–1	75	950 ± 40	Beta 143601
326–1	83	950 ± 30	Beta 143602
321–1	89	960 ± 30	Beta 143607
1335–4	94	1180 ± 40	Beta 178923
1301–4	105	1170 ± 40	Beta 178924
325–1	118	1130 ± 40	Beta 143604
322–1	123	1150 ± 40	Beta 143603
329–1	142	1100 ± 30	Beta 143605
4144 ditch	113–119	1110 ± 40	Beta 178927

Note: All samples except 514–3 were charcoal fragments directly recovered from the excavation profiles.

(figure 9.5). The Hatahara site also lies on a peninsula, but this landform and site itself are much larger than the Lago Grande site and its particular setting. Topographic mapping indicates that the Hatahara site covers 16 hectares (figure 9.6). Hatahara is a multicomponent site, with remains from three major ceramic complexes represented in its composite stratigraphy. The oldest strata are related to the Manacapuru phase, above which are locally superimposed a Paredão phase occupation across much of the site and a Guarita phase occupation toward the surface of the site over a comparably large or even larger area (figure 9.7).

Hatahara is a large, dense ADE site presently covered by grass and a productive papaya plantation (Petersen, Neves, and Heckenberger 2001:figure 3.1). Beyond the site proper, the soil is a clay-rich yellow Oxisol, with low fertility and high acidity, which we believe represents the original soil of the site before the formation of ADE. The Hatahara site also exhibits a horseshoe or circular shape plan, like Lago Grande (figure 9.6). Fourteen radiocarbon dates place the primary occupation at Hatahara within two periods, between the seventh century AD and the end of the first millennium AD, although older dates are also known (table 9.3).

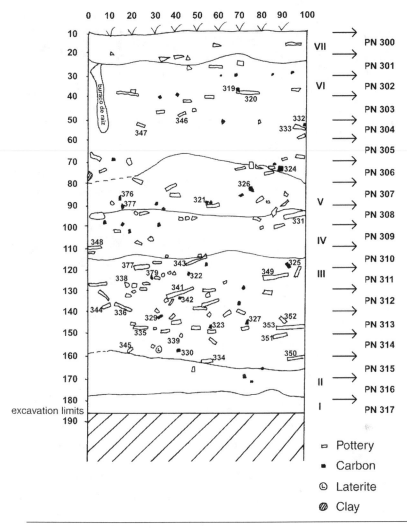

FIGURE 9.4 Stratigraphic profile of Excavation Unit 1 at the Lago Grande site in the CAP study area.

The Hatahara site has exceptional conditions of preservation for carbonized plant remains, or "charcoal" (which defines ADE in part) and burned and unburned bones. Preservation of bone remains, unlike charcoal, is rather unusual for open-air sites within Amazonia. Well-defined pit features have yielded bones of fish, turtles, and terrestrial mammals. Favorable soil conditions also have been conductive for the preservation of human remains at Hatahara. Eleven secondary human burials, some of them together, were excavated from

Hatahara site

FIGURE 9.5 Aerial view of the Hatahara site and the Solimões River in the CAP study area.

one burial mound, Mound 1 (Machado 2004; Neves et al. 2004; Petersen, Neves, and Heckenberger 2001:figure 3.2). These burials seem to have been placed simultaneously at the bottom of this mound, which was built using pottery sherds and soil. All of the mounds are generally 1 to 2 meters high, minimally cover several hundred square meters, and have an oblong shape and flat top. At least one larger funerary mound is also known at the Açutuba site, along with several other smaller ones.

Mound 1 at the Hatahara site, currently under most intensive study, was built with ADE and sherds from all three ceramic complexes of the site, but with a much higher concentration of Paredão phase sherds (Machado 2004). Mound 2 has a similar stratigraphy, but its ceramics have yet to be studied, and burials have yet to be detected. In total, 10 mounds have been defined at the Hatahara site, although whether all of them contain burials is unclear. The horizontal, parallel placement of many Paredão phase ceramics within several mounds, the large size of some sherds, and the lack of visible perturbations in the stratigraphy confirm that the mounds were built primarily in one or a few episodes (figure 9.8). However, the placement of later Guarita phase ceramics on each mound suggests later ceremonial additions to the mounds and thus some form of continuity over time.

Clearly, mound construction at the Hatahara site was done intensively in one or more episodes, which involved the movement of several tons of earth and included both contemporaneous and earlier ceramics. Such work demanded

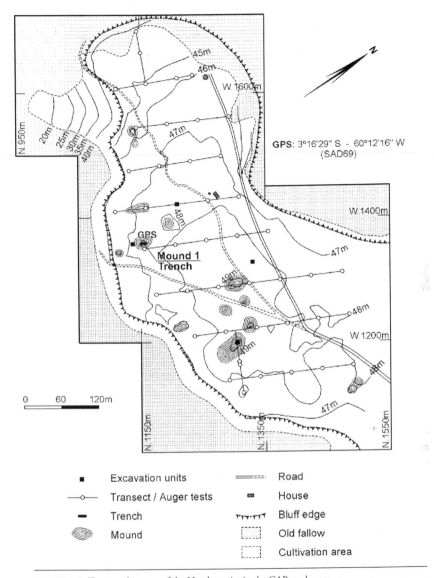

FIGURE 9.6 Topographic map of the Hatahara site in the CAP study area.

some degree of labor mobilization, coordination, and logistics, although the mounds at the Hatahara site may not have been built simultaneously. Notably, mound construction was somewhat of a wasteful practice in that it used fertile ADE soil, which might have been used instead for farming. This situation parallels mound building in the Illinois River valley (United States) by the Hopewell culture, where sod was used as fill material (Van Nest et al. 2001:646). It is unclear

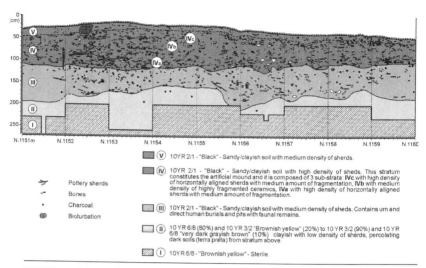

FIGURE 9.7 Stratigraphic profile of Mound 1 at the Hatahara site.

TABLE 9.3 Hatahara Site: Revised Radiocarbon Sequence, Noncalibrated (1 Sigma Standard Deviation)

Sample Number	Depth (cm)	Date (Years BP)	Lab Number
361	30–40	1010 ± 80	Beta 143583
1879	58	980 ± 40	Beta 143585
1880	60	960 ± 30	Beta 143586
1881	80	1000 ± 40	Beta 143588
505	84	1000 ± 40	Beta 143589
10688	108	1030 ± 40	Beta 178914
10737	123	1060 ± 40	Beta 178917
10714	146	910 ± 40	Beta 178915
1892	155	960 ± 40	Beta 143595
1855	160–170	1070 ± 70	Beta 143596
10740	172	940 ± 40	Beta 178918
1869	180–190	1080 ± 40	Beta 143598
10729	180	1180 ± 40	Beta 178916
1873	192	1300 ± 40	Beta 143599

Note: This is a selected list.

FIGURE 9.8 View of Mound 1 profile at the Hatahara site.

whether ADE employed as a building material in the mounds at Hatahara was collected from inside or outside the site. A number of subtle low depressions within the site, possible borrow areas for fill, support the first alternative.

In any case, the use of ADE as a building material in the mounds at the Hatahara site reveals an interesting cycle in the life history of anthropogenic soils. That cycle began with the human transformation of yellow Latosols into ADE. It probably was a long-term process, taking many years of the intentional and unintentional accumulation of organic debris, including large amounts of charcoal, by sedentary populations (Glaser and Woods 2004; Lehmann et al. 2003; Petersen, Neves, and Heckenberger 2001; Woods and McCann 1999). Once formed at the Hatahara site, ADE was partially recycled as material for mound building, suggesting that the inhabitants had no shortage of fertile soil. Finally, with the abandonment of the settlement by the beginning of the second millennium AD, the ADE was not farmed for many centuries until the onset of the Colonial period. The intact mound stratigraphy shows little or no root disturbance of the horizontally placed sherds, suggesting that high forest never grew on the site after its abandonment. Therefore, the occupation of Hatahara and similar sites in the CAP area created a relatively permanent landscape mosaic on the river bluff areas, comprising islands of secondary growth that were surrounded by larger areas of high forest.

SYNTHESIS AND DISCUSSION

We now return to some of our initial statements and discuss them in light of the preliminary data from the Lago Grande and Hatahara sites. The chronology of occupation of both sites shows that they were settled continuously over many decades or a few centuries at the end of the first millennium AD and thereafter. The relatively long span of occupation created the conditions for the development of ADE at both sites and at many others in the CAP area. According to Denevan (1992b), pre-Columbian forms of agriculture in interfluvial uplands within Amazonia were likely associated with a recurrent pattern of continuous long-term cultivation of the same plots, given the technological difficulties involved in constantly opening new farm plots in areas of high forest with stone axes. Recent analysis and experimental work with flaked stone axes found in CAP sites confirms Denevan's hypothesis, demonstrating that these axes would have been useful only for felling trees with small to medium-size diameters (Costa 2002).

The chronology of occupation of both the Lago Grande site and the Hatahara site is also consistent with Denevan's hypothesis because it documents long-term, more or less continuous occupation of these settlements. These settlements were supported by stable agricultural systems combining intensive agriculture with short fallow periods on bluffs above the peak of floods, with less predictable but highly productive short-term cultivation in the floodplain potentially relevant as well. Floodplain cultivation was not necessary, however, considering that comparably large settlements existed along the Negro River, as at the Açutuba site, where the floodplain was unsuitable for farming.

Recent ethnographic studies of farming among Tikuna and *caboclo* (mixed-blood peasants) settlements in várzea settings also support our argument (Murrieta and Dufour 2004; Shorr 2000). In Campo Alegre, a large Tikuna settlement of 1,300 people on the upper Solimões River in Brazil, Shorr discovered that farmers allocate the fertile soils of the várzea for a few select crops such as watermelon that are not necessities (2000:81). Among the Tikuna, manioc is the staple crop, and it is grown on high ground above the floodplain. Shorr notes that these choices represent a primary emphasis on highly predictable, although less productive, manioc farming on infertile upland soils, given the destructive flood patterns of the Solimões River (2000:85). Likewise, Murrieta and Dufour (2004) demonstrate that in the dietary patterns of several caboclo settlements on Ituqui Island, downstream from Santarém, manioc provides the largest source of calories, floodplain crops are unimportant, and fish are the largest source of protein for each surveyed household. Interestingly, however, fish provide the second largest input of calories after manioc within each household. Together with the archaeological evidence, these contemporary examples suggest that white-water floodplains, though fertile, were probably never a primary breadbasket for pre-Columbian people in central Amazonia due to the uncertainties of floodplain cultivation there, a point previously made by Meggers (1996). Mixed strategies based on manioc agriculture in nonfloodplain settings, complemented by opportunistic floodplain agriculture and the capture of fish and animal protein from floodplain settings, provided an effective economic base for those communities (Heckenberger 1998).

In any case, the development of ADE, settlements, mounds, and long-term short-fallow cultivation in central Amazonia surely transformed the bluffs and their adjacent uplands into a landscape mosaic toward the beginning of the first millennium AD, if not earlier. This pattern also prevailed in other parts of tropical lowland South America. For instance, early-sixteenth-century descriptions of the coastal Atlantic Forest (Mata Atlântica) at the outskirts of Rio de Janeiro in eastern Brazil show a mosaic of vegetation similar to that found in intensive and short-fallow cultivation (Dean 1995).

In the central Amazon basin, excavation profiles opened over the past decade display an orderly, parallel layering of ceramic sherds and other cultural remains that indicate these materials did not undergo postdepositional mechanical mixing over the centuries (Heckenberger, Petersen, and Neves 1999; Neves 2001, 2003a, 2003b). Where forests exist, one of the major sources of postdepositional disturbance of archaeological sites is root perturbation. The low incidence of such mechanical mixing indicates that, at least in some major CAP sites, settlement abandonment was not necessarily followed by reforestation of tall trees with deeper roots. Instead, secondary-growth forests may have developed on abandoned settlements. Therefore, besides ADE, pre-Columbian occupations in central Amazonia seemingly left long-lasting influences in the patterns of plant distribution on a regional scale.

Landscape changes are always tied on some scale to the societies and social processes that generated them. Therefore, it remains for us to delineate the structure and functioning of the social systems that generated the late pre-Columbian landscape history of central Amazonia. The classification of prehistoric societies in Amazonia has promoted considerable debate recently: Were they stratified, hierarchical chiefdoms *or* egalitarian, heterarchical societies *or* both? (See Drennan 1995; Heckenberger, Petersen, and

Neves 1999; Meggers 1993–95; Roosevelt 1987, 1993, 2000.) Research elsewhere in Amazonia shows that there is no simple answer to this important question.

In the lower Tapajós and the upper Solimões (Amazon) rivers, hierarchical, full-fledged, chiefdomlike societies during the sixteenth century AD have been documented based on interpretation of the earliest historic accounts (e.g., Meggers 1993–95, 1996; Myers 1992; Porro 1994). In other areas such as Marajó Island and possibly the upper Xingu River basin, comparably hierarchical forms of social organization collapsed well before the beginning of local European colonization; thus, they were not documented at all historically. This pattern indicates that the late political history of Amazonia was characterized by the successive rise, fall, and disintegration of chiefdoms in different places over many centuries, sometimes at rather different tempos and for various reasons (Petersen et al. forthcoming).

The data from central Amazonia match this general pattern as well, as indicated by the occupation and abandonment of local settlements well before the arrival of Europeans and other intruders. Settlement occupation and abandonment in Amazonia have been traditionally explained as adaptive strategies due to limited or unpredictable resources (e.g., Meggers 1996). This explanation may account for some instances of settlement abandonment, but most likely grossly simplifies what were more complex processes. The CAP data suggest that agricultural systems were fairly stable in central Amazonia (and elsewhere regionally) during the first and early second millennia AD and thus cannot necessarily be invoked to explain why particular settlements were abandoned. Based on this premise, we propose here that political forces as much as adaptive forces were key factors leading toward settlement abandonment in pre-Columbian Amazonia. In this perspective, settlement abandonment can be understood as the correlate of population fission, the outcome of internal processes of political conflict and fission of communities.

Such conflicts would be expected to emerge from a constant tension between, on the one hand, centralizing centripetal hierarchical ideologies—verified in the archaeological record in labor mobilization for mound-building activities—and, on the other hand, centrifugal, fragmentary, and decentralized household-based productive units. The latter inference about local kin-based productive units in the CAP area is conjectural and based on the fact that there are no unequivocal signs of labor mobilization for intensive agriculture. Instead, agriculture was likely a small-scale, kin-based endeavor. In central Amazonia, this economic basis would have created the conditions for village fissioning or abandonment or both during late pre-Columbian times in the face of insurmountable internal political conflict (Turner 1979:165–166).

The pattern of alternating centralization and decentralization in social relations is also verifiable among contemporary indigenous societies, albeit on a different scale than presumably pertained during pre-Columbian times. Among the present-day Tukanoans of the northwestern Amazon, who arguably represent the most visibly hierarchical societies of lowland South America, two complementary models of social organization coexist. One is hierarchical and based on unilineal descent, exogamy, virilocality, and agnatic residential groups, whereas the other is egalitarian and based on endogamy and the consanguineal character of residential and territorial groups (Hugh-Jones 1995:237). Among other contemporary societies of the Amazon, such as the Parakanã, the same pattern of alternation between more and less hierarchical modes of social organization has also been documented (Fausto 2001:533). In this sense, Amazonian archaeology should as much address the processes of past social collapse as it presently looks for the emergence of complexity (Petersen et al. forthcoming).

We reject traditional explanations of settlement abandonment based on environmental limitations within Amazonia. Quite to the contrary, our argument assumes that critical ecological resources were sufficiently abundant to permit long-term occupation in large settlements and cumulative landscape transformation from at least the beginning of the first millennium AD onward, if not earlier. The specifics of these hypotheses remain to be tested, but the archaeological record of central Amazonia documents a rich history of landscape transformation over several millennia. We are confident that this history can be understood only in light of the indigenous people who transformed nature and in turn were transformed by it long ago.

ACKNOWLEDGMENTS

Recent CAP fieldwork has been funded by grants from the Fundação de Amparo à Pesquisa do Estado de São Paulo (FAPESP Grants 02/02953-0 and 99/92925-7). The William T. Hillman Foundation, the Wenner-Gren Foundation, the University of Maine (Farmington), the University of Vermont, the Carnegie Museum of Natural History, and the University of São Paulo provided earlier funding.

Marcos Castro provided the digital mapping of the Lago Grande, Hatahara, and other sites. We thank Fernando Costa, Patrícia Bayod Donatti, Luiz Fernando Erig Lima, and Juliana Salles Machado for letting us use data from their graduate projects. The senior author also extends his thanks to William Balée, Clark Erickson, the staff at Tulane University, and the participants in the Symposium on Neotropical Historical Ecology for providing a wonderful environment for open and relaxed presentations and the fruitful discussion of ideas. William Balée, Clark Erickson, and Denise Schaan provided useful criticism that helped clarify this chapter in many ways. The interpretations are the authors' sole responsibility, however.

REFERENCES

Balée, W. 1988. Indigenous adaptation to Amazonian palm forests. *Principes* 32:47–54.

——. 1994. *Footprints of the Forest: Ka'apor Ethnobotany—the Historical Ecology of Plant Utilization by an Amazonian People.* New York: Columbia University Press.

Barse, W. 1995. El período arcaico en el Orinoco y su contexto en el norte de Sud América. In I. Cavelier and S. Mora, eds., *Ambito y ocupaciones tempranas de la América tropical,* 99–113. Bogotá: Fundación Erigaie, Instituto Colombiano de Antropologia.

——. 2003. Holocene climate and human occupation in the Orinoco. In J. Mercader, ed., *Under the Canopy: The Archaeology of Tropical Rain Forests,* 249–270. New Brunswick, N.J.: Rutgers University Press.

Boomert, A. 2000. *Trinidad, Tobago, and the Lower Orinoco Interaction Sphere.* Alkmaar, Netherlands: Cairi.

Brochado, J., and D. W. Lathrap. 1982. Chronologies in the New World: Amazonia. Unpublished manuscript.

Cavelier, I., C. Rodríguez, L. F. Herrera, G. Morcote, and S. Mora. 1995. No solo de caza vive el hombre: Ocupación del bosque Amazónico, Holoceno Temprano. In I. Cavelier and S. Mora, eds., *Ambito y ocupaciones tempranas de la América tropical,* 27–44. Bogotá: Fundación Erigaie, Instituto Colombiano de Antropología.

Cleary, D. 2001. Science and the representation of nature in Amazônia: From La Condamine through da Cunha to Anna Roosevelt. In I. Vieira, J. M. da Silva, D. Oren, and M. D'Incao, eds., *Diversidade biológica e cultural da Amazônia,* 273–296. Belém, Brazil: Museu Paraense Emílio Goeldi.

Colinvaux, P. 2001. Paradigm lost: Pleistocene environments of the Amazon basin (continued forest cover in perpetual flux, part 2). *Review of Archaeology* 22 (1): 20–31.

Colinvaux, P., P. de Oliveira, and M. Bush. 2000. Amazonian and neotropical plant communities on glacial time-scales: The failure of the aridity and refuge hypotheses. *Quaternary Science Reviews* 19:141–169.

Costa, F. W. S. 2002. Análise das indústrias líticas da área de confluência dos rios Negro e Solimões. Dissertação de mestrado, Programa de Pós-Graduação em Arqueologia do Museu de Arqueologia e Etnologia da Universidade de São Paulo.

Daly, D. C., and J. D. Mitchell. 2000. Lowland vegetation of tropical South America: An overview. In D. Lentz, ed., *Imperfect Balance: Landscape Transformations in the Precolumbian Americas,* 391–453. New York: Columbia University Press.

Dean, W. 1995. *With Broadax and Firebrand: The Destruction of the Brazilian Atlantic Forest.* Berkeley: University of California Press.

Denevan, W. 1992a. The pristine myth: The landscape of the Americas in 1492. *Annals of the Association of American Geographers* 82 (3): 369–385.

——. 1992b. Stone vs. metal axes: The ambiguity of shifting cultivation in prehistoric Amazonia. *Journal of the Steward Anthropological Society* 20:153–165.

Denevan, W. 1996. A bluff model of riverine settlement in prehistoric Amazonia. *Annals of the Association of American Geographers* 86 (4): 654–681.

Denevan, W. 2001. *Cultivated Landscapes of Native Amazonia and the Andes.* Oxford: Oxford University Press.

Donatti, P. B. 2003. A arqueología da margem norte do Lago Grande, Iranduba, Amazonas. Dissertação de mestrado, Programa de Pós-Graduação em Arqueologia do Museu de Arqueologia e Etnologia da Universidade de São Paulo.

Drennan, R. 1995. Chiefdoms in northern South America. *Journal of World Prehistory* 9 (3): 301–340.

Erickson, C. L. 2000a. An artificial landscape-scale fishery in the Bolivian Amazon. *Nature* 408:190–193.

———. 2000b. Los caminos prehispánicos de la Amazonia boliviana. In L. Herrera and M. Cardale de Schrimpff, eds., *Caminos precolombinos: Las vias, los ingenieros y los viajeros,* 15–42. Bogotá: Instituto Colombiano de Antropología y Historia.

———. 2003. Historical ecology and future explorations. In J. Lehmann, D. Kern, B. Glaser, and W. I. Woods, eds., *Amazonian Dark Earths: Origins, Properties, Management,* 455–500. Dordrecht, Netherlands: Kluwer Academic.

Fausto, C. 2001. *Inimigos fiéis: História, guerra e xamanismo na Amazônia.* São Paulo: Editora da Universidade de São Paulo.

Glaser, B., and W. I. Woods, eds. 2004. *Amazonian Dark Earths: Explorations in Space and Time.* Berlin: Springer-Verlag.

Guapindaia, V. 2001. Encountering the ancestors: The Maracá urns. In C. McEwan, C. Barreto, and E. Neves, eds., *Unknown Amazon: Culture in Nature in Ancient Brazil,* 156–173. London: British Museum Press.

Heckenberger, M. J. 1998. Manioc agriculture and sedentism in Amazonia: The Upper Xingu example. *Antiquity* 72: 663–648.

Heckenberger, M. J., A. Kuikuro, U. T. Kuikuro, J. C. Russell, M. Schmidt, C. Fausto, and B. Franchetto. 2003. Amazonia 1492: Pristine forest or cultural parkland? *Science* 301:1710–1714.

Heckenberger, M. J., E. G. Neves, and J. B. Petersen. 1998. De onde surgem os modelos? Considerações sobre a origem e expansão dos Tupi. *Revista de Antropologia* 41:69–96.

Heckenberger, M. J., J. B. Petersen, and E. G. Neves. 1999. Village size and permanence in Amazonia: Two archaeological examples from Brazil. *Latin American Antiquity* 10 (4): 353–376.

Hilbert, P. P. 1968. *Archaeologische untersuchungem am mittlerem Amazonas.* Berlin: Dietrich Reimer Verlag.

Hugh-Jones, S. 1995. Inside-out and back-to-front: The androgynous house in the northwest Amazon. In J. Carsten and S. Hugh-Jones, eds., *About the House: Lévi-Strauss and Beyond,* 226–252. Cambridge, U.K.: Cambridge University Press.

Ingold, T. 2000. *The Perception of the Environment: Essays in Livelihood, Dwelling, and Skill.* London: Routledge.

Klinge, H. 1965. Podzol soils in the Amazon basin. *Journal of Soil Science* 16 (1): 95–103.

Lathrap, D. W. 1968. The hunting economies of the tropical forest zone of South America: An attempt at historical perspective. In R. B. Lee and I. DeVore, eds., *Man the Hunter,* 23–29. Chicago: Aldine.

———. 1977. Our father the cayman, our mother the gourd: Spinden revisited, or a unitary model for the emergence of agriculture in the New World. In C. Reed, ed., *Origins of Agriculture,* 713–751. The Hague: Mouton.

Lehmann, J., D. Kern, B. Glaser, and W. I. Woods, eds. 2003. *Amazonian Dark Earths: Origin, Properties, Management.* Dordrecht, Netherlands: Kluwer Academic.

Lima, L. F. E. 2003. Levantamento arqueológico das áreas de interflúvio na área de confluência dos rios Negro e Solimões. Dissertação de mestrado, Programa de Pós-Graduação em Arqueologia do Museu de Arqueologia e Etnologia, Universidade de São Paulo.

Machado, J. S. 2004. A formação de montículos artificiais: Um estudo de caso no sítio Hatahara, Amazonas. Unpublished report submitted to the Fundação de Amparo à Pesquisa do Estado de São Paulo.

Magalhães, M. 1994. *Arqueologia de Carajás: A presença pré-histórica do homem na Amazônia.* Rio de Janeiro: Companhia Vale do Rio Doce.

Meggers, B. J. 1993–95. Amazonia on the eve of European contact: Ethnohistorical, ecological, and anthropological perspectives. *Revista de Arqueología Americana* 8:91–115.

———. 1996. *Amazonia: Man and Culture in a Counterfeit Paradise.* 2d ed. Chicago: Aldine.

Meggers, B. J., and J. Danon. 1988. Identification and implications of a hiatus in the archaeological sequence on Marajó Island, Brazil. *Journal of the Washington Academy of Sciences* 78 (3): 245–253.

Meggers, B. J., and C. Evans. 1983. Lowland South America and the Antilles. In J. Jennings, ed., *Ancient South Americans,* 287–335. San Francisco: W. H. Freeman.

Meggers, B. J., and E. Miller. 2003. Hunter-gatherers in Amazonia during the Pleistocene-Holocene transition. In J. Mercader, ed., *Under the Canopy: The Archaeology of Tropical Rain Forests,* 291–316. New Brunswick, N.J.: Rutgers University Press.

Mercader, J. 2003. Introduction: The Paleolithic settlement of rain forests. In J. Mercader, ed., *Under the Canopy: The Archaeology of Tropical Rain Forests,* 1–31. New Brunswick, N.J.: Rutgers University Press.

Miller, E., et al.(others not specified). 1992. *Arqueologia nos empreendimentos hidrelétricos da Eletronorte: Resultados preliminares.* Brasília: Eletronorte.

Mora, S., L. F. Herrera, I. Cavelier, and C. Rodríguez. 1991. *Cultivars, Anthropic Soils, and Stability: A Preliminary Report of Archaeological Research in Araracuara, Colombian Amazon.* Latin American Archaeology Reports no. 2. Pittsburgh: University of Pittsburgh.

Morcote-Ríos, G., and R. Bernal. 2001. Remains of palms (Palmae) at archaeological sites in the New World: A review. *Botanical Review* 67 (3): 309–350.

Murrieta, R., and D. Dufour. 2004. Fish and *farinha:* Protein and energy consumption in Amazonian rural communities on Ituqui Island, Brazil. *Ecology of Food and Nutrition* 43:231–255.

Myers, T. P. 1992. The expansion and collapse of the Omagua. *Journal of the Steward Anthropological Society* 20 (1–2): 129–152.

Nelson, B., and A. de Oliveira. 2001. Área botânica. In J. Capobianco, A. Veríssimo, A. Moreira, D. Sawyer, I. Santos, and L. Pinto, eds., *Biodiversidade na Amazônia Brasileira,* 132–153. São Paulo: Editora Estação Liberdade, Instituto Socioambiental.

Neves, E. G. 1995. Village fissioning in Amazonia: A critique of monocausal determinism. *Revista do Museu de Arqueología e Etnologia da USP* 5:195–209.

———. 2000. Levantamento arqueológico da área de confluência dos rios Negro e Solimões, Estado do Amazonas. Unpublished report submitted to the Fundação de Amparo à Pesquisa do Estado de São Paulo (FAPESP).

———. 2001. Indigenous historical trajectories in the upper Rio Negro basin. In C. McEwan, C. Barreto, and E. Neves, eds., *Unknown Amazon: Culture in Nature in Ancient Brazil,* 266–285. London: British Museum Press.

———. 2004. Levantamento arqueológico da área de confluência dos rios Negro e Solimões, Estado do Amazonas: Continuidade das escavações, análise da composição química e montagem de um sistema de informações geográficas. Unpublished report submitted to the Fundação de Amparo à Pesquisa do Estado de São Paulo (FAPESP)

Neves, E. G., J. B. Petersen, R. N. Bartone, and C. A. da Silva. 2003. Historical and socio-cultural origins of Amazonian Dark Earths. In J. Lehmann, D. Kern, B. Glaser, and W. I. Woods, eds., *Amazonian Dark Earths: Origins, Properties, Management,* 29–50. Dordrecht, Netherlands: Kluwer Academic.

Neves, E. G., J. B. Petersen, R. N. Bartone, and M. J. Heckenberger. 2004. The timing of *terra preta* formation in the central Amazon: Archaeological data from three sites. In B. Glaser and W. I. Woods, eds., *Amazonian Dark Earths: Explorations in Space and Time,* 125–134. Berlin: Springer Verlag.

Nimuendajú, C. 1981. *Mapa etno-histórico do Brasil e regiõs adjacentes.* Brasilia: IBGE.

Nordenskiöld, E. 1930. *L'archéologie du bassin de l'Amazone.* Ars Americana, vol. 1. Paris: Ars Americana.

Oliver, J. 2001. The archaeology of forest foraging and agricultural production in Amazonia. In C. McEwan, C. Barreto, and E. G. Neves, eds., *Unknown Amazon: Culture in Nature in Ancient Brazil,* 50–85. London: British Museum Press.

Petersen, J. B., M. J. Heckenberger, and E. G. Neves. 2003. A prehistoric ceramic sequence from the central Amazon and its relationship to the Caribbean. In L. Alofs and R. Dijkoff, eds., *Proceedings of the 19th International Congress for Caribbean Archaeology,* 250–259. Aruba: Archaeological Museum of Aruba.

Petersen, J. B., M. J. Heckenberger, E. G. Neves, J. G. Crock, and R. N. Bartone. Forthcoming. Collapse among Amerindian complex societies in Amazonia and the insular Caribbean: Endogenous or exogenous factors? In R. Reycraft and J. Railey, eds., *I Fall to Pieces: Global Perspectives on the Collapse and Transformation of Complex Societies.* New York: Plenum Press.

Petersen, J. B., E. G. Neves, R. N. Bartone, and M. A. Arroyo-Kalin. 2004. An overview of Amerindian cultural chronology in the central Amazon. Paper presented at the annual meeting of the Society for American Archaeology, Montreal, March 31–April 4.

Petersen, J. B., E. G. Neves, and M. J. Heckenberger. 2001. Gift from the past: *Terra preta* and prehistoric Amerindian occupation in Amazonia. In C. McEwan, C. Barreto, and E. G. Neves, eds., *Unknown Amazon: Culture in Nature in Ancient Brazil,* 86–105. London: British Museum Press.

Piperno, D., and D. Pearsall. 1998. *The Origins of Agriculture in the Lowland Neotropics.* San Diego: Academic Press.

Politis, G. 1996. *Nukak.* Bogotá, Colombia: Instituto Amazónico de Investigaciones Científicas (SINCHI).

——. 2001. Foragers of the Amazon: The last survivors or the first to succeed? In C. McEwan, C. Barreto, and E. G. Neves, eds., *Unknown Amazon: Culture in Nature in Ancient Brazil,* 26–49. London: British Museum Press.

Porro, A. 1994. Social organization and political power in the Amazon floodplain: The ethnohistorical sources. In A. Roosevelt, ed., *Amazonian Indians from Prehistory to the Present,* 79–94. Tucson: University of Arizona Press.

Posey, D. A. 1985. Indigenous management of tropical forest ecosystems: The case of the Kayapó Indians of the Brazilian Amazon. *Agroforestry Systems* 3:139–158.

——. 1986. Manejo de floresta secundária, capoeiras, campos e cerrados (Kayapó). In B. Ribeiro, coord., *Etnobiologia,* vol. 1 of *Suma etnológica Brasileira,* 173–185. Petrópolis, Brazil: Vozes/FINEP.

Prous, A. 1992. *Arqueologia brasileira.* Brasília: Editora da Universidade de Brasília.

Raffles, H. 2002. *In Amazonia: A Natural History.* Princeton, N.J.: Princeton University Press.

Ramos, A., P. Silverwood-Cope, and A.G. Oliveira. 1980. Patrões e clientes: Relações inter-tribais no alto rio Negro. In A. Ramos, ed., *Hierarquia e simbiose,* 135–182. São Paulo: HUCITEC.

Roosevelt, A.C. 1987. Chiefdoms in the Amazon and Orinoco. In R. Drennan and C. Uribe, eds., *Chiefdoms in the Americas,* 153–185. Lanham, Md.: University Presses of America.

———. 1991. *Moundbuilders of the Amazon: Geophysical Archaeology on Marajó Island, Brazil.* San Diego: Academic Press.

———. 1993. The rise and fall of the Amazonian chiefdoms. *L'Homme* 33 (2–4): 255–282.

———. 1995. Early pottery in the Amazon: Twenty years of scholarly obscurity. In W.K. Barnett and J. Hoopes, eds., *The Emergence of Pottery: Technology and Innovation in Ancient Societies,* 115–131. Washington, D.C.: Smithsonian Institution Press.

———. 1999. Archaeology of South American hunters and gatherers. In R.B. Lee and R. Daly, eds., *The Cambridge Encyclopedia of Hunters and Gatherers,* 86–91. Cambridge, U.K.: Cambridge University Press.

———. 2000. The lower Amazon: A dynamic human habitat. In D. Lentz, ed., *Imperfect Balance: Landscape Transformations in the Precolumbian Americas,* 455–491. New York: Columbia University Press.

Roosevelt, A.C., J. Douglas, and L. Brown. 2002. The migrations and adaptations of the first Americans: Clovis and pre-Clovis viewed from South America. In N. Jablonski, ed., *The First Americans: The Pleistocene Colonization of the New World,* 159–235. Memoirs of the California Academy of Science no. 27. San Francisco: California Academy of Science.

Roosevelt, A.C., R. Housley, I.M. Imazio da Silveira, S. Maranca, and R. Johnson. 1991. Eighth millennium pottery from a prehistoric shell midden in the Brazilian Amazon. *Science* 254:1621–1624.

Roosevelt, A.C., M. Lima Costa, C. Lopes Machado, M. Micab, N. Mercier, H. Valadas, J. Feathers, W. Barnett, M. Imazio da Silveira, A. Henderson, J. Silva, B. Chernoff, D. Reese, J.A. Holman, N. Toth, and K. Schick. 1996. Paleoindian cave dwellers in the Amazon: The peopling of the Americas. *Science* 272:373–384.

Sauer, C.O. 1969. *Seeds, Spades, Hearths, and Herds: The Domestication of Animals and Foodstuffs.* 2d ed. Cambridge, Mass.: MIT Press.

Shorr, N. 2000. Early utilization of flood-recession soils as a response to the intensification of fishing and upland agriculture: Resource-use dynamics in a large Tikuna community. *Human Ecology* 28 (2): 73–107.

Silva, C.A. da. 2003. Sítios arqueológicos no Município de Manaus. Manuscrito depositado no Museu Amazônico, UFAM.

Silverwood-Cope, P. 1990. *Os Makú: Povo caçador do noroeste da Amazônia.* Brasília: Editora UnB.

Simões, M. 1974. Contribuição à arqueologia dos arredores do baixo rio Negro. In *Programa Nacional de Pesquisas Arqueológicas,* 5:165–200. Publicações Avulsas no. 26. Belém, Brazil: Museu Paraense Emílio Goeldi 26.

———. 1981. Coletores-pescadores ceramistas do litoral do Salgado (Pará). *Boletim do Museu Paraense Emílio Goeldi* 78:1–32.

Simões, M., and A. Kalkmann. 1987. Pesquisas arqueológicas no médio rio Negro (Amazonas). *Revista de Arqueologia* 4 (1): 83–116.

Stahl, P.W. 2002. Paradigms in paradise: Revising standard Amazonian prehistory. *Review of Archaeology* 23 (2): 39–51.

Sternberg, H. 1998. *O homem e a água na várzea do Careiro.* 2d ed. Belém, Brazil: Museu Paraense Emílio Goeldi.

Turner, T. 1979. The Gê and Bororo societies as dialectical systems: A general model. In D. Maybury-Lewis, ed., *Dialectical Societies: The Gê and Bororo of Central Brazil,* 147–178. Cambridge, Mass.: Harvard University Press.

Van der Hammen, T., and H. Hooghiemstra. 2000. Neogene and quaternary history of vegetation, climate, and plant diversity in Amazonia. *Quaternary Science Reviews* 19:725–742.

Van Nest, J., D. K. Charles, J. E. Buikstra, and D. L. Asch. 2001. Sod blocks in Illinois Hopewell mounds. *American Antiquity* 66 (4): 633–650.

Whitehead, N. L. 1994. The ancient Amerindian polities of the Amazon, the Orinoco, and the Atlantic Coast: A preliminary analysis of their passage from antiquity to extinction. In A. Roosevelt, ed., *Amazonian Indians from Prehistory to the Present,* 33–53. Tucson: University of Arizona Press.

Woods, W. I., and J. M. McCann. 1999. The anthropogenic origin and persistence of Amazonian Dark Earths. *Yearbook of the Conference of Latin American Geographers* 25: 7–14.

Wright, R. M. 1999. Destruction, resistance, and transformation—southern, coastal, and northern Brazil (1580–1890). In F. Salomon and S. B. Schwartz, eds., *The Cambridge History of the Native Peoples of the Americas,* 287–381. Cambridge, U.K.: Cambridge University Press.

10

HISTORY, ECOLOGY, AND ALTERITY

Visualizing Polity in Ancient Amazonia

MICHAEL HECKENBERGER

DIVERSITY, COMPLEXITY, HYBRIDIZATION, and *globalization* are a few of the terms that today define human ecological research. Largely gone are assumptions of equilibrium and homeostatic self-regulation, mechanistic feedback loops, and closed social and ecological systems that dominated ecological anthropology from the mid- to late twentieth century. In their place, many researchers today talk of sociocultural self-organization and scalar iteration, uncertainty and negotiation, and dynamic change. In Amazonia, specifically, recent studies indicate that much of what is commonly considered as untouched nature is not entirely natural after all, but the by-product of dynamic human-environment interactions. Many parts of the Amazon, in fact, might be better seen as "cultural forests" (Balée 1989) or even forest gardens and managed parklands, gone to fallow over the past 500 years, rather than the outcome of autonomous natural processes of biological diversity. Hence, the "myths" of pristine nature and primitive culture, long taken for granted by natural scientists and historians alike, are giving way to a new vision of historical ecology that foregrounds the symbolic, social, and political dimensions of human ecology (Balée 1998; Biersaack 1999; Crumley 1991; Denevan 1992; Descola 1992).

Within the general perspective of historical ecology, taken here to mean "the complex historical relationship between human beings and the biosphere" (Balée and Erickson, introduction, this volume), this chapter focuses on questions of deep history and sociopolitical organization in Amazonia. Sociopolitical organization is obviously a critical dimension of human ecology: how human groups ranging from small, dispersed, and generally autonomous social groupings to large, densely settled, regional populations create and "map onto" landscapes. In Amazonia, archaeological research demonstrates that landscapes were altered gradually by millennia of human occupations and substantially by large, settled, agricultural populations in late prehistory, around AD 1000–1500 (Erickson, chapter 8, this volume; Heckenberger, Petersen, and Neves 1999; Heckenberger et al. 2003; Roosevelt 1999).

In Amazonia, it has been difficult to address questions of cultural variability and the diversity of landscape because little is known about the long-term history of the region; and, as is often the case, an absence of robust historical knowledge is often taken as a lack of history at all—in other words, the history is "naturalized" into an imagery of pristine wilderness and primitiveness. Amazonia is an imaginary place, which came into existence in Western cartography and culture history only recently. The idea of "Amazonia" as a relatively homogeneous ecological and cultural area crystallized in the mid-1900s, particularly within a cultural ecological model of a "tropical forest culture" prominent in regional anthropology over the past 50 years. Even if Amazonia is minimally defined as the Amazon drainage basin itself (largely restricted to northern Brazil, with some 30 percent in Bolivia, Peru, Ecuador, Columbia, and the Guianas), its variability is considerable and arguably no less so than that of other maximal "culture areas" (e.g., the Eastern Woodlands, temperate Europe, or Oceania).

In light of the grand historical schemes of Andean prehistory, the rise and fall of states and empires, Amazonia has always paled by comparison: history, like topography, is typically viewed as flat in the lowlands. Although Amazonia is more diffuse and more poorly known than the Andes, however, there is no a priori reason to assume that its history is less complex than that of other world areas of similar proportions in terms of the numbers of actual human bodies and the changes that transpired relative to them. Nothing as large and powerful as the Inka Empire (around AD 1476–1533) ever emerged in the lowlands, but archaeology and ethnohistory raise the possibility of ancient Amazonian social formations similar in scale to many chiefdoms and kingdoms in other parts of the world. In most areas and during most time periods, however, little empirical work informs the analysis. The question of what constitutes social complexity in Amazonia, precipitated by the increasingly common claims of large, regional social formations and parallel claims of human-induced environmental change, is rarely addressed. Even in the few places where regional specialists have long agreed that complex societies were present in 1492, perhaps for more than a millennium—such as Marajó Island, Santarém, the middle and upper Amazon, or in lowland Bolivia—little is known about their political, economic, or ideological systems. In particular, if Amazonian settlements in prehistory approached the size of towns or small cities elsewhere in the ancient world (with populations in the low to middle thousands), as many specialists now think, it is important to consider the variable, long-term effects of large, settled, complex societies on diverse Amazon landscape or, even, whether we might think in terms of an urban ecology[1] for parts of pre-Columbian Amazonia.

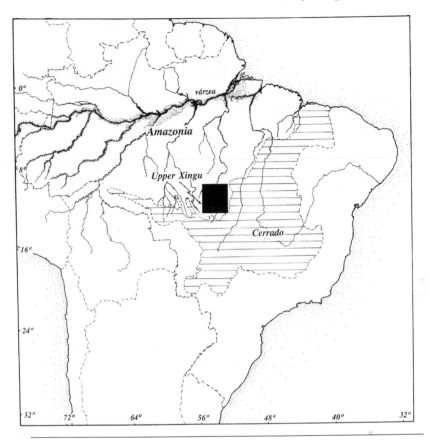

FIGURE 10.1 The Upper Xingu region in Amazonia.

The idea of urban ecology is ultimately tied to what is meant by *urban*, a term many may feel is inappropriate to Amazonia. One thing is certain: the classic definitions of cities, states, and urbanism in, for instance, Mesopotamia, the Mediterranean, or Europe must be dramatically altered or abandoned altogether if they are to be useful for understanding Native American polities. To consider this question, what constitutes social complexity, urbanity, or statehood—or not—in Amazonia, this chapter addresses issues of polity and sociopolitical change in terms of the long-term or deep history of the southern Amazon and Upper Xingu region (figure 10.1), particularly with respect to reconsideration of a settled/mobile or riverine/upland dichotomy in light of distinctive modalities of social valuation, ranking, and exchange. It sketches in broadest outline these nested histories, but first a brief consideration of what is meant by history and historical ecology is merited.

HISTORICAL FACTS AND PROBLEMS IN AMAZONIA

History, as taken here, is not limited to documentary analysis or literary criticism, but involves examination of other residues and traces inscribed on the skin of the land and in human bodies. There are two ways to "do" history in a place like Amazonia. The first, the more common by far, involves deducing it from the things we know or think we know about the present (or recent past): quantified assemblages of contemporary linguistic, ecological, and cultural "populations" and what happens to them in recent times "before our eyes" or as deduced from our (Western) historical experience. The second way to do history, relatively rare in the Amazon, involves addressing things from the past, collected and interpreted in the present, of course, but made in and of the past nonetheless. For many historical anthropologists, there is really no other way to do history, even though we might think about it from the narrow vantage point of the present.

In Amazonia, the scaffolding upon which historical interpretation stands is weak, not because it is rickety, which assumes wear and tear, but because it is based on so few well-documented case studies.[2] From a scientific standpoint, Amazonia is plagued by problems of sampling and incompatibility. The heralded new syntheses and revisionist arguments still have the ring of a "Whig view of history." Research since the mid-1900s provides robust bodies of evidence about archaeology, linguistics, ethnohistory, ethnography, and current land use in a few areas, but many areas remain poorly documented. To document them requires direct historical linkages—in other words, demonstrated continuity within specific sociohistorical contexts. The transdisciplinary language necessary to integrate diverse perspectives and scales, at any rate, is poorly developed because Amazonian history generally lacks well-documented chronologies or specific narratives of the deep past.

From a historical point of view, little-known places such as Amazonia are especially problematical. There are few carefully documented historical cases or well-formulated historical problems regarding such places, from any viewpoint. Furthermore, the necessary analytical mechanisms for bridging scales, perspectives, or disciplines are poorly developed, except through the distanced viewpoints of presentism and uniformitarianism, as well as through outright speculation. Vansina notes a similar problem of "voids and blinders"—for example, the major gaps of historical knowledge about equatorial Africa and the biases of its authors—when he states that traditional viewpoints argue that "environment determines history and the unlucky peoples here have no history because they never changed" (1990:3). Amazonia, like Africa, "[even though] as large as the United States east of the Mississippi, almost as large as Western Europe ... remains *terra incognita* for the historian ... [because] peoples living there 'were too busy surviving in such a hostile environment' to change" (Vansina 1990:3).

Familiar images of Amazonia are typically imbued with a unique naturalistic tone: Garden of Eden or Green Hell. It is a river and forest world of immense proportions and most commonly seen by outsiders as filled to excess with water, plants, biting insects, and toothy, dangerous animals—and, of course, with occasional "savage" or "archaic" people, more or less frozen in time in a seemingly timeless primordial forest. In fact, in the Western imaginary the Amazon is the epitome of mystery and danger, brimming not only with life but also with death, pestilence, and disease, all woven together in an endless maze of hostile, dark, and hopeless places. Schama, in his book *Landscape and Memory*, devotes a few pages to Amazonia, specifically his evocative summary of Ralegh's description of a "creeper-strangled, monster-bloated, erotically lubricated, filmy, floating world," an image, Schama continues, that "passed down time like driftwood, as the myth of thwarted imperial penetration, fetching up again in the imaginations of Alexander von Humboldt, Joseph Conrad, John Huston, and many many more pilots of delusion" (1995:312). It is not a landscape, but an antilandscape—that is to say, untouched or pure nature. Furthermore, Amazonia is largely known from the banks of the major rivers; in other words, landscapes are known from a distance, from passing through them, around them, or, lately, over them, but not from immediate personal experience with them, entering in them or entering into relations with them, through what Ingold (1993) calls a "dwelling perspective."

Regardless of what kind of wilderness Amazonia is seen to be, idyllic or infernal, there was seldom any doubt that the people natural scientists ran into were our "contemporary ancestors." By the twentieth century, the idea of finding lost cities or civilizations hidden in the jungle, a subtext that runs parallel to the idea of the primitive and pristine, became viewed as the idle dream of deluded explorers: there would be no discoveries like the great tropical centers of Meso-america, Peru, or Southeast Asia. It is no surprise, then, that we have yet really to consider what kind of "state" we might find or even look for in Amazonia because we assume ahead of time that it did not exist. This viewpoint, of course, supposes that we know what we are looking for with regard to non-Western civilizations—political power based on economic exploitation (surplus) and direct coercion, promulgated on writing, state religion, and the economic and administrative control of central places, cities, and capitals—but we know a priori that Amazonians did not have these!

Indeed, there are no pyramids in Amazonia, no stone masonry, not even a durable standing structure of any kind from many parts of the area, but does this mean there was no state? Likewise, there is no abstract system of writing or numeration. But does this mean that there was no developed cartography, geometry, or architecture? Can we assume that the powerful could not inscribe their wills on others both locally or within regional theaters of power or that there was no elite ideology based on some form of genealogical history—in other words, that there was no political economy? Further, can we be certain that, in terms of actual human bodies and the changes that transpired in their lives, there were no technoeconomic or demographic trends that rivaled those that are elsewhere associated with "civil society"? Recent developments in historical ecology have cast doubt on these assumed absences. Thus, it is no longer tenable to assume any isomorphism between ancient and modern Amazonian peoples without clear historical support.

Even areas most isolated from direct colonial activities were not insulated from the major historical flows set in motion by sixteenth-century European colonialism. The historical ecology of Amazonia over the past 500 years, like that of the Americas in general, clearly shows us that indigenous peoples, although not as fragile and doomed as once suggested, have a history that has trended toward the "vanishing point." It is the history of what Dobyns (1983) aptly calls a "widowed land." By 1750, when la Condamine traveled the Amazon River—roughly two centuries after Europeans first described the river as home to diverse, densely settled, and regional societies and to an equally rich environment—he found the floodplains desolate, with only a few haphazard indigenous settlements. Gone were the great chiefly polities that thronged these floodplains, and in their place polyethnic social formations (often inflections of already polyethnic precolonial polities) had emerged and become increasingly more influenced by European- and African-derived practices and logics. This latter point is particularly critical here, not in suggesting that postcolonial indigenous groups were any less "traditional" or "indigenous" than their forebears, but in casting doubt on viewpoints that frame the problem in these terms.

Writing generally comes late to the Amazon, and what exists is spotty. Thus, writing indigenous history, at least deep history, requires that archaeology provide our clues. The question turns on issues of archaeological visibility, such as settlement plans, constructions, material culture, and the landscape itself. It revolves around what today are often called materiality and spatiality. It requires a definition of *personhood* that is not based strictly on sight—how people see or ontologically perceive social "others," or how they traffic with one another in direct exchange—but instead on how they resolve into groups (larger "persons"), how they move through time and space, and how we can recognize these distributions in visible material residues and their associations.

THE AMAZON CHRONICLES: DEEP HISTORY

The great antiquity and diversity of cultural and biological patterns in Amazonia that have recently come to light raise a host of new historical problems and diversity of opinion on regional culture history. There is nonetheless some consensus among specialists regarding the antiquity of: (1) early tropical forest foragers, around 8500–4500 BC; (2) early horticulturalists, around 5000–1000 BC; (3) the spread of forest farmers through much of the region, associated with several large linguistic diaspora (notably, Arawak and Tupí-Guaraní, among others), around 1000 BC–AD 500; and (4) the development of fairly large, integrated regional societies, around AD 500–1500 (Heckenberger 2005; Lathrap 1970; Moseley and Heckenberger 2005; Roosevelt 1989, 1999). Four main periods of cultural development, after the emergence of tropical forest farmers around 2000–1000 BC, are suggested: (1) the Formative period, around 1000 BC–AD 500; (2) the Regional Development period, AD 500–1250; (3) the Pre-Colonial Climax or "Classic" period of peer communities organized around major monumental centers, AD 1000 to 1250–1650; and (4) the European World System period, after around AD 1500–1650. The periodization highlights dominant general developments and innovations, notably the "rise" of rank, polity, and world systems (native and exogenously centered).

FORMATIVE PERIOD

Many regional specialists agree that local development in several parts of Amazonia, specifically the development of forest farming, was followed by significant migration and diffusion (e.g., see Lathrap 1970). Notably, there was an "expansion" of Arawakan populations along river and coastal corridors as well as concomitant (largely) upland migrations of Tupian, Cariban, and Gê speakers, as well as several smaller linguistic families (Panoan, Tukanoan). For the first time, significant differences between riverine (*várzea,* or floodplain) and upland *(terra firme)* trajectories appeared, in terms of not only subsistence and scale, but also basic patterns of sociality. From a linguistic point of view, this process began between 2000 and 1000 BC, when proto-Arawak speakers began to disperse from greater northwest Amazonia. Tupí and in particular Tupí-Guaraní peoples also began to disperse from an origin in southwest Amazonia about this time, and Carib (from northeastern Amazonia) and Gê (from the southeastern peripheries of Amazonia and south) peoples also may have begun to disperse as well (Urban 1992).

The Arawak diaspora best represents the Formative period, as it is defined elsewhere (e.g., Ford 1969), particularly insofar as the Arawakan diaspora involves a developed agricultural economy, in some cases based on manioc and fishing, architectural elaboration and monumentality, and social hierarchy. Some continuity in Arawakan-speaking groups can be suggested in several areas based not only on linguistic distributions, but also on material culture (notably a ceramic technology, a "sloping horizon," generally called Saladoid-Barrancoid or Amazonian Barrancoid—which extends from circa 500 BC to the present) and a common (circular plaza) settlement pattern (Heckenberger 2002). The Arawak peoples expanded across much of riverine and maritime areas between 500 BC and AD 500.

Material culture affinities have also been suggested among some Tupí-Guaraní speakers (the coastal Tupi-Guarani tradition) but, as Viveiros de Castro (1992) notes, the unity of Tupí-Guaraní speakers and perhaps the "equally metamorphic" Carib-speaking peoples lies not so much in technology, demographic organization, economics, or sociopolitics, all of which vary significantly within the language family, but instead in cosmology. Even if the exact language and material culture correlations cannot be entirely worked out, the language distributions and ethnological comparisons clearly indicate a more riverine, settled orientation among Arawak and related peoples.

The principle triggers of the major diaspora varied significantly through time and space, depending on local ecological and social conditions and cultural predispositions ("tradition"). Thus, the causes of each of the diaspora cannot be reduced to an underlying evolutionary or structural force, but instead resulted from the combined influence of local human-environment interaction and sociopolitical relations within and between communities. Furthermore, each of the diaspora was uniquely influenced by specific conditions at its inception and by what we might loosely call the weight of "tradition" or "history"—in other words, the inertia produced through the reproduction of culturally typical schemas and institutional orientations in dynamic sociohistorical contexts.

The distinctive sociopolitical, economic, and ideological orientations of riverine Arawak as opposed to those of upland Amazonian (Tupian and Carib) and central Brazilian (Gê) groups are of particular note here. Proto-Arawak language already contained words for manioc and other domesticated plants, ceramics, and major river fauna (Payne 1991). Linguistic and archaeological evidence also suggests the earliest expressions of the unique sociosymbolic ensemble of hereditary chieftainship typical of ethnographic Arawak peoples about this time, notably a (linguistically marked) dichotomy between older (senior) and younger (junior) siblings, over time expressed as a founder's ideology and an exclusive (sacred) central plaza.

In essence, the Arawak peoples, among native peoples in lowland South America, were (and are) uniquely hierarchical (Heckenberger 2002, 2005; Hill and Santos-Granero 2002). Within this hierarchical social structure, political competition, based at least in part on heredity and ancestry—namely, a founder's ideology rooted in notions of primogeniture—was critical as political agents (cadet lines) split from parent settlements and founded new settlements (and a new line of local ancestors). The elaboration of such an ideology was typically manifest in settlements and landscapes, particularly as an extension of ancestrality, because placing ancestors in local history typically required first fixing them in place. Almost invariably, such notions of hierarchy and ancestrality are also typically tied to patterns of sociality that are regional by nature. Parenthetically, the opposite model might be suggested for Tupí-Guaraní speakers, a model that, as noted by several authors (Fausto 1992; Viveiros de Castro 1992), lacks the hierarchical division of society based on hereditary rank—what Clastres (1987) calls the "One." Tupí-Guaraní speakers also rarely "fix" themselves in landscapes in the permanent way that Arawak speakers generally do and are often characteristically mobile and atomistic. This distinction underscores the cultural specificity of certain social logics, in this case social hierarchy, and the fact that some social and ecological practices and strategies are rooted in deep cultural histories.

REGIONAL DEVELOPMENT PERIOD

By around AD 1, the largest diaspora, focused on speakers of the Arawak and Tupí-Guaraní languages, had reached their near maximal extent (although significant population movements occurred later as well). The Arawak diaspora in particular bears resemblance to other major tropical diaspora in Africa (Niger-Congo/Bantu languages) and in the Pacific (Oceanic Austronesian languages) in terms of a tendency to "ramify" early in the diaspora (Firth 1936; Kirch 1984). The resemblance would lie not in similar ecological conditions or agricultural innovation, which in each case were present for thousands of years before the major movements, but instead in the hierarchical model of sociality and "founder's ideology." It was the development of this general social logic that increasingly separated Arawak peoples occupying river and coastal areas from their upland neighbors.

This idea resonates in some respects with Carneiro's (1970) idea of circumscription, specifically the common tendency for political competition to stimulate population movement where possible and for political groups to "stratify" in places where it is not. There is one important caveat: scarcity, in this case, is defined first and foremost by social and symbolic capital (i.e., prestige within a ranked social structure), not by economic capital. The Formative period, or early diaspora, was characterized by ramification among Arawakan peoples and possibly among other social groups, whereas the subsequent millennium (once again giving or taking 500 or so years depending on location), the Regional Development period, was more commonly characterized by hybridization, internal differentiation, and often significant stratification (regional hierarchies) in emerging multilingual and polyethnic regional social systems.

During the Regional Development period, beginning by around AD 500 or earlier in various areas, fairly large and settled communities emerged and became nodes in regionally integrated settlement systems. This period, like the Formative period, is a "center of gravity" concept because many groups followed different trajectories. Following the logic of this chapter, this period was the first time in Amazonian history that large regional social formations organized around primary centers appeared in the basin (although earlier cases are probable). During this period, regional societies were established often through dynamic processes of ethnogenesis along diverse pathways of social complexity.

As noted, the spread of early agriculturalists was the hallmark of the Neolithic in various parts of the world, among which Indo-Europeans, Austronesian (Oceanic), and Niger-Congo (Bantu) cases are the best described. These processes of population movements in Amazonia, perhaps more than in any other place, were accompanied by or prompted extremely varied and complex processes of ethnogenesis, as newcomers (migrants related to the major diaspora) and autochthones developed into hybrid regional social systems. Ethnogenesis and hybridity, in fact, were perhaps the more notable aspects of the period in many areas after around AD 500 and were critical in the development of many of the pluriethnic and multilingual societies known from around AD 1550 forward. The development and reconstitution of hybrid, regional social systems did not erase the diverse lines of history apparent from earlier periods, specifically the unique characteristics of distinctive, antecedent sociohistorical communities. This period was, however, one of cultural divergence (regionalism) across the broad region and of the coalescence of groups within regions. In the Formative period, the alterity between dense, settled populations and more mobile and "predatory" upland groups was firmly established in a variety of places across Amazonia and to some degree provided the symbolic and social "glue" through which macroregional systems of interaction were created and maintained (Heckenberger 2005). I take up the question of social identity again later in specific terms of relations (alterities) between the Xinguanos (settled, regional, and accommodating) and their neighbors (mobile, atomistic, and predatory). In the centuries before and after 1492, processes of regional interaction and integration, including the indirect supraregional forces of fluctuating native world systems, lay at the root of change as much as, if not more than, local adaptation to the natural environment. This pattern culminated in the world-system dynamics of Inka and Euro-American political and economic domination, among others.

CLASSIC PERIOD

By around AD 1000, regional polities had reached their maximal proportions in terms of size, relative power, and social and political connections in nested hierarchies. During this climax or Classic period of regional sociopolitical systems, the largest settlements and most powerful regional polities dominated much of riverine Amazonia. In most contexts, articulation peaked within regional political economies that incorporated diverse ideologies, social identities, and modes of production. The earliest expression of a small Classic period polity was perhaps on Marajó island, around AD 400–600, but the trend was widespread by AD 1000. Exact population figures are generally lacking, and large settlements such as Santarém at the mouth of the Tapajós River, Açutuba on the lower Negro River, and the galactic clusters of settlements in the Upper Xingu region likely held thousands of persons, thus placing these settlements well in the range of towns and small cities as defined elsewhere. These settlements formed part of larger integrated communities at the regional scale, with populations that likely numbered well into the thousands or even tens of thousands. The sociopolitical organization of most of these societies is still poorly known; however, archaeological and ethnohistorical evidence does suggest that by 1492 some were organized as minimally to moderately stratified societies described as chiefdoms and kingdoms throughout the world.

The Amazonian Polychrome Ceramic tradition—minimally including Marajoara, Miracanguera, Guarita, and Santarém complexes, as well as western Amazon complexes such as Zebu, Nefueri, Caimito, Napo—extended the length of the Amazon River and along several major tributaries by around AD 1000 (see Neves and Petersen, this volume). This suggests not only cultural sharing but broad sociopolitical articulation through prestige-goods exchange at supraregional levels. Widespread change in ceramic technologies and settlement patterns associated with these and other pre-Columbian polities, such as the Upper Xingu region indicate fundamental changes in local and regional political economies. Major polities of this ancient regime would include the Tapajós and related chiefdoms, perhaps centered on a capital at Santarém, the late Marajoara chiefdoms, the Middle Amazon chiefdoms including regional centers like Açutuba, and the ethnohistorical Omagua and related (Kokama) chiefdoms, as well as, in fact, the entire network of polities participating in broad exchange networks of the Polychrome Ceramic tradition (Heckenberger, Petersen, and Neves 1999; Roosevelt 1999). In the southern Amazon, a whole different theater of complex social formations was present, including the Arawakan-speaking Baure, Pareci, and Xinguano peoples, among others (e.g., heterarchically organized confederations of related Gê- and Tupian-speaking groups).

Ethnogenetic processes and continued population movements were common, although in some areas regional hierarchies and central places became generally larger and more fixed through time (Hill 1996). Regional political economies continued to be based on the fundamental alterity between settled groups, usually major river people, and less-permanent upland or hinterland groups. World systems emanating from the Andes were likely critical forces in many places and times. Given what we know of other parts of the ancient world, it seems almost impossible that a geopolitical entity as large and powerful as the Inka Empire had little effect on Amazonia. Suffice it to say that Europe's expansion into Amazonia demonstrates how profound and widespread the influence of such world systems could be.

EUROPEAN WORLD SYSTEM PERIOD

The period after around AD 1500 is marked by general demographic decline and the political and economic collapse of most large polities of the ancient regime, followed by the colonization of the region by nonindigenous peoples and the reconstitution of indigenous regional social formations, usually part indigenous and part foreign in the long run, although occasionally societies persisted that were more or less entirely one or the other. Change came quickly as disease raged across large areas, and virtually all people were affected. By the mid-1500s, cultural change occurred throughout the hemisphere, and indigenous peoples from the mightiest to the most modest were either incorporated into the world system of early modern Europe or died. By around 1750, the major flood-plain societies had ceased to exist and in their place were dispersed indigenous and newcomer populations, including small boomtowns for the extraction of canela, cacao, gold, rubber, wood, and other things. There is no doubt that European colonialism was the principal catalyst of change during this period, which commonly occurred in the absence of Europeans. If ethnogenesis and cultural "hybridity" were already important elements of pre-Columbian social formations, as suggested earlier, the rate of change was accelerated after 1492, and most indigenous peoples were reborn as "mixed-blood" peoples, hybrids of locals and newcomers, throughout the Amazon. By the same token, many "traditions" were undoubtedly invented during this time. Nonetheless, the marks of a genuinely indigenous heritage have persisted in the landscape and in human bodies, some of which, like the peoples of the Upper Xingu region, are the cultural survivors of the ancient polities of Amazonia.

ECOLOGY AND ALTERITY: THE RIVER/UPLAND DICHOTOMY REVISITED

The deep history of Amazonia, in particular the recognition of complex social formations in a variety of areas, requires new anthropological perspectives on human-environment interactions that are not derived entirely from twentieth-century ethnography. Several issues in particular merit reconsideration. First, contrary to orthodox cultural materialist views (ecofunctionalism), the emergence and development of Amazonian complex societies were not simply predictable outcomes of the operation of impersonal forces of demography and technoeconomics in the context of highly general ecological parameters (i.e., circumscription of agricultural land or subsistence resources). Second, social complexity was not a concomitant or an outcome of increasing economic and administrative centralization (i.e., as predicted by classic central-place theory). Third, the explicit or implicit idea of a generalized Amazonian viewpoint that generally lacked or denied political authority—in other words, that did not include any form of coercive or institutionalized power, even the relatively small-scale power ("from within") associated with the political institution of the chieftaincy—is inappropriate or at least problematical for describing some Amazonian peoples, notably the large polities of late prehistoric times.

Most regional ethnologists, including those focusing on deep histories, agree that uniquely Amazonian perspectives, practices, and sociocultural tendencies must guide discussions of sociality and ecology. The problem is that the approaches that have been most successful at revealing these features—ethnography with little time depth—also often deny complexity (i.e., social hierarchy, political economy, and regional integration) and create the image of Amazonians as small, simple, and politically homogeneous throughout the region. In other words, the almost exclusive basis for understanding Amazonian political economies has been ethnographic data, which assumes that the past is more or less isomorphic with the present. This view is no longer tenable, as demonstrated in numerous cases where the distant past was quite different, larger, and more complex than the present. Furthermore, even societies that were more or less similar in size and configuration to ethnographically described groups (i.e., with low-density, dispersed settlements; small territories; egalitarianism; and community political autonomy) were enmeshed in regional political economies that included large, settled, regional, and hierarchical polities. As now recognized in many parts of the ancient world, such as Africa, Oceania, and much of the Americas, the concentration of symbolic and social capital or prestige broadly defined, rather than economic centralization (exploitation) and direct coercion, was the primary basis of political power in early complex societies. In Amazonia, such a prestige-goods economy was largely a question of producing and reproducing persons rather than things, although significant economic goods were sometimes involved.

Setting aside questions of variability and extent of political economy in Amazonia, a reconsideration of the long-noted division between riverine (várzea) and upland (terra firme) peoples, usually characterized by contrasts in size and permanence of settlement and relative economic wealth and power, is merited (Steward 1949). The idea that rivers support bigger and more settled folk and that uplands support more mobile, smaller, and more dispersed populations, often predatory on their settled neighbors, is a leitmotif of cultural development in many parts of the world. In few places, however, have the evolutionary implications of such a view been emphasized as much as in Amazonia. There is widespread agreement regarding the relative ecological parameters of riverine versus upland cultural adaptations in Amazonia (Carneiro 1995; Denevan 2001; Lathrap, Gebhart-Sayer and Mester 1985; Meggers 1996; Roosevelt 1989, 1999).

Ecological explanations have dominated regional culture history since the 1950s, and by the 1970s Robert Carneiro and Donald Lathrap had proposed their now classic ecological models of differential population growth and pressure leading to inevitable competition and conflict (Carneiro 1995; Roosevelt 1991). Among ecofunctionalists, Lathrap was the most keenly aware of the histories of specific culture and language groups, but, like Carneiro, he viewed the issue of demographic growth and economic development as primary causal agents operating in a somewhat nonproblematic way. In other words, one gets the impression that, regardless of cultural or temporal specificity, a cultural group situated within the várzea would thrive and grow; but the same group in the upland would remain the same, in homeostatic equilibrium with nature. Whereas Steward, Carneiro, and Lathrap saw the dichotomy as a shallow gradient from more to less productive lands, Anna Roosevelt (1991, 1994) made the distinction singular in guiding cultural evolution. Specifically, following elements of Meggers and Evans' discussion of agricultural limitations in Amazonia, Roosevelt suggested that restricted distributions of high-fertility soils for the intensification of cereal crops made the várzea a unique setting that increased economic intensification (Lathrap and Carneiro favored the idea of riverine resources and manioc agriculture [Carneiro 1985; Lathrap, Gebhart-Sayer, and Mester 1985]).

Regardless, the consensus among ecological determinists was that the political apparatus of the state, here in the nascent form of chiefdoms, had emerged predictably in the rich floodplains, but predictably not in uplands. An alternative view considered the rise of social hierarchy and political economy as a corruption rather than an advancement of "primitive" or "archaic" society (Clastres 1987). Nonetheless, most authors agreed that the threshold was defined in terms of demographic and economic potential and that the riverine chiefdoms were an unusual mutation of a more general Amazonian pattern typified by small, autonomous communities as known from the uplands ethnographically (Descola 1996:330). However, since these views were presented, there has been little discussion among cultural historians of the internal variability in either ideal type in terms of the distinctive sociopolitical identities of riverine settled agriculturalists and other smaller and more mobile people in the uplands. Furthermore, the degree to which this dichotomy structured regional systems—specifically the social division (alterity) between groups that tended to reproduce riverine or upland strategies, which depended as much on cultural and political choice as on ecological, economic, or demographic potential—has been inadequately explored.

More recently, Roosevelt (1999) has suggested that hierarchy, defined in terms of economic alienation and direct political coercion, appeared rarely in Amazonia and quite late in the centuries just prior to 1492. She proposes that the majority of ancient Amazonian complex societies were heterarchical. Although perhaps extending the evolutionary ladder slightly and recognizing diverse pathways of early complexity, Roosevelt sees hierarchy as emerging relatively unproblematically from heterarchy. Hierarchy also remains a question of demographic, economic, and technological scale and degrees of centralization, tied to ecological potentials rather than resulting from cultural choice or the internal inertia of sociopolitical systems.

Alternatively, I have proposed that hierarchy, as a system of social values, is deeply rooted in some Amazonian groups (most notably Arawak and related groups) and is hallmarked by a social ideology of hereditary ranking, including titular chiefs and a bifurcation of society into elite and commoner ranks (Heckenberger 2003, 2005). As social value, hierarchy need not be tied to pronounced economic centralization and exploitation or political coercion and certainly need not be tied to the bureaucratic apparatus of the state (e.g., standing armies, police, and true economic capital). This view again raises the issue of the riverine/upland dichotomy, but in a new way as a reflection of social alterity between contemporary groups participating in broad regional political economies. From early in the Formative period, around 1000–2000 BC or earlier, the dichotomy existed between settled riverine and maritime populations, commonly Arawak-speaking and related peoples, and their more mobile upland neighbors, commonly Carib and Tupian peoples, among others. These distinctive social identities were transformed and inflected in each new setting as immigrants and autochthones forged alliances and new identities, but in diverse portions of the region a pattern emerged time and again between settled, regional riverine societies and mobile, upland, small-scale societies. Early on, this general pattern produced contrastive strategies of economic, social, and political actions. In other words, these two contrasting identities are not only dichotomous endpoints in a continuum of Amazonian social forms or technoeconomies, but also alter egos in regional systems of interaction.

What stands out here is the internal differentiation and regional nature of these political economies, as revealed, for instance, in material culture and settlement pattern. In settled areas, social alterity was thus inflected in two directions: horizontally (between more "settled," regional social formations and more atomistic, mobile formations) and vertically (between upper and lower ranks in settled, regional societies). In other words, we can note a continuum of social identities that vary in terms of the degree to which social reproduction corresponds or is defined by what Giddens (1984) has coined an existential contradiction, the definition of humanity and society based on the contours of nature. All societies are characterized by an "existential contradiction," but in some societies (in Amazonia and in general) there is also an internal distinction between upper and lower segments of societies, glossed by Giddens (1984) as a "structural contradiction." Thus, societies and regions can be compared based on the form of relations between society and the "outside" and by the way a society defines its "inside." In societies with a notable structural contradiction, which divide into upper and lower segments based on real or fictive genealogy, regional integration based on elite interaction is also common. In some places, such as along the Amazon River, this division was reflected in appearance of prestige goods, such as elite ceramics (polychrome and other highly decorated fine wares) and fine objects of shell, wood, and stone, as well as utilitarian goods. The division was also reflected in the use of domestic and ceremonial space, particularly with the appearance of increasingly monumental and exclusive central plazas.

In settled areas, the primary rituals of symbolic and social reproduction do not hinge upon the disintegrating act of ritual cannibalism and predation, transforming ontological others to incorporate them as social relations, which characterize the practices of the Tupí-Guaraní peoples to the north. Instead, they hinge upon reconstitutive and performative ritual acts of (re)placing ancestors, founders, whereby living persons take on or come to stand for the histories of families, houses, communities, and even small regions, such as the Upper Xingu. In the latter, the political cum ritual process of making chiefs and turning them into ancestors is the critical context of social reproduction and is embodied not only in living chiefly persons, but in certain places and things owned or controlled by them (Heckenberger 2005).

Although all Amazonian societies define social identity vis-à-vis an existential contradiction (i.e., Amazonian animistic systems, sensu Descola [1996], or what Viveiros de Castro [1998] calls "Amerindian perspectivism"), a structural contradiction is quite restricted in the region. Societies that lack a structural contradiction altogether are typically smaller, more mobile, and atomistic societies, wherein "humanity" ends at the boundary of the community and the incorporation of outsiders is the basis of social reproduction (Clastres 1987; Fausto 1992, 1999; Viveiros de Castro 1992). Societies with some degree of a structural contradiction are typically more settled (rooted in place), comprise large regional social formations, and are sociosymbolically reproduced through "replacing" ancestors and replicating affinal alliances. It is important to recognize, however, that large social formations, sedentism, and social complexity are not formulaic in Amazonia or elsewhere and may well be based on nonheirarchical as well as hierarchical (structural contradiction) forms of sociopolitical integration. Nonetheless, the emergence and transformation of such a structural contradiction are critical not only because the contradiction is part of the internal development of regional political economies, but also because it is a major dimension of regional political economies. Within broad regions, pervasive alterities, which in part correspond to broadly defined ecologies, notably the river/upland dichotomy, are critical factors of social relations and change. It is also important to remember that such distinctions—mobile or settled, atomistic or regional, heterarchical or hierarchical—are not simply additive, one emerging predictably from the other, but rather different contemporaneous orientations of social complexity in dynamic regional political landscapes.

THE SOUTHERN AMAZON

Discussions of social complexity have in general been narrowly focused on the archaeology and ethnohistory of the Amazon River, accentuating the importance of rich alluvial floodplains. This focus has helped perpetuate the geographic determinist and ecofunctional explanations of culture change mentioned earlier and has downplayed the cultural specificity of certain cultural characteristics, such as settled, regional, and hierarchical social formations among Arawak speakers throughout much of Amazonia. One area that has only recently entered into debates is the southern periphery of the Amazon forests, transitional to the savannas of the Bolivian llanos and the scrub forests of central Brazil—an area densely occupied by complex societies in 1492 (Denevan 1966, 2001; Erickson 2000 and chapter 8, this volume; Métraux 1942). Early European explorers in the region (1550–1750) noted the sophisticated, powerful, and "civilized" cultures in the headwater areas of the Guaporé (Baure peoples), Tapajós (Pareci), and Xingu (Xinguano) rivers. These cultures were numerous and densely settled, and they had a high level of engineering in planned plaza villages with ceremonial houses ("temples"), idols, and an elaborate ritual life (Métraux 1942; Schmidt 1917). Later scholars classified these cultures as chiefdoms (Oberg 1955; Steward and Faron 1959).

In the southern Amazon, the first expeditions (around 1690–1720) to the Baure and Pareci noted large circular plaza villages integrated into large regional social formations (Block 1994; Denevan 1966). In the 1720s, Pires de Campo (1862), for instance, noted among Pareci peoples numerous circular plaza villages (averaging 30 to 40 houses each) linked with wide, straight roads. In one day's march, his expedition, the first in this area, encountered 12 such communities. The southern Amazon chiefdoms had already been dramatically altered by the deleterious effects of European expansionism throughout the region.

Many authors have noted common characteristics among Arawak peoples that are seen as hallmarks of Amazonian chiefdoms, such as social hierarchy and regional organization, sedentism, fairly intensive agriculture, dense populations, and developed technology, including navigation skills. Max Schmidt's (1914, 1917) discussions of the Arawak, based on his research in the southern Amazon (1900–1910), were the first synthetic treatment to highlight these characteristics. Lathrap's (1970) model of an Amazonian "Neolithic Revolution" builds on Schmidt's views regarding the riverine expansion of Arawak peoples as early root crop agriculturalists and navigators. Schmidt attributed the directionality of the Arawak diaspora to the reproduction of a general settled riverine (and later maritime) orientation. However, he emphasized certain sociopolitical features—specifically the splitting or budding off of high-ranking individuals ("lords"), actually cadet lines (i.e., the subordinate kindred of secondary chiefs), which became grafted upon (over) autochthonous groups—as the most critical mechanisms of dispersal of Arawak peoples, rather than technological innovation (i.e., swidden or "developed tropical forest agriculture") as stressed by Lathrap.

Steward and Faron also noted commonalities between complex societies in "Circum-Caribbean" areas and those of the southern Amazon (1959:2), but were emphatic that the cultural authorship of the complex features were not Arawak (287–289). They felt that the advanced nature of the "theocratic chiefdoms of eastern Bolivia" resulted from their proximity to high civilization in the Andes (implicating diffusion) and from a "high subsistence level, which permitted the development of dense populations, large communities, and some social and economic specialization" (252–253). Clastres (1987) also noted the correlation between Arawak peoples and the chieftaincy in northern Amazonia, but also considered the unique conditions of Andean influence (by diffusion) as the critical factor in the Arawak peoples' development. Schmidt, we might note, also made a connection, specifically between the Arawak and the highland Andean culture Tiwanaku, but in terms of common origins and migration rather than in terms of trait diffusion.

The southern Amazon is a case in point of the type of regional social and ecological orientations described earlier: large settled populations on major rivers (commonly Arawak and people socially related to them) who were surrounded by diverse "others," notably including macro-Tupian peoples to the north and macro-Gê to the south in the uplands (in western Bolivia, adjacent to the Andes, the situation is more complex). Today, the underlying Arawak identity of many of the major chiefly social formations as settled agriculturalists characterized by hereditary social hierarchies and supralocal integration is more obvious due to better linguistic, ethnohistoric, and archaeological data. As the early diffusionists noted, there are also remarkable similarities in Arawak material culture. As discussed earlier, these settled Arawak groups were also invariably enmeshed in diverse social and political relations with other more mobile, predatory peoples, which formed the basis of broad regional social and politicoeconomic systems. There is also a basic commonality among the southern Arawak peoples with respect to spatial organization, notably settlements with a central plaza. This traditional feature of southern Arawak can be traced back two millennia in other parts of the Arawak diaspora (Caribbean and perhaps central Amazon), making it the oldest spatial organizational feature in the tropical lowlands that can be relatively confidently attributed to any specific cultural-linguistic group (Heckenberger 2005).

Among Arawak chiefly societies throughout the southern Amazon, central plazas express hierarchical distinctions between center people (adult men, chiefs, and ancestors) and others. It is the graphic spatial representation of the twofold social alterity noted previously (between inside and outside and between upper and lower segments of society). On the one hand, it reflects the

existential contradiction typical of all Amazonian peoples, whereby social identity (self and otherness) is defined and reproduced through interaction with
the outside, the contours of nature and supernature, including social outsiders (foreigners) and ontological others. On the other hand, it also represents
a structural contradiction, whereby society is internally divided in two ways,
which Lévi-Strauss called dialectic and concentric (hierarchical). The former
is well represented in the dualistic cultural groupings (age grades, associations,
and gender-based distinctions) instantiated and reproduced through both
pervasive and situational spatial orientations to plazas. The southern Arawak
are unique in the degree to which they emphasize the latter distinction, the
concentric or hierarchical orientations around up/down, senior/junior, sacred/
profane, chiefly/nonchiefly.

The Xinguano nation provides a particularly clear idea of such a hierarchi-
cally divided society. Xinguanos (and the southern Arawak peoples in general)
lack age grades and the warrior and naming societies found in surrounding
(in particular Gê) societies and instead are organized around actual and fictive ties to founding ancestors, ties reproduced through formal and informal
discourse and onomatology, spatial relations between persons and things, and
bodily disposition and adornment. This pattern of cosmological authentication
of persons is tied to hereditary patterns of leadership, and individuals higher in
the ranking are metaphorically constructed as ancestors in ritual (i.e., as par-
ents and grandparents). The pattern has crystallized over time into an ethos of
aristocracy (the definition of an elite group or caste), and it generally applies to
all southern Arawak and arguably was already in place in nascent form in early
diaspora communities of the Formative period.

Early Arawak peoples (proto-Xinguano) occupied the Upper Xingu region by
at least AD 800–900, although, based on radiocarbon dates, it is proposed that
colonization could have been as early as AD 1–500. The Upper Xingu region is the
southeastern endpoint of the Arawak diaspora (it is also the southeastern extent
of the linguistic diaspora of Cariban-speaking peoples, who were established
in the region by AD 1500 at the latest). The precedence of the Arawaks can be
suggested based on ethnological comparisons across the southern Amazon. The
primary features that characterize Arawaks, Xinguanos, Pareci, and Baure are
notably the plaza ritual complex, the chieftaincy, spatial organization (including
not only community layouts, but also communal structures and roads, among
other things), and material culture (hammocks, flutes, masks, idols, houses,
fishing equipment, hammocks, among others). Other groups share certain of
these features (e.g., Mundurucu, Tapirapé, Kayapó, Suyá, Karajá, and Bororo),
but as a constellation the features are highly variable in these others groups but
nearly ubiquitous among southern Arawak. In the Upper Xingu region, the
words used for most of these features—in particular for the structures, objects,
and songs related to the plaza ritual complex—are Arawak.

It is possible to suggest, based on ethnological and linguistic comparisons, that these features emerged in one form or another in proto-Arawakan populations. At least, we can note that plazas, social hierarchy based on birth order (primogeniture), and the political cum ritual process of chief or ancestor making are not unusual in Amazonia despite some programmatic assertions to the contrary (see, e.g., Meggers 1996). In historic times, most of these hierchical societies spoke diverse languages, but when historical lines can be traced over more than a century, it turns out that the root of these plurilingual societies is typically Arawak (i.e., the earliest ancestors appear to be Arawakan speakers). The plazas themselves were part and parcel of Arawakan speakers during the Formative period, although central public and sacred space mutated substantially through time from the small circular plaza villages of early times (such as those known from the southern Caribbean, around 500 BC to AD 500–1000 [Heckenberger 2002]). In addition, as noted earlier, numerous non-Arawakan-speaking groups adopted circular plaza orientation (for example, the northern Gê, Bororo, Mundurucu, Tapirapé). In late prehistoric and early colonial times, public centers were massive, awesome, and highly restricted from most of the populous.

The relation between plazas and hierarchy may be fortuitous, but clearly the centric spatial grammar orients and to some degree defines relations between persons, including divisions between young and old, male and female, and notably the hierarchical distinction between chief-ancestor-sacred and commoner-affine-profane in historic-period Arawak societies. This structural contradiction, at least in nascent form, was characteristic of proto-Arawak and part and parcel of the diaspora. In other words, not only were early diaspora communities settled riverine farmers, but they also included a system of institutional (kin group) social ranking. In other words, the correlation between features commonly seen in historically known Arawak peoples, particularly in the southern Amazon, is most importantly the central plaza ritual complex, which includes circular and other shaped plazas, and the masks, idols, ceremonial structures, flutes, and esoteric knowledge tied to the plaza ritual complex.

In the Upper Xingu region, clear continuity of many elements of this typical southern Arawak social and ecological orientation can be demonstrated between AD 1000 and the present based on basic village organization (circular plaza villages), regional settlement locations, and material culture (a late variant of an Amazonian Barrancoid pottery style [Dole 1961–62; Lathrap 1970]). This does not mean that Xinguano material culture, language, and lifestyle remained unchanged; in fact, many elements of it changed dramatically. What is particularly important is that plaza ritual complex; the "temple-idol-priest" complex, as it has been called (Steward and Faron 1959), grew in scale and elaboration over time. The plaza (ring) village was in place in a rudimentary form since initial colonization of the Upper Xingu region , but evolved into an integrated pattern of major and minor centers organized in galactic (multicentric) clusters.

Galactic clusters, which range from a few to eight or more settlements, were organized as peer communities in a regional polity (Heckenberger 2005; Heckenberger et al. 2003). In the Upper Xingu region, these settlements were linked through an integrated system of curbed roads some 10–30 meters wide and connected to major causeways within villages some 20–50 meters wide and arranged radially from central plazas and occasionally connected to smaller circular enclosures or secondary plazas in and around villages. This integrated plan of settlements in turn articulated with elaborate systems of ditches, ramparts, canoe ports, bathing paths, reservoirs, bridges, and other constructed features. These constructed public works neatly partitioned the entire landscape. A late prehistoric configuration of settlements is suggested based on the large, contemporaneous earthworks dated to between AD 1250 and AD 1650 (based on 20 radiocarbon dates in excellent stratigraphic context from ten 10-by-1-meter excavation trenches and the horizontal integration of the major earthworks).

In the Upper Xingu region, monumentalism takes on an essentially lateral character, consisting of complex regional archaeological features from defining the space of individual settlements to the organization of regional galactic settlement clusters. The modern Xinguano (Kuikuru) village at Ipatse Lake has similar sociopolitical divisions linking place to persons now and over time through key concepts, such as that in which "house" and "person" are tied together, wherein chiefly persons are always thought of in terms of the dual bodies, of the center, and of the cardinal places of the central plaza ring of houses, where they compose and decompose kindred around themselves (figures 10.2 and 10.3). Person, house, village, region, and world can be meaningfully dissected and compared because they are ordered according to an overarching cultural logic that organized space and sociality into right-left, north-south, center-periphery, and up-down.

The regional galactic configuration (AD 1250–1650) of the Xinguano polity is unique in Amazonia. This configuration, which divides and partitions space according to the central plaza village and radial road network, creates an exquisitely segmented cartography, a unique landscape that is not only a road map but a history (see Seeger 1976, 1981, for the neighboring Suya). Like the galactic polity described by Tambiah (1985), the form and content of the central plaza village and the network of central plaza villages are at the intersection of the topographic, the sociopolitical, and the cosmological—a pattern that on the ground represents universal principles of dialectic and quadra-partition encompassed within the overarching hierarchy of ancestrality. In terms of historical ecology, the regional organization of galactic clusters within the broader Xinguano peer polity creates a patterned signature or "footprint" on the forested landscape. The fairly precise orientation of the clusters—conditioned by local principles of "ethnophysics" (astronomy, mathematics, calendrics, and engineering), ecological zonation, and sociopolitical relation—creates a gridlike

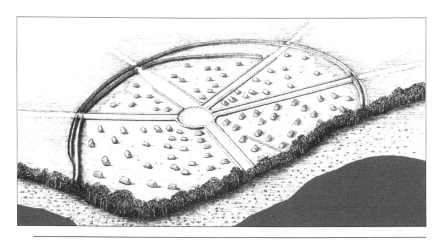

FIGURE 10.2 Reconstructed prehistoric village of Kuhikugu.

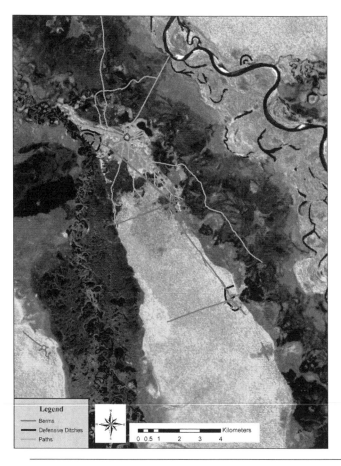

FIGURE 10.3 The Ipatse Cluster showing primary village linked by curbed roads.

or latticelike organization across the region. In other words, the anthropogenic landscape is organized into centers or nodes of varying size linked by corridors and satellites, an organization that profoundly influences the distribution of vegetation and faunal communities and reflect (map) the long-term history of galactic clusters in the region.

SKIN OF THE LAND AND BODY OF THE STATE: AMAZON URBANISM?

The archaeology and history of the Xinguanos and the southern Arawak peoples provide one example of what an Amazonian polity might have looked like around 1492. More than a confederacy, but less than an autocracy, the polity defies neat definition, at least according to what we are used to calling civil society, cities, and states. Nonetheless, we are forced by this and other cases, such as the societies along the Amazon River (Neves and Petersen, chapter 9, this volume; Roosevelt 1999), to reconsider the question of Amazonian urbanism. Most specialists agree that no "lost cities" await discovery in the Amazon, at least recognizable cities in the traditional sense. The question remains as to whether there is any form of Amazonian sociality that might deserve to be considered as a form of urbanism. Urbanism usually implies an urban/rural or city/countryside dichotomy, which in turn obviously has a profound effect on ecology: the skin of the land, the state of the human body(ies), and the body of the state (including the small-size form commonly called chiefdoms). Thus, the question of social complexity and urbanism is not simply about the definition of cities or towns, chiefdoms or states, but also about the elucidation of long-term trajectories of dynamic human-environment interactions, in particular the historical ecology of large, settled populations or social formations that elsewhere would be large, complicated, and powerful enough to be called "complex societies."

Graham (chapter 2, this volume) discusses this topic with respect to the unique tropical ecology of the more obviously urban ancient Maya. In Amazonia, the question must be framed differently because few settlements would classify as urban, even if one accepts Flannery's (1994) proposal of a minimum of 5,000 persons for New World cities. At present, Santarém, which may cover 25 square kilometers, or roughly the size of the better-known sites of Cahokia, Chan Chan, or Teotihuacan (Roosevelt 1999), is the only site large enough to be a city. Even if Amazonian settlements rivaled the size of what are considered towns and cities elsewhere, they share little in common with European feudal towns or with Mediterranean and Mesopotamian city-states. I believe the important question is not if cities or states existed in Amazonia, as if there were simple definitions of cities and states that serve equally across time and space (for example, based on the presence or absence of pyramids, surplus, writing, taxation, and the like). Rather than assuming we know what we are looking for, generally based in large part on Western historical experience, we might instead ask: Would we recognize an Amazonian complex society if we saw it? How do we insert the idea of Amazonian towns or cities (or what anywhere else would be called such) into the flat and somewhat featureless sociospatial landscape of the ubiquitous Amerindian village? The question is not what we call the large, regional social formations that we now know existed in various parts of Amazonia, but how they were organized, and answering this question requires in-depth field research into the historical ecology of discrete regions.

Visualizing polity in Amazonia or alternate (non-Western) forms of urbanism or urban ecology in general depends on our ability to define the precise relationships between human bodies, polities (i.e., the institutional settings of social and political relationships), and the nonhuman biosphere. This ability requires attention to the actual flows and arenas of power as they were reflected in social relations, institutions, and the movements of human social bodies in the past, as well as to the dramatic alterations of landscapes over many generations or even over thousands of years (see also Erickson, chapter 8, and Neves and Petersen, chapter 9, this volume). Today, researchers interested in the question of historical ecology (long-term human-environment interactions) largely agree that a substantial part of Amazonia is cultural forest, perhaps more than Balée's (1989) initial estimate of 12 percent of the region. Most researchers also agree that in some areas this forest was the result of a large-scale "domestication" of the landscape.

The domestication of nature has been a remarkably fruitful idea in Amazonia with regard both to the symbolic constructions of nature and "supernature" in Amazonian animistic systems (Descola 1992, 1996; Viveiros de Castro 1992, 1998) and to ecological questions of agricultural development, settlement, and technology (Denevan 1966, 2001; Erickson 1995 and chapter 8, this volume; Lathrap 1970). In considerations of the past societies that most obviously transformed the landscapes around them, the symbolic, social, and historical construction of political power in these domesticated contexts remains poorly framed. What might the "skin of the land" have looked like in relation to the large polities that dominated much of Amazonia, not only in terms of economic potential or improvements, but also in terms of the values and norms of specific cultural groups? Indigenous groups are situated or situate themselves in place in highly variable ways. Thus, nature cannot be seen as unproblematic, simply a matter of better classification and cataloging before it yields up its secrets, but rather must be seen as historical, symbolic, social, and, above all else, contingent on the dynamic relation between humans and the nonhuman environment. The big surprise that comes from the emerging perspective of the Amazonian past is that the remarkable biodiversity of the region corresponds to an equally impressive and unexpected diversity in the ethnosphere, or the sphere of human cultural systems.

When we return to the question of urban ecology, we have to ask what can be learned from this cursory reconstruction of deep Amazonian history. By 1500, complex societies were common along much of the Amazon River, its major navigable tributaries, and other areas. They were organized into broad political economies, which included chiefly warfare and prestige-goods exchange.

This modality of power associated with chiefly societies was tied in many cases to a core of genealogically defined high-ranking elite vested with specialized knowledge and privilege related to plaza ritual and ancestor veneration. Such a structure was not the only avenue of political complexity in the lowlands, but it was the core of political power in chiefly societies. Indeed, statehood throughout much of the Americas was characterized by creation of similar enclosures or containers of power, including temples, plazas, and boundary markers that concentrated ritual and political activities and in turn the generation and reproduction of social and symbolic capital. This is perhaps what Steward and Faron had in mind when they coined the term *theocratic chiefdoms* of the southern Amazon, where "the men's house takes on state level significance" (1959:24). Lathrap (1985) describes just such a calculus of political power inflected by the spatial geometries of central plazas.

The galactic organization of polity—as described in the Upper Xingu of southern Amazonia, whereby linked sacred centers were ranked in regional networks according to their size and the ranked importance of the ancestors buried there—suggests that political power was not based on control and centralization of basic food resources and economic administration of community life (contra classic central place models). Here we might note the simultaneously decentered (in basic economic and administrative terms) and centric (in spatial, sociopolitical, and ritual terms) system. The Xinguano political configuration of diverse nodes of ritual cum political centralization (plazas) "thickening" into higher and higher (more inclusive) levels of social ranking and integration and, in a general sense, into more powerful centers is reminiscent of Geertz's (1980) model of the Balinese "theater state" and Tambiah's (1985) reformulation of it in the Southeast Asian galactic polity. In other words, there is little evidence of a pervasive center-periphery dichotomy in terms of basic subsistence or administration in the Upper Xingu, but rather the concentration of public spectacle and political action. Specifically, the dynamic formulation and reformulation of the social relations of integration revolves around major chiefs in the context of chiefly initiations when the sons of chiefs are "put up" as future chiefs, inheriting the symbols of the chieftaincy and becoming living ancestors, and when the great chiefs in turn become true ancestors in the spectacular mortuary feasts.

Such a sociopolitical organization, although based principally on ritual and prestige building—in other words, on social and symbolic factors rather than on technoeconomic control—is no less conditioning of the distribution of human bodies' space and thus of the nature of human-environment interactions. It is possible, if not likely, that in terms of human bodies, these polities were as large as those commonly defined by towns and small cities. Further, their impacts on local and regional ecology, on transformations of the "skin of the land," were no less pronounced than in other areas of complex societies in 1492 or before.

Even today Amazonia ill-fits traditional models of urbanism. Browder and Godfrey (1997) describe a pluralistic theory of "disarticulated urbanism" in the Brazilian Amazon. In many respects, this concept may be inappropriate for indigenous social systems, at least for the pre-Columbian systems described here. These authors' subject is contemporary urbanism, development, and globalization of the Brazilian Amazon, specifically the changing nature of frontier development through a transition from a hierarchical extractive-mercantile (urban-rural) economy in colonial times to a more disarticulated and heterogeneous contemporary frontier within a mosaic of dispersed urbanized settings (towns and cities) interspersed with an equally varied and dynamic frontier. Here again we are faced with the tyranny of terminology, using the terms of our past and present to apprehend other pasts and presents. Nonetheless, the first three (of nine) criteria for Browder and Godfrey's (1997) pluralistic theory are relevant here: (1) "the Amazon is a heterogeneous social space" (360); (2) "the configuration of settlement systems in Amazonia is irregular and polymorphous, disarticulated from any single master principal of spatial organization" (362); and (3) urban growth is "functionally disarticulated from . . . agricultural development" in many parts of Amazonia (361–362). These ideas of heterogeneity, the critique of central place models, and in particular the independence of ritual and political action from economic exploitation or outright coercion (a "power from" versus a "power over") do have significant implications for understanding pre-Columbian polities in Amazonia.

In 1492, Amazonia was a patchwork of loosely articulated regional social networks that included relatively large regional politicoeconomic centers (chiefdoms and kingdoms), usually but not always in major river settings (Whitehead 1994). These small polities were engaged in peer-polity relationships with similar centers within contiguous regions, forming regional cultures sharing some cultural and social values and practices. These polities were engaged in relations not only with other settled peoples, but with more mobile peoples in diverse hinterlands. Even the small, egalitarian communities that have been so well documented over the past century were not autonomous, but rather embedded within broad social networks of communities. What is important to emphasize here is that the organization of polity will directly effect dynamic human-environment relations and that alternative patterns or pathways of social complexity are not only determined by ecology, but determinant of it. The principal lesson of historical ecology in Amazonia, as elsewhere, is that nature—whether primordial natural settings, intrinsic human nature, or the unique and shared nature of Amazonian peoples—can no longer be assumed. The job ahead of us is to go beyond this generic observation to reveal the dimensions of long-term change and variability within and between discrete regions in Amazonia—for instance, how long-term and even pervasive social identities and alterities have conditioned how landscapes came into being and changed over time.

NOTES

1. The idea of "concentric zonation" may have some relevance, but the specifics of urban ecology as originally defined by the Chicago school of urban sociology are not critical here.
2. See Whitehead (1998) for a related discussion of historiography in Amazonia in Balée's (1998) overview of historical ecology (also see Neves and Petersen, this volume).

REFERENCES

Balée, W. 1989. The culture of Amazonian forests. In D. A. Posey and W. Balée, eds., *Resource Management in Amazonia: Indigenous and Folk Strategies*, 1–21. Advances in Economic Botany no. 7. Bronx: New York Botanical Garden.

——. 1998. Introduction. In W. Balée, ed., *Advances in Historical Ecology*, 1–10. New York: Columbia University Press.

Biersaack, A. 1999. The new ecological anthropologies. *American Anthropologist* 10:1–22.

Block, D. 1994. *Mission Culture on the Upper Amazon: Native Tradition, Jesuit Enterprise, and Secular Policy in Moxos, 1660–1880*. Lincoln: University of Nebraska Press.

Browder, J. O., and B. J. Godfrey. 1997. *Rainforest Cities: Urbanization, Development, and Globalization of the Brazilian Amazon*. New York: Columbia University Press.

Carneiro, R. L. 1970. Theory of the origin of the state. *Science* 169:733–738.

——. 1985. Slash-and-burn cultivation among the Kuikuru and its implications for cultural development in the Amazon basin. In P. Lyon, ed., *Native South Americans: Ethnology of the Least Known Continent*, 73–91. Prospect Heights, Ill.: Waveland Press.

——. 1995. The history of ecological interpretations of Amazonia: Does Roosevelt have it right? In L. E. Sponsel, ed., *Indigenous Peoples and the Future of Amazonia: An Ecological Anthropology of an Endangered World*, 45–65. Tucson: University of Arizona Press.

Clastres, P. 1987. *Society Against the State: Essays in Political Anthropology*. New York: Zone.

Crumley, C. L. 1991. Historical ecology: A multidimensional ecological orientation. In C. L. Crumley, ed., *Historical Ecology: Cultural Knowledge and Changing Landscapes*, 1–16. Santa Fe: SAR Press.

Denevan, W. 1966. *The Aboriginal Cultural Geography of the Llanos de Mojos of Bolivia*. Ibero-americana no. 48. Berkeley: University of California Press.

——. 1992. The pristine myth: The landscape of the Americas in 1492. *Annals of the Association of American Geographers* 82:369–385.

——. 2001. *Cultivated Landscapes of Native Amazonia and the Andes*. Oxford: Oxford University Press.

Descola, P. 1992. Societies of nature and the nature of society. In A. Kuper, ed., *Conceptualizing Society*, 107–127. New York: Routledge.

——. 1996. *In the Society of Nature: A Native Ecology in Amazonia*. Cambridge, U.K.: Cambridge University Press.

Dobyns, H. 1983. *Their Numbers Became Thinned*. Knoxville: University of Tennessee Press.

Dole, G. E. 1961–62. A preliminary consideration of the prehistory of the upper Xingu basin. *Revista do Museu Paulista* 13:399–423.

Erickson, C. L. 1995. Archaeological methods for the study of landscapes of the Llanos de Mojos in the Bolivian Amazon. In Peter Stahl, ed., *Archaeology in the Lowland American Tropics: Current Analytical Methods and Applications,* 66–95. Cambridge, U.K.: Cambridge University Press.

———. 2000. An artificial landscape-scale fishery in the Bolivian Amazon. *Nature* 408:190–193.

Fausto, C. 1992. Fragmentos de história e cultura Tupinambá: Da etnologia como instrumento crítico de conhecimento etno-historico. In M. Carneiro da Cunha, ed., *História dos índios do Brasil,* 381–396. São Paulo: Companhia das Letras.

———. 1999. Of enemies and pets: Warfare and shamanism in Amazonia. *American Ethnologist* 26:933–956.

Firth, R. 1936. *We the Tikopia.* London: George Allen and Unwin.

Flannery, K. V. 1994. Childe the evolutionist: A perspective from nuclear America. In David Harris, ed., *The Archaeology of V. Gordon Childe,* 101–120. Chicago: University of Chicago Press.

Ford, J. A. 1969. *A Comparison of Formative Cultures in the Americas: Diffusion or the Psychic Unity of Man.* Washington, D.C.: Smithsonian Institution Press.

Geertz, C. 1980. *Negara.* Princeton, N.J.: Princeton University Press.

Giddens, A. 1984. *The Constitution of Society.* Berkeley: University of California Press.

Heckenberger, M. J. 2002. Rethinking the Arawakan diaspora: Hierarchy, regionality, and the Amazonian formative. In J. Hill and F. Santos-Granero, eds., *Comparative Arawak Histories: Rethinking Culture Area and Language Group in Amazonia,* 99–122. Urbana: University of Illinois Press.

———. 2003. The enigma of the great cities: Body and state in Amazonia. *Tipiti: Journal of the Society for the Anthropology of Lowland South America* 1:27–58.

———. 2005. *The Ecology of Power: Culture, Place, and Personhood in the Southern Amazon, AD 1000–2000.* New York: Routledge.

Heckenberger, M. J., A. Kuikuru, U. T. Kuikuru, J. C. Russell, M. Schmidt, C. Fausto, and B. Franchetto. 2003. Amazonia 1492: Pristine forest or cultural parkland? *Science* 1302:1710–1714.

Heckenberger, M. J., J. B. Petersen, and E. G. Neves. 1999. Village permanence in Amazonia: Two archaeological examples from Brazil. *Latin American Antiquity* 10:353–376.

Hill, J. D. 1996. *Ethnogenesis in the Americas.* Urbana: University of Illinois Press.

Hill, J. D., and F. Santos-Granero, eds. 2002. *Comparative Arawakan Histories: Rethinking Language Family and Culture Area in Amazonia.* Urbana: University of Illinois Press.

Ingold, T. 1993. The temporality of landscape. *World Archaeology* 25:152–174.

Kirch, P. V. 1984. *The Evolution of Polynesian Chiefdoms.* Cambridge, U.K.: Cambridge University Press.

Lathrap, D. W. 1970. *The Upper Amazon.* London: Praeger.

———. 1985. Jaws: The control of power in the early nuclear American ceremonial center. In C. Donnan, ed., *Early Andean Ceremonial Centers,* 241–267. Washington, D.C.: Dumbarton Oaks.

Lathrap, D. W., A. Gebhart-Sayer, and A. M. Mester. 1985. Roots of the Shipibo art style: Three waves at Imariacocha, or there were "Incas" before the "Inca." *Journal of Latin American Lore* 11: 31–119.

Meggers, B. J. 1996. *Amazonia: Man and Culture in a Counterfeit Paradise.* Washington, D.C.: Smithsonian Institution Press.

Métraux, A. 1942. *The Native Tribes of Eastern Bolivia and Western Matto Grosso.* Bureau of American Ethnology Bulletin no. 134. Washington, D.C.: Smithsonian Institution Press.

Moseley, M. E., and M. J. Heckenberger. 2005. From village to empire in South America. In C. Scarre, ed., *The Human Past,* 640–677. London: Thames and Hudson.

Oberg, K. 1955. Types of social structure among the lowland tribes of South and Central America. *American Anthropologist* 57:472–488.

Payne, D. L. 1991. A classification of Maipuran (Arawakian) languages based on shared lexical retentions. In D. C. Derbyshire and G. K. Pullum, eds., *Handbook of Amazonian Languages,* 355–499. The Hague: Mouton.

Pires de Campos, A. 1862. Breve noticia que dá o Capitão Antonio Pires de Campos, do gentio barbaro que há na derrota viagem das minas do Cuyabá e seu reconcavo … até o dia 20 maio de 1723. *Revistra Trimensal do Instituto Historico e Geographico e Ethnografico do Brasil* (Rio de Janeiro) 25:437–458.

Roosevelt, A. C. 1989. Natural resource management before the Conquest: Beyond ethnographic projection. In D. A. Posey and W. Balée, eds., *Resource Management in Amazonia: Indigenous and Folk Societies,* 30–61. Advances in Economic Botany no. 7. Bronx: New York Botanical Garden.

——. 1991. Moundbuilders of the Amazon: Geophysical Archaeology on Marajo Island. San Diego: Academic Press.

——. 1994. Amazonian anthropology: Strategy for a new synthesis. In A. C. Roosevelt, ed., *Amazonian Indians from Prehistory to Present: Anthropological Perspectives,* 1–29. Tucson: University of Arizona Press.

——. 1999. The development of prehistoric complex societies: Amazonia, a tropical forest. In E. A. Bacus and L. J. Lucero, eds., *Complex Societies in the Ancient Tropical World,* 13–34. Archaeological Papers of the American Anthropological Association no. 9. Washington, D.C.: American Anthropological Association.

Schama, S. 1995. *Landscape and Memory.* New York: Vintage.

Schmidt, M. 1914. *Die Paressi-Kabishi.* Leipzig: Baessler Archiv IV Heft 475.

——. 1917. *Die Aruaken: Ein Beitrag zum Problem der Kulturverbrietung.* Leipzig: Viet.

Seeger, A. 1976. Fixed points on arcs in circles: The temporal, processual aspect of Suyá space and society. *Proceedings of the International Congress of Americanists* 2:341–359.

——. 1981. *Nature and Society in Central Brazil: The Suyá Indians of Mato Grosso, Brazil.* Cambridge, Mass.: Harvard University Press.

Steward, J. H. 1949. South American cultures: An interpretive summary. In *Comparative Ethnology of South American Indians,* 669–772, vol. 5 of J. H. Steward, ed., *Handbook of South American Indians.* Bureau of American Ethnology Bulletin no. 143. Washington, D.C.: U.S. Government Printing Office.

Steward, J. H., and L. Faron. 1959. *Native Peoples of South America.* New York: McGraw-Hill.

Tambiah, S. 1985. *Culture, Thought, and Social Action.* Cambridge, Mass.: Harvard University Press.

Urban, G. 1992. A história da cultura brasileira segundo as línguas nativas. In M. Carneiro da Cunha, ed., *História dos índios no Brasil,* 87–102. São Paulo: Companhia das Letras, SMC, FAPESP.

Vansina, J. 1990. *Paths in the Rainforest: Towards a History of Political Tradition in Equatorial Africa.* Madison: University of Wisconsin Press.

Viveiros de Castro, E. B. 1992. *From the Enemy's Point of View: Humanity and Divinity in an Amazonian Society.* Chicago: University of Chicago Press.

——. 1998. Cosmological deixis and Amerindian perspectivism. *Journal of the Royal Anthropological Institute* 4 (3): 469–488.

Whitehead, N. L. 1994. Ancient Amerindian polities of the Amazon, Orinoco, and Atlantic Coast: A preliminary analysis of their passage from antiquity to extinction. In A. Roosevelt, ed., *Amazonian Indians from Prehistory to the Present: Anthropological Perspectives,* 33–53. Tucson: University of Arizona Press.

———. 1998. Ecological history and historical ecology: Diachronic modeling versus historical explanation. In W. Balée, ed., *Advances in Historical Ecology,* 30–41. New York: Columbia University Press.

11

BETWEEN THE SHIP AND THE BULLDOZER

Historical Ecology of Guajá Subsistence, Sociality, and Symbolism After 1500

LORETTA A. CORMIER

CONTEMPORARY FORAGERS HAVE been a subject of intense interest since the origins of anthropology, when they were situated on the lowest rung of a cultural evolutionary ladder. The perspective of historical ecology offers an approach whereby the subsistence strategy of any culture is explicable only through understanding the dynamics of human-ecological interaction through time. In early Amazonianist ecological anthropology, the approach of cultural ecology employed the notion of "limiting factors," which suggested that a poverty of a local resource, such as poor soils (Meggers 1957, 1973) or insufficient protein (Good 1987; Gross 1975; Harris 1974), affects population size, degree of sedentism, cultural behaviors and values, or general cultural development.

Although environmental constraints most certainly can affect the subsistence strategies that are possible in a given area, limiting factors alone as an explanation is problematic for understanding Amazonian foraging peoples. The handful of foragers or near-foragers in Amazonia are situated among agriculturalists practicing shifting cultivation. When peoples inhabiting the same ecological zones exploit the same resources through different modes and imbue the same environment with differing symbolic meanings, such monocausal interpretations do not provide sufficient explanation. Moreover, when we move beyond data from the ethnographic present, many contributions to this volume make clear that the archaeological record demonstrates historical differences in the way the landscape has been used over time.

As Balée and Erickson aptly describe in the introduction to this volume, "landscape" represents the collision of nature and culture, with historical ecology being inherently an anthropogenic research program. In this chapter, approaching landscape as a sphere of interaction represents an attempt to connect the dots between the recent ethnography of the Guajá[1] foragers of Maranhão, Brazil (figure 11.1) and past historical and ecological events that may be responsible for a probable shift from agriculture to foraging after European colonization. If painted with the broadest of brushes, the well-known events of colonization affecting the Guajá are similar to those undergone by numerous other Amazonian indigenous peoples. However, individual cultures were not affected homogeneously. Responses varied from isolation to assimilation to resistance and even to extinction. Furthermore, indirect consequences of European contact included both ecological change and indigenous interethnic political change.

The Guajá provide an example of the importance of historical ecology for understanding culture because their way of life as authentic hunter-gatherers is, quite remarkably, predicated on past indigenous agricultural forest modification.

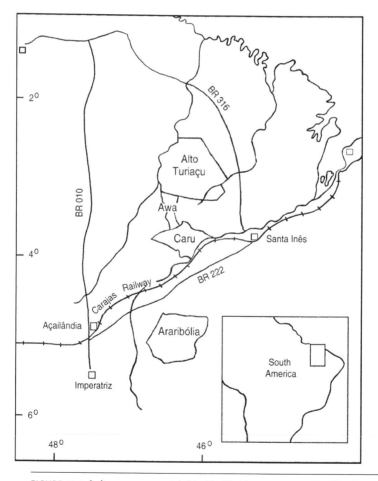

FIGURE 11.1 Indigenous reserves inhabited by Guajá groups, eastern Brazilian Amazon.

In the specific case of the Guajá, their mode of subsistence, for at least the past 150 years, can be understood as an indirect interaction with agriculturalists through exploitation of anthropogenic forests. Balée (1988, 1989, 1992, 1994) has described the Guajá as relying heavily on the babaçu palm *(Orbignya phalerata,* or *Attalea speciosa),* which is a rapid colonizer of old fallow fields (also see Clement, chapter 6, this volume). Thus, the Guajá have lived in the wake of agriculturalists, with indigenous modification of the forest being a prerequisite for their way of life.

In this chapter, I address the diversity of Amazonian foraging and the historical ecology of Guajá foraging. First, I critically evaluate the term *foraging* itself. Second, I examine several case studies and models in order to demonstrate the diversity of foraging in Amazonia. Third, I explore the historical ecology of the Guajá in depth to make the case that their foraging is a postcolonial adaptation. Finally, I discuss Guajá foraging in a social and symbolic context to establish the interconnections between their subsistence strategy and other domains of their culture.

THE FORAGING LIFE

The term *hunter-gatherer* or *forager* is not completely transparent; it has been used to describe a wide range of subsistence strategies. In the recent *Cambridge Encyclopedia of Hunters and Gatherers* (Lee and Daly 1999:3), the term *hunter-gatherer* includes continuous hunter-gatherers, respecialized former agriculturalists, groups who make limited use of domesticates, and those who grow no domesticated plants but gain substantial calories from them through trade with agriculturalists. Further, basing the classification of *forager* on subsistence strategy alone (specifically, those who lack or produce limited domesticated foods) can also be problematic when considering societies such as the Pacific Northwest Coast chiefdoms whose only domesticate was the dog (Kelly 1995:293–295).[2]

The terms *hunter-gatherer, forager,* and *trekker* are sometimes used interchangeably, but they are not completely equivalent. The difference between the terms *hunter-gatherer* and *forager* is largely a historical one. Although both refer to those who do not cultivate domesticated plants or raise domesticated animals, the term *forager* is increasingly coming into favor over the term *hunter-gatherer* in order to avoid prioritizing hunting over gathering (Kelly 1995:xiv). The critique derives in part from both ethnographic evidence that gathering provides more calories in some groups than does hunting, as Lee (1968) argued in an early work about the !Kung people of southern Africa and others. Second, the term *hunter-gatherer* has tied into the feminist critique that "man the hunter" implies a privileging over "woman the gatherer" in both prehistory and contemporary societies (e.g., see Dahlberg 1981). It should be noted that in either hunter-gatherer or gatherer-hunter societies, hunting and gathering does not always represent a strict division of labor along gender lines (Cormier 2003a; Estioko-Griffin and Griffin 1981).

An additional problem with the term *hunter-gatherer* is that it does not reflect the role that fishing may play in the diet. In most Amazonian groups, fishing provides a significant source of calories (Beckerman 1983; Chernela 1989, 1993:87–109; Gragson 1992). Although one may still sometimes see the cumbersome term *hunter-gatherer-fishers* in the literature, the term *forager* encompasses the three subsistence strategies while prioritizing none. Finally, the term *trekker* refers to groups that spend six months or more per year away from a village site to which they ultimately return (Balée 1994:210). Thus, they may use domesticated plants, but spend at least half of the year subsisting through foraging activities.

One problematic use of the term *forager* is pairing it, as in the loaded descriptive term *traditional foragers*. At worst, one might argue that the term *traditional* has been evolutionary and ecosystemic anthropologists' ethnocentric precept for these seemingly ecologically noble savages living in a pristine wilderness. However, I have made the same mistake in my own work. Elsewhere (Cormier 2002a), I erroneously describe the Guajá as "traditional foragers." The context for that description was to create a contrast between a presumably longstanding precontact way of life and recent postcontact changes to their lifestyle, which have involved limited incorporation of domesticates. However, describing such differences with the term *traditional* might be misinterpreted not just to suggest the recent change between their precontact and postcontact way of life, but to impy that long-term cultural continuity and even stasis existed until the European conquest.

From historical records, the Guajá have been documented as foragers for at least 150 years in the approximate vicinity that they inhabit today, the region drained by the Pindaré, Turiaçu, and Gurupi rivers (Gomes 1985). Over the past twenty-five years, their culture has begun to undergo dramatic changes. Their foraging grounds are being reduced and circumscribed due to development; they have undergone depopulation, primarily from infectious disease; and they are beginning to adopt agriculture and Western material items into their culture. They might best be described as a *postforaging society*, a term becoming more frequently used (Barac 1999; Headland 2004; Trigger 1999). The term reflects the inadequacy of the term *foragers* or *hunter-gatherers* for contemporary Amazonian peoples, who to varied degrees are socially, politically, economically, and ecologically affected by Western societies.

Describing the Guajá as foragers also involves a degree of classification by means of the ethnographic present, given that many are now beginning to sow and harvest domesticates. The term *forager* might be considered an anthropological prototype (Rosch 1978). Although perhaps no group conforms exactly to any culture type, foraging peoples can broadly be described according to a suite of common features, including family and band social organization, a high degree of individualism with relative egalitarianism for same-sex, same-age members, a tendency toward weak expression of gender hierarchy, a high degree of nomadism with a seasonal concentration and dispersion pattern, a common-property regime, and little to no raising of domesticated plants and animals (Lee and Daly 1999:3–5). The Guajá have essentially conformed to this description in the recent past.

WHY DO FORAGERS FORAGE?

I discuss in depth later the arguments for the validity of the agricultural regression hypothesis as explanation for Guajá foraging, but in brief the hypothesis suggests that the Guajá were farmers before they were foragers and lost agricultural knowledge due to the cultural upheaval following European colonization. However, the specific case of the Guajá should not necessarily lead to the conclusion that all Amazonian foragers have undergone an identical process. Several case studies have posed alternate explanations: symbiosis with agriculturalists, cultural allopatry, political choice, and precolonial foraging.

Perhaps the best-known pattern of tropical forest foraging is found among African and Southeast Asian groups involved in extensive trade relations with agricultural peoples who give them crops in exchange for forest products. Headland and Reid borrow the biological term *symbiosis* to describe the relationship through an analogy to mutualism, commensalism, cooperation, and similar relationships among interacting populations (1989:45). Among the African Efe (Bailey and Peacock 1988:101) and Mbuti (Ichikawa 1999:211), more than 60 percent of their calories come from trade with neighboring farmers. In northwest Amazonia, a similar relationship can be found with the foraging Maku and the agricultural Tukanoans. The Maku trade wild game, baskets, and labor to the Tukanoans for the crops, in particular bitter manioc, on which they are heavily dependent to meet their caloric needs (Jackson 1983; Milton 1984; Silverwood-Cope 1972). Jackson has gone so far as to describe this interaction as a master-servant relationship (1983:154).

These forager-farmer relationships were the basis of the so-called wild yam question of the 1980s (e.g., see Bailey et al. 1989; Headland 1987). Specifically, the question was whether noncultivated species, in particular yams and palms, that thrive in edge environments and light gaps created from treefalls and natural fires, could provide a quantity of carbohydrate foods sufficient to substitute for domesticated plants among foragers. Although tropical forests have a high degree of biodiversity, many plants contain toxins, rendering them inedible for humans, and those that are edible may be thinly scattered, widely dispersed, or located high in the canopy out of human reach (Hutterer 1983:179). Due to a slowed rate of digestion and an enlarged colon, which allow for detoxification and fermentation of the cellulose in leaves, terrestrial gorillas are able to eat plants that would be toxic for humans (Fleagle 1988:243). Tropical forest plants (with the exception of palms) do not concentrate energy in seeds and nuts that are rich in fats and carbohydrates the way that plants do in climates producing seasonal stress (Hutterer 1983:179). The central debate concerned whether it was possible at all for independent foragers to obtain sufficient wild plant foods in the tropical forests and to survive without relying on trade relations with food producers. In Amazonia, the answer is yes. Although the Maku fit the model of symbiosis with agriculturalists, other Amazonian foragers deviate from the African pattern of institutionalized relationships with food producers.

Milton (1991) has developed another model of Amazonian foraging that also involves shifting the level of analysis from a single culture to interethnic relations within a region. She has used the allopatry/sympatry distinction from nonhuman primatology in exploring dietary differences among four Amazonian tropical forest groups. Although the model is not specific to hunter-gatherers, it does have implications for dietary differences among neighboring groups. Milton characterizes tropical forest groups as largely allopatric due to territorial exclusion and often hostile interethnic relations, with accompanying differences in resource exploitation in the groups she studied. However, she found that environmental availability of these resources alone could not explain these differences. She argues that food selection is in part related to general interethnic cultural differentiation, similar to differences in features such as body ornamentation.

Milton's model raises the question of the role of culture choice in food selection. Rival (1998, 2002) has made an explicit argument for the importance of intentionality in understanding subsistence strategies. She argues that cultural choice has been underestimated in explanations of Amazonian foraging and describes the Huaorani of Ecuador as making limited use of domesticates, not because they cannot cultivate crops, but because they prefer to concentrate on other types of food. She disagrees with the assumption that foragers in Amazonia are regressed, deculturated, or devolved agriculturalists. For the Huaorani, she argues that their trekking and foraging is a political choice rather than an example of cultural loss, for it is linked to their cultural identity on multiple levels. In her view, Huaorani foraging is deeply embedded in cultural structures of meaning, specifically the predator-prey relation. It is not merely a subsistence activity, but a symbolic embodiment whereby the Huaorani take the position of prey in flight. Rival therefore does not view Huaorani foraging as a postcolonial phenomenon, but as a consistent structure into which European colonists took the role of the non-Huaorani predators.

Rival's point, I believe, is well taken. If a choice in subsistence strategy exists, not all societies will make the same choice. Although the environmental availability of a resource is logically antecedent to its exploitation, environmental availability alone cannot always predict the degree to which a society will exploit a given resource. One case in point is food choice among the Matisgenka of the Madre de Dios River in Peru. Howlers are the most abundant mammal, yet the Matisgenka take howlers roughly 10 times less frequently than the similar-size woolly and spider monkeys (Shepard 2002). Another example is the food choice of a largely foraging subgroup of the Parakanã[3] of the Xingu River basin in Brazil, who, like the Guajá, are Tupí-Guaraní speakers described by Milton (1991) as relying heavily on babaçu palms as a subsistence item. However, for the Parakanã group Milton described, monkeys are taboo.

Perhaps it should be self-evident that cultural approaches to food are always embedded in complex symbolic structures, no less so for Americans in our geometry of crossing and uncrossing of knife and fork in order to eat a meal properly. In my home state of Alabama, it is bread, not rice, that disappears from the grocery store shelves in tornado season although both are equally available. However, no matter how ecologically constrained a culture's approach to food may be, environmental determinism alone is an insufficient explanation.

Finally, it is important to raise the question of whether or not independent Amazonian foragers could have existed in the midst of farmers prior to European colonization. That is to say, could foragers have exploited old fallow fields or similar habitats created by nonanthropogenic processes such as light gaps created by treefalls or naturally occurring fires? The work of Stearman on the Sirionó and the Yuquí addresses this possibility.

The Sirionó and the Yuquí, like the Guajá, lack indigenous fire-making technology, which Stearman (1984) explains as due to their "deculturation"; interestingly, she posits that deculturation began in precolonial times, with European contact exacerbating the process. Stearman (1984, 1989) advocates Métraux's (1942, 1963) hypothesis that the Sirionó-Yuquí represent remnants of a Guaraní intrusion into Bolivia (see also Balée 2001). Stearman argues that their dispersal and isolation into hostile territory might have made it difficult for them to establish and maintain agriculture and could have led to greater dependence on game and wild plant foods (1984:641). Further, like the Guajá (Balée 1995, 1999), the Yuquí (Stearman 1984) and the Sirionó (Balée 2001) retain words in their vocabulary for domesticates they do not cultivate.[4]

Comparing and contrasting these cases with the Guajá, one can conclude that the Guajá are unlike the Maku because the Guajá have existed as foragers in the absence of trade relations with agriculturalists. However, one might argue that Guajá exploitation of the anthropogenic forest entails an analogous relationship of dependency on food producers, albeit an indirect relationship because they exploit forest types created in the wake of agriculturalists.

Milton's explanation of regional interethnic differentiation in resource exploitation and Rival's explanation of political choice do have important implications for the Guajá. As discussed in the final section of this chapter, Guajá foraging, regardless of its ultimate origins, is not merely a subsistence strategy, but central to understanding their social relations and cosmological beliefs. However, it is important for the purposes here to draw a salient contrast between the Guajá and the Huaorani: unlike the Guajá (until recently), the Huaorani do make limited use of domesticates. This is not to suggest that "foraging" should be treated as a hermeneutically sealed category. Ethnological comparisons of foragers, near foragers, trekkers, and even postforaging societies are most certainly valid and worthwhile. However, the point should be raised because the core question addressed here is narrow and involves understanding how the Guajá, surrounded by farmers, did not possess the knowledge of how to cultivate domesticates.

The Yuquí provide the closest analogy to the Guajá, with their absence of fire-making technology and lack of domesticates, yet they maintain Tupí cognates for domesticates in their language, suggesting culture loss. If Stearman's hypothesis is correct, it demonstrates an example of precontact agricultural loss in an Amazonian group. Although such a scenario would also be theoretically possible for the Guajá, I argue in the next section that evidence from Guajá historical ecology suggests that foraging was adopted in a postcolonial context.

WHY DO THE GUAJÁ FORAGE?

The myriad facets of colonization—including assimilation, displacement, disease, warfare, enslavement, and depopulation—have been rightly addressed as forming a complex watershed event affecting Amazonian cultural and ecological adaptation. Although Amazonian history clearly did not begin in 1500, it can perhaps be considered a historical case analogous to punctuated equilibrium because of the dramatic and rapid changes that occurred within a relatively short time span (Balée 1995). In my view, the evidence suggests that it was in this context that the Guajá likely made a transition from agriculture to foraging. This statement does not suggest that all Amazonian foragers forage as a result of colonialism. Rather, it describes the historical ecological trajectory of one group.

The consequences of European contact were likely as much a factor in the Guajá subsistence mode as the presence of the anthropogenic forest. Although little is known about the early history of the Guajá, they are members of a Tupí-Guaraní linguistic group that is hypothesized to have dispersed in several waves from the region of what is now Rondônia in southwest Amazonia (Jensen 1999; Rodrigues 1984–85). The Guajá speak a language classified as the southern group of Tupí-Guaraní Subgroup VIII, which includes five groups in addition to the Guajá: the Anambé, Ka'apor, Tukunyapé, Turiwára, and Amanayé. Evidence suggests that many speakers of southern Tupí-Guaraní Subgroup VIII, who have been documented in the vicinity of the Tocantins River basin at the time of contact, migrated eastward in the wake of European colonization (Cormier 2003b).

In 1767, Noronha (1856) described two of the groups, the Turiwára and the Amanayé, as being on the Tocantins River, and Capuchin missionary catechism reports in the late 1800s mentioned them as being to the east on the Capim River (APEP 1864, 1873, 1875, 1877). Balée has provided ethnobotanical and oral history evidence indicating that the Ka'apor migrated from the Tocantins River to their present location to the east in the vicinity of the Guajá (1994:25, 30–35). The Anambé today remain in the vicinity of the Tocantins River (CEDI/PETI 1990:66). Noronha also referred to a group called the "Uayá" on the Tocantins in 1767, which Balée (1994:25) and Nimuendaju (1948:135) hypothesize may have been the Guajá. Even if the Guajá were not the Uayá, it is likely that they were on the Tocantins River before European colonization. The first definite description of the Guajá would place them in the same eastward migratory trajectory as the other members of the southern Tupí-Guaraní Subgroup VIII, specifically an 1853 report describing them in Maranhão, along the affluents of the Caru and Gurupi rivers (Gomes 1985).

The indigenous groups along the Tocantins River were subject to compulsory labor through the mission towns of the Jesuits for about 100 years between 1655 and 1757, and later, when the Jesuits were expelled, they were subject to similar, yet more severe, compulsory labor and virtual enslavement under the General Company for the Commerce of Brazil (Companhia Geral do Comércio do Grão Pará e Maranhão), which was founded in 1755 (Azevedo 1930; Brito 1991; Cruz 1963; Sweet 1975). The Portuguese enslaved some groups as a whole along the Tocantins River . One recorded slaving expedition in 1653 described more than 500 native people taken from the area of the Tocantins River (Sweet 1975:124–125). Another practice was to erect a cross in a village and upon returning several months later to declare the entire village slaves if the cross had not been cared for properly (Sweet 1975:120).

It is possible that the Guajá were one of the groups enslaved wholesale, and this enslavement might explain their probable loss of agriculture. Escaped slaves might have difficulty reestablishing themselves as agriculturalists. Sweet describes a 1730 expedition that recovered more than 100 slaves who had been surviving by raiding plantations (1975:164). Flight and displacement were also consequences of attempts to escape from being captured as slaves. Sweet describes a mid-1650s Jesuit priest's report that a "hundred-league" stretch of the Tocantins River had been abandoned in flight from fear of the slavers (1975:126). Evidence suggests that the Guajá were exposed in some way to these Portuguese slavers. The name for one of the Guajá divinities, *kapitã,* probably derives from the term given to Portuguese labor bosses, who were called *capitão.* According to Balée, the Ka'apor today use a cognate of *capitão* to refer to their headmen (1994:168). In addition, as previously mentioned, Balée (1992, 1994, 1999) has described linguistic artifacts of agriculture in the Guajá language, especially regarding words for certain widespread domesticates.

Concurrent with enslavement were the effects of disease. The devastating effects of Old World infectious disease have been documented for multiple indigenous Amazonian populations. As is well known, infectious diseases often spread from indigenous group to indigenous group, even without direct contact with Europeans. It is also worth mentioning that epidemic disease was often the impetus for flight of native Amazonians from their work posts under the General Company for the Commerce of Brazil (Fonseca de Castro 1996). Wagley described a similar process of a Tapirapé village disbanding in 1895 due to a smallpox epidemic ([1940] 1974:374). More recently, in 1981, Fundação Nacional do Índio (FUNAI) recorded the flight of many Guajá into the forest after a respiratory virus was introduced (Parise 1988).

Forline (1997), an ethnographer of the Guajá, suggests that the Cabanagem Civil War (1835–1841) may have been the key factor behind Guajá migration. The conflict between Portuguese loyalists to the Crown and native-born Brazilians seeking an independent government in Grão Pará left between 20,000 and 30,000 dead (Cleary 1998). It was the bloodiest war in Amazonian history. Given that the Guajá were first documented in Maranhão in the decade after the Cabanagem War, it is quite possible that this event was ultimately the impetus for their migration. Certainly, any one event or combination of events may have been responsible, including epidemic disease, slavery, or warfare (whether it involved fighting against European settlers, waging interethnic warfare under conditions of destabilized politics, or being caught in the midst of the Cabanagem Civil War).

Although migration by itself does not necessarily suggest radical culture change, a forced displacement in the case of the Guajá might have made it difficult for them to reestablish agriculture, as the example of escaped slaves surviving through raiding suggests. In addition, according to Balée's (1988, 1994) oral history, the Ka'apor (who today share a reserve with the Guajá) would not have allowed the Guajá to keep permanent settlements. The Guajá may have encountered similar difficulty in attempting to establish territory in other areas, particularly if they were a depopulated group, less able to compete. One might think that they at least would have been able to be involved in part-time maize production with its low start-up costs (Balée 1992, 1994). Considering this, the loss of knowledge of agriculture altogether, coupled with a current lack of indigenous fire-making technology, might have resulted from the actual enslavement of the Guajá. As Sweet (1975) confirmed, sometimes entire villages were taken as a whole. The Guajá may have been one of these groups.

Erikson (2002) describes an interesting ecological change among the Matis in response to relatively recent effects of depopulation of the group due to infectious disease. The Matis decreased their reliance on agriculture and increased their reliance on hunting. His description of the preference of hunting over agriculture is reminiscent of Rival's position that cultural choice is critical to understanding subsistence strategies. It is also an example of a contemporary group that increased hunting behavior because changes in their population density made it possible to decrease reliance on agriculture. This view generally supports Beckerman's (1978, 1979) earlier hypothesis that prior to contact most protein may have been obtained from plant foods and fish, with heavy exploitation of animal protein becoming practical only after the depopulation of native peoples.

The Matis adaptation ties into Denevan's (1992a) argument that it is important to understand early European colonization and native demographic collapse as a major ecological event (see also Heckenberger et al. 2003; Meggers 2003). Perhaps there has been a tendency to view ecological impact in terms of the active forces of European displacement and replacement of indigenous peoples through disease, warfare, large-scale settlement, and deforestation. But in the historical interstices between the ships of the past and the bulldozers of today, depopulation itself must have had ecological consequences. As we begin to understand more clearly how densely populated Amazonia was before contact and the multiple ways in which indigenous Amazonian peoples were agents in environmental modification (see chapters 5, 7, 8, 9, and 10, this volume), then sudden, massive disappearance of these active ecological agents necessarily would have had environmental effects. The sheer loss of numbers of people across varied landscapes would have created shifts in the ecological pressure on flora and fauna simply from the absence of these human agents.

One such ecological change is evident in Clement's (1999a, 1999b) description of a generalized process of what might also be described as a type of agricultural regression in Amazonia. Clement argues that one result of the massive depopulation of Amazonia was the loss of genetic diversity among domesticated plant species. Because domesticated species depend on human management, these species in abandoned fields would have died out, regressed to the wild genotype, or, at the very least, suffered a population contraction. Further, the genetic diversity within a given species is a function of both the ecological variables of a localized habitat and the human cultural knowledge of managing the species, which often includes experimentation. Although a domesticated species might still exist, the genetic variation within the species and human knowledge of management of genetic varieties are lost.

Denevan (1992b,1998, and chapter 5, this volume) also argues that patterns of shifting cultivation observed in contemporary Amazonian groups may not be a long-standing tradition, but rather emerged with the introduction of steel tools. Thus, he suggests that the introduction of the technology of the iron and steel ax fostered a pattern of short-term use and abandonment of sequential garden sites rather than long-term management of single garden sites. This view would be consistent with Beckerman's (1978) argument that escape from European predation and infectious disease should be considered as contributing the two most important selective pressures on settlement size and degree of sedentism in Amazonia. European contact may thus well have created both the need for increased mobility and short-term use of garden sites and the technological means to facilitate such a mode of exploitation.

Although old fallow fields certainly existed before European contact (and similar environmental conditions could exist in naturally occurring light gaps),[5] if Denevan is correct, then the proliferation of abandoned old fallow fields that the Guajá exploit probably derived from ecological changes occurring since 1500. To be clear, Devevan is not suggesting a postcontact intensification of agriculture; he means quite the opposite (see Denevan, chapter 5, this volume). As contributions to this volume suggest, the archaeological record indicates a pattern of precolonial, highly managed forms of intensified agriculture. Denevan is describing a depopulated landscape with farmers now engaging in a pattern of short-term usage and abandonment of agricultural fields. The pattern would create a decrease in areas of intense, managed agricultural zones and an increase in old fallow fields resulting from short-term usage by more mobile groups. This short-term usage and high mobility should be understood not merely as due to the steel ax facilitating rapid field preparation, but in the historical context of factors provoking flight (i.e., slavery, warfare, disease). In this light, although a forager adaptation to old fallow forest, such as seen with the Guajá, could have been possible before contact, their patch-to-patch movement from old fallow field to old fallow field would seem to be a much more feasible strategy after 1500. In addition, the sudden, massive negative space where people used to be between the initial waves of depopulation and the subsequent large-scale European settlement should at least be considered as a factor influencing the human-environment relationships and indigenous peoples' subsistence strategies. The complex of historical ecology of early contact may well have been responsible for the Guajá loss of agriculture and after contact may have facilitated their ability to adapt as foragers exploiting old fallow fields.

FORAGING AND THE FOREST SIBLINGS

Much of my work has focused on the relationship of the Guajá to local monkeys, which I believe provides an example of their adaptation to the anthropogenic forest. When I first went to work among the Guajá in 1996, my research centered on understanding two seemingly contradictory relations: monkeys were a key source of food, and numerous monkeys were kept as pets. The central question involved understanding how the Guajá conceptualization of monkeys could encompass monkeys' being both hunted as prey and nurtured as pets. In Western society, companion animals are typically avoided as food. In brief, what I found was that monkeys comprise part of a symbolic cannibalistic complex. They are considered to be nearly human and for that reason are actually desired, not eschewed, as food. A brief review of my main findings (Cormier 2002a, 2002b, 2003b) provides an ethnographic background.

The Guajá are beginning to live continuously in association with domesticated plants; however, they still exhibit an ecological strategy of seasonal concentration and dispersion whereby they tend toward heavy exploitation of fish in the dry season and dispersal and increase of their trekking behavior in the wet season, with monkeys being the most frequently used animal food (along with fish and game) (Cormier 2002a, 2003b). Guajá ethnobotanical knowledge is an important hunting strategy, for they know far more plants that are edible for game than are edible for humans. More than 80 percent of the plants they identify have at least one use as food for an animal they hunt, and more than half of those are plants eaten by monkeys (Cormier 2002b). In addition, the anthropogenic habitats in which the Guajá caloric staple, the babaçu palm, thrives is an environment that also attracts monkeys. Balée identified 34 species of plants that are dominant colonizers of old fallow forests in the Guajá region (1994:133). Of these old fallow species, the Guajá described to me approximately 35 percent of them as plants eaten by monkeys (Cormier 2002b). Thus, given that such a high percentage of old fallow disturbance indicators are plants for local monkeys, the degree of Guajá emphasis on monkeys as a dietary item might also be considered another aspect of their adaptation to the anthropogenic forest.

The social role of monkeys among the Guajá is related to their general notions of Guajá kinship with plants and animals and to the specific functions of monkeys in their culture. These beliefs conform well to recent reformulations of Amazonian animism (e.g., see Descola 1996; Viveiros de Castro 2001). The term *animism* minimally applies to religious systems that include the belief in souls and spirits, particularly when such an attribute is assigned to plants, animals, and landscape features. In brief, Descola broadens the anthropomorphic connotation of animism in the Amazonian context to include the extension of within-group kinship relationships outward to nonhuman beings. In a similar vein, Viveiros de Castro's GUT theory argues that affinity, rather than consanguinity, is the generic mode of relatedness in Amazonia. Moreover, Viveiros de Castro's (1998, 1999) notion of multinatural perspectivism is a linked concept, for it posits a prevalent animistic theme in Amazonian thought that human beings and animals share an essential worldview and animals are considered to be nonhuman "persons."

For the Guajá, the personhood of plants, animals, and supernatural beings is expressed in part through the kinship terminology they use to classify them. Most are referred to by the same term the Guajá use for the matrilateral sibling (*harəpiana*). Relationships among men and relationships among women are fairly egalitarian. They have no headman or designated shaman. Males do have some authority over women, but it is weakly expressed. The Guajá reckon descent through males, but also display the notion of partible paternity, recently elaborated for Amazonian groups more generally by Beckerman and Valentine (2002). They describe a widespread Amazonian belief that the fetuses are made from the buildup of semen; thus, in native theory, it is possible for a child to have more than one "biological" father.

For the Guajá, such plural paternity is viewed largely as required for the successful completion of a pregnancy. My lack of children was a concern, and it was suggested to me that the reason my husband and I had not conceived while in the field was that I had only one sexual partner. One informant somewhat wearily explained that his wife was pregnant and that he had to *tanta, tanta, tanta* (literally "bang, bang, bang"), and he openly identified his wife's other sexual partners. That is not to say that no sexual jealousies ever existed. On several occasions, male informants expressed unhappiness with their wives' relationships with other men. However, the feeling did not seem to be so much a jealousy specific to the sexual relationship, but a more generalized displeasure over a wife's preference to spend time with another man.

The Guajá practice both polygyny and polyandry. However, in the group where I did my fieldwork, polyandrous relationships were temporary and typically involved a transitional state between two marital partners. In other words, the cases of polyandry I observed were the means by which Guajá women divorced their husbands. The woman would begin to refer to a new man as her husband (*iménə*). For a time, she would be the wife of both men and then eventually cease to use the *iménə* term for her first spouse.

Households are based on a mother, the children she has borne, and her husband, who may or may not be one of the fathers to any given child borne by the woman. Thus, children are always raised with their matrilateral—in essence, affinal—siblings and have at least some patrilateral, consanguineal siblings outside the household. The relationship of the Guajá to nonhuman beings is based on the close, coresidential relationship of matrilateral siblings raised in the same household.

The matrilateral siblingship forms the primary basis for Guajá relationships to nonhuman beings. In general, most communities of plants, animals, and spiritual beings are referred to as *harəpiana* matrilateral siblings with the Guajá people as a whole. However, there are at least three significant exceptions. One is the Guajá use of the patrilateral, consanguineal kin term (*harəpihará*) to refer to their spirit siblings. A second is the use of the patrilateral sibling term in reference to howler monkeys. And a third is the designation of pet monkeys, which are incorporated more directly into the local kinship system, as the "child" (*mɵmɵrə*) of their human adoptive mothers.

Monkeys are a substantial presence in the Guajá community, with approximately 40 monkey pets kept at any given time in a village of just more than 100 people.[6] Monkeys serve several purposes in that community. One is simply as companion animals, similar to the role of our dogs and cats. They also serve a role in the enculturation of boys as hunters and of girls as child caretakers. Boys are given toy bows and arrows as toddlers and begin to play hunter with monkey pets.[7] Girls as young as toddlers begin to spend some time carrying monkeys on their heads and are referred to as the monkeys' "mothers." But, more commonly, infant monkeys are the primary responsibility of adult women who nurture them as they would children. They are played with, sung to, bathed, breast-fed, and fed premasticated foods directly from the women's mouths. Once a monkey is brought into the household and designated as the child of a woman, it is never eaten as food.

Pollock (2002) has described the notion of multiple maternity among the Arauan-speaking Kulina of western Amazonia, whereby women who breast-feed their sisters' children are viewed as additional mothers because their contribution of breast milk involves a sharing of substance needed to sustain a child, analogous to men's contribution of semen in plural paternity. This is similar to Conklin's (2001) description of "consanguinealization of affines" among the Wari', whereby exchange of semen, breast milk, and even sweat creates relations of consanguinity through shared substance. Even though I did not identify this type of consanguinealization among the Guajá, women's nurturance of monkeys and their designation as mothers of monkeys do seem to be counterparts of male plural paternity, particularly given the quasi-human status of monkeys in Guajá beliefs. The mothers of monkeys are the females who nurture them, rather than exclusively the beings that bore them.

With regard to more general relations to the nonhuman world, although most relations are viewed as affinal, the patrilateral sibling term is used to describe special relationships with spirit siblings (*haima*). Each individual is named for a plant, animal, divinity, or landscape feature with which he or she shares in a collective spiritual nature. Thus, an individual named Tatu is a spirit patrilateral sibling of the armadillo (*tatu*) and of any other individuals named Tatu. At a higher node of reckoning, although most human and non-human populations are viewed as metamatrilateral siblings, some populations (both human and nonhuman) are believed to be more closely related and are considered to be patrilateral siblings. For example, the tapir and the white-lipped peccary are termed matrilateral siblings, but the white-lipped peccary and the collared peccary are patrilateral siblings. The Guajá people stand in this patrilateral relationship only with howler monkeys, the principal monkey species hunted for food. The Guajá explain the metalevel of shared consanguinity through their creation myth. Howler monkeys were at one time human beings who were transformed into monkeys by their mythical creator divinity or culture hero to provide food for the Guajá. In the Guajá sacred sky home, the Yu divinity hunts and controls howlers. He is a *haima* spiritual sibling with the yu palm (*Astrocaryum gynacanthum* Mart.), which the Guajá identify as an important howler food. Thus, Yu eats howlers and howlers eat yu. Parallel equations are found in Guajá descriptions of consumption and consanguinity among other forms of life, which can be interpreted as a symbolic cannibalism theme in their cosmology (Cormier 2003b).

DISCUSSION AND CONCLUSION

Although not all questions about Guajá historical ecology have been answered here, two points do seem clear. First, their subsistence strategy is intricately integrated with their social and cosmological orders. Ecology, kinship, and worldview are interdependent and mutually reinforcing so that environmental exploitation, reckoning of social relations in the human and nonhuman world, and symbolic constructions form a coherent whole. However, this interdependence does not suggest stasis. Guajá hunting and gathering have involved exploitation of anthropogenic old fallow forest not only where their caloric staple, the babaçu palm, thrives, but where the evidence suggests that monkeys are attracted to the same habitat. What is less transparent is the relative weight to be accorded to the variables of (1) the direct effects of depopulation and displacement due to European colonization; (2) the indirect consequences of changes in the human-ecological matrix through shifts in the human, floral, and faunal densities; (3) changes in interethnic politics due to shifts in territory from both depopulation and competition for territory among displaced groups; and (4) the role of cultural tradition, cultural identity, and cultural choice in food selection itself.

In the case of the Guajá, all these variables have likely come into play, but I would infer from the foregoing data that the most likely determinant of their foraging adaptation is culture loss in the wake of early contact, epidemic disease, warfare, and either flight from slavers or actual enslavement. The evidence suggests that they were part of the Tupí-Guaraní Subgroup VIII on the Tocantins River that migrated eastward between the mid-1700s and the mid-1800s. Given that the members of this subgroup, with the exception of the Guajá, are currently agriculturalists, taken together with Balée's evidence that the Guajá have linguistic artifacts of agriculture in their language, it seems likely that the Guajá were themselves at one time agriculturalists.

Denevan's (1992a, 1992b, 1998) work furnishes a likely chain of events. First, he describes a floral and faunal rebound in the early stages of colonization after the radical depopulation of indigenous peoples, yet before the large-scale settlement of Amazonia by Europeans. Second, he posits that short-term shifting cultivation and garden abandonment form a post-1500 pattern related to the introduction of the steel axe. For the Guajá, the general depopulation of native Amazonians might have created shifts in the human-ecological landscape that might have initially reduced hunting pressure on natural resources and made hunting and gathering a more feasible option. This option was coupled with a likely increase in the number of abandoned old fallow fields and other abandoned settlements available for Guajá patch-to-patch nomadic exploitation. Balée's (1988, 1994) research with the Ka'apor supplies direct historical evidence for interethnic political factors that would have prevented Guajá bands from later establishing territory, settlements, and agriculture.

Whether or not the contemporary Guajá are foragers by choice is exceedingly difficult to ascertain. Without a doubt, they valorize their way of life (Cormier 2003b). However, a culture's self-evaluation is probably the least reliable of gauges, for it may be as much an ethnocentric justification as it is an expression of free will. But, clearly, simple environmental determinism or even adaptation to post-1500 anthropogenic forests cannot by itself be a complete explanation for understanding Guajá culture as a whole. Guajá social and symbolic behaviors as well as constructions related to monkeys are unique to their culture. Though this configuration is unique, however, no single feature is altogether distinctive insofar as numerous Amazonian analogues exist in the primacy of siblingship, animistic inclusion of nonhuman beings in constructions of sociality, plural paternity, and symbolic cannibalism.

As a final word, I would argue that *agricultural regression* and *cultural devolution* are not synonyms. The term *agricultural regression* suggests a loss of a particular type of cultural knowledge. However, the term *cultural devolution* might be misread to suggest unilinear evolutionism—that a group has taken a step backward on an inevitable chain of cultural development. For the Guajá, the argument that they have lost agricultural knowledge is not meant to suggest that they themselves are devolved. Although their subsistence strategy does represent agricultural regression if it is understood as loss of knowledge they once had, agricultural regression alone is not implied in this change. Their way of life is a contemporary adaptation to the cultural and ecological exigencies they have faced over time and across varied landscapes of eastern Amazonia. If they are in fact former agriculturalists adapting to a post-1500 anthropogenic forest, which strongly seems to be the case, their subsistence strategy is best understood as a cultural innovation in response to changing historical, social, and ecological conditions.

The program of historical ecology seeks to understand landscape as a dynamic sphere encompassing the interplay of interrelations among human cultures, the environments they inhabit, and the means by which they mutually influence one another over time. The Guajá's cultural features can be understood as involving cultural continuity, affinities with larger Amazonian symbolic constructs, interregional Amazonian relations, and Western contact. Similarly, their ecological environment is as much a product of history and culture as it is a product of nature, as is their way of life in eastern Amazonia. When the Guajá were first documented in the historical record, they were observed as a foraging people who exploited old fallow fields where their staple carbohydrate, the babaçu palm, thrived. The same old fallow fields also differentially concentrate the plants that the Guajá identify as key subsistence items for monkeys, which are in turn a key source of game for the Guajá. The complex role of monkeys in Guajá kinship and cosmology may well have its roots in Guajá foraging. However, it should not be characterized as a sui generis cultural invention, even if, as I argue, it is an innovative response of former agriculturalists. Rather, it is best understood as a local ecosymbolic instantiation of the Guajá's place in history, culture, and nature in Amazonia.

NOTES

1. According to Nimuendaju, the name Guajá is a derivation of an earlier Brazilian term *gwazá,* which itself is likely a corruption of the term *wazaizara,* used by the Tembé and Guajajára for the Guajá, which means "feather ornament owner" (1948:135). Although the name Guajá is not pejorative, it is not the people's autodesignation. They call themselves Awá or Awá-te, "true people." I have continued to use Guajá to avoid confusion with other groups who also call themselves *awa.*

2. The Kwakiutl, Coastal Salish, and Haida participated in a dog-eating ceremony. However, dogs were also used for hunting and kept as pets, but not used for food outside of the ritual (Schwartz 1997:34–36).

3. The largely foraging Parakanã Subgroup is worth mentioning because Balée (1995) has observed that although geographically separated, many Amazonian foragers are Tupí-Guaraní speakers (the Guajá, Hetá, Aché, Avá-Canoerio, Yuquí). A similar point was raised by participants at the 2002 Symposium on Neotropical Historical Ecology at Tulane University. Although not addressed here, the potential significance of this point does warrant further consideration.

4. Stearman (1984) points out that although the Yuquí sometimes obtained domesticates by crop raiding, their words for these domesticates were unlikely to be borrowed because there were no other Tupí-speaking peoples nearby.

5. See Stearman 1995 for a description of how "patch dynamics" influenced Yuquí foraging.

6. I identified a total of 90 monkeys during 15 months of fieldwork, but many of them died and were replaced.

7. Although not directly related to the argument at hand, it should be noted that these practice shots are "play" for the boys, but not for the monkeys.

REFERENCES

Arquivo Público do Estado do Pará (APEP). 1864. Letter to the president of the province of Pará. In "Patentes e principais de índios 1863–1888," unpublished document, Belém, Brazil.

———. 1873. Cathechese e civilisação de índios. Unpublished document, Belém, Brazil.

———. 1875. Letter to the president of the province of Pará. In "Patentes e principais de índios 1863–1888," unpublished document, Belém, Brazil.

———. 1877. Letter to the president of the province of Pará. In "Patentes e principais de índios. 1863–1888," unpublished document, Belém, Brazil.

Azevedo, J. L. 1930. *Os Jesuítas no Grão Pará: Suas missões e a colonização*. Coimbra, Portugal: Imprensa da Universidade.

Bailey, R. C., G. Head, M. Jenike, B. Own, R. Rechtman, and E. Zechenter. 1989. Hunting and gathering in the tropical rain forest: Is it possible? *American Anthropologist* 91:59–82.

Bailey, R. C., and N. R. Peacock. 1988. Efe pygmies of northeast Zaire: Subsistence strategies in the Ituri Forest. In I. de Garine and G. A. Harrison, eds., *Coping with Uncertainty in the Food Supply*, 88–117. Oxford: Clarendon Press.

Balée, W. 1988. Indigenous adaptation to Amazonian palm forests. *Principes* 32:47–54.

———. 1989. The culture of Amazonian forests. In D. A. Posey and W. Balée, eds., *Resource Management in Amazonia: Indigenous and Folk Strategies*, 1–21. Advances in Economic Botany no. 7. Bronx: New York Botanical Garden.

———. 1992. People of the fallow: A historical ecology of foraging in lowland South America. In K. H. Redford and C. Padoch, eds., *Conservation of Neotropical Forests: Building on Traditional Resource Use*, 35–57. New York: Columbia University Press.

———. 1994. *Footprints of the Forest: Ka'apor Ethnobotany—the Historical Ecology of Plant Utilization by an Amazonian People*. New York: Columbia University Press.

———. 1995. Historical ecology of Amazonia. In L. E. Sponsel, ed., *Indigenous Peoples and the Future of Amazonia: An Ecological Anthropology of an Endangered World*, 97–110. Tucson: University of Arizona Press.

———. 1999. Mode of production and ethnobotanical vocabulary: A controlled comparison of Guajá and Ka'apor (eastern Amazonian Brazil). In T. L. Gragson and B. G. Blount, eds., *Ethnoecology: Knowledge, Resources, and Rights*, 24–40. Athens: University of Georgia Press.

———. 2001. Environment, culture, and Sirionó plant names. In L. Maffi, ed., *On Biocultural Diversity: Linking Language, Knowledge, and the Environment*, 298–310. Washington, D.C.: Smithsonian Institution Press.

Barac, V. 1999. From primitive to pop: Foraging and post-foraging hunter-gatherer music. In R. B. Lee and R. Daly, eds., *The Cambridge Encyclopedia of Hunters and Gatherers*, 434–440. Cambridge, U.K.: Cambridge University Press.

Beckerman, S. 1978. Reply to Eric Barry Ross: Food taboos, diet, and hunting strategy: The adaptation to animals in Amazonian cultural ecology. *Current Anthropology* 19:17–19.

———. 1979. The abundance of protein in Amazonia: A reply to Gross. *American Anthropologist* 81:533–560.

———. 1983. Carpe diem: An optimal foraging approach to Barí fishing. In R. B. Hames and W. T. Vickers, eds., *Adaptive Responses of Native Amazonians*, 269–299. New York: Academic Press.

Beckerman, S., and P. Valentine, eds. 2002. *Cultures of Multiple Fathers: The Theory and Practice of Partible Paternity in Lowland South America*. Gainesville: University of Florida Press.

Brito, C. M. 1991. *Índios das "Corporações": Trabalho compulsório no Grão-Pará nos esquemas do diretório.* Belém, Brazil: Arquivo Público do Estado do Pará.

Centro Ecumênico de Documentação e Informação / Projeto Estudo sobre Terras Indígenas no Brasil (CEDI/PETI). 1990. *Terras indígenas no Brasil.* Rio de Janeiro: Museu Nacional.

Chernela, J. M. 1989. Managing the rivers of hunger: The Tukano of Brazil. In D. A. Posey and W. Balée, eds., *Resource Management in Amazonia: Indigenous and Folk Strategies,* 238–248. Advances in Economic Botany no. 7. Bronx: New York Botanical Garden.

———. 1993. *The Wanano Indians of the Brazilian Amazon: A Sense of Space.* Austin: University of Texas Press.

Cleary, D. 1998. "Lost altogether to the civilised world": Race and the Cabanagem in northern Brazil, 1750–1850. *Comparative Studies in Society and History* 40:109–135.

Clement, C. R. 1999a. 1492 and the loss of Amazonian crop genetic resources. I. The relation between domestication and human population decline. *Economic Botany* 53: 188–202.

———. 1999b. 1492 and the loss of Amazonian crop resources. II. Biogeography at contact. *Economic Botany* 53:203–216.

Conklin, B. A. 2001. *Consuming Grief: Compassionate Cannibalism in an Amazonian Society.* Austin: University of Texas Press.

Cormier, L. A. 2002a. Monkey as food, monkey as child: Guajá symbolic cannibalism. In A. Fuentes and L. D. Wolfe, eds., *Primates Face to Face: The Conservation Implications of Human-Nonhuman Primate Interconnections,* 63–84. Cambridge, U.K.: Cambridge University Press.

———. 2002b. Monkey ethnobotany: Preserving biocultural diversity in Amazonia. In J. R. Stepp, F. S. Wyndam, and R. Zarger, eds., *Ethnobiology and Biocultural Diversity: Proceedings of the Seventh International Congress of Ethnobiology,* 313–325. Athens: University of Georgia Press.

———. 2003a. Decolonizing history: Ritual transformation of the past by the Guajá Indians of eastern Amazonia. In N. L. Whitehead, ed., *History and Historicities: New Perspectives in Amazonia,* 123–139. Lincoln: University of Nebraska Press.

———. 2003b. *Kinship with Monkeys: The Guajá Foragers of Eastern Amazonia.* New York: Columbia University Press.

Cruz, E. 1963. *História do Pará.* Vol. 1. Belém, Brazil: Universidade Federal do Pará.

Dahlberg, F., ed. 1981. *Woman the Gatherer.* New Haven, Conn.: Yale University Press.

Denevan, W. M. 1992a. The pristine myth: The landscape of the Americas in 1492. *Annals of the Association of American Geographers* 82:369–385.

———. 1992b. Stone vs. metal axes: The ambiguity of shifting cultivation in prehistoric Amazonia. *Journal of the Steward Anthropological Society* 20:153–165.

———. 1998. Comments on prehistoric agriculture in Amazonia. *Culture and Agriculture* 20: 53–59.

Descola, P. 1996. Constructing natures: Symbolic ecology and social practice. In P. Descola and G. Pálsson, eds., *Nature and Society: Anthropologica Perspectives,* 82–102. New York: Routledge.

Erikson, P. 2002. Several fathers in one's cap: Polyandrous conception among the Panoan Matis (Amazonas, Brazil). In S. Beckerman and P. Valentine, eds., *Cultures of Multiple Fathers: The Theory and Practice of Partible Paternity in South America,* 123–136. Gainesville: University of Florida Press.

Estioko-Griffin, A. A., and P. B. Griffin. 1981. Woman the hunter: The Agta. In F. Dahlberg, ed., *Woman the Gatherer,* 121–151. New Haven, Conn.: Yale University Press.

Fleagle, J. G. 1988. *Primate Adaptation and Evolution.* San Diego: Academic Press.

Fonseca de Castro, A. 1996. Manuscritos sobre a Amazônia colonial: Repertorio referente a mâo-de-obra indígena do fundo secretaria do governo (colônia e império). *Anais do Arquivo Público do Pará* 2 (1): 9–121.

Forline, L. C. 1997. The persistence and cultural transformation of the Guajá Indians: Foragers of Maranhão State, Brazil. Ph.D. diss., University of Florida.

Gomes, M. P. 1985. *Relatório antropológico sobre a Área Indígena Guajá (Awá-Gurupi).* Setembro, Brazil: Ministério da Justiça, Fundação Nacional do Índio.

Good, K. 1987. Limiting factors in Amazonian ecology. In M. Harris and E. Ross, eds., *Food and Evolution: Toward a Theory of Human Food Habits,* 407–421. Philadelphia: Temple University Press.

Gragson, T. L. 1992. Fishing the waters of Amazonia: Native subsistence economies in a tropical rain forest. *American Anthropologist* 94:428–440.

Gross, D. 1975. Protein capture and cultural development in the Amazon basin. *American Anthropologist* 77:526–549.

Harris, M. 1974. *Cows, Pigs, Wars, and Witches: The Riddles of Culture.* New York: Random House.

Headland, T. N. 1987. The wild yam question: How well could independent hunter-gatherers live in a tropical rain forest ecosystem? *Human Ecology* 15:463–491.

——. 2004. Thirty endangered languages in the Philippines. Paper presented at the annual meeting of the American Anthropological Association, New Orleans, November 20–24.

Headland, T. N., and L. A. Reid. 1989. Hunter-gatherers and their neighbors from prehistory to the present. *Current Anthropology* 30:43–66.

Heckenberger, M. J., A. Kuikuro, U. T. Kuikuro, J. C. Russell, M. Schmidt, C. Fausto, and B. Franchetto. 2003. Amazonia 1492: Pristine forest or cultural parkland? *Science* 301:1710–1714.

Hutterer, K. L. 1983. The natural and cultural history of Southeast Asian agriculture: Ecological and evolutionary considerations. *Anthropos* 78:169–212.

Ichikawa, M. 1999. The Mbuti of northern Congo. In R. B. Lee and R. Daly, eds., *The Cambridge Encyclopedia of Hunter and Gatherers,* 210–214. Cambridge, U.K.: Cambridge University Press.

Jackson, J. 1983. *The Fish People: Linguistic Exogamy and Tukanoan Identity in Northwest Amazonia.* Cambridge, U.K.: Cambridge University Press.

Jensen, C. 1999. Tupí-Guaraní. In R. M. Dixon and A. Aikhenvald, eds., *The Amazonian Languages,* 125–163. Cambridge, U.K.: Cambridge University Press.

Kelly, R. L. 1995. *The Foraging Spectrum: Diversity in Hunter-Gatherer Lifeways.* Washington, D.C.: Smithsonian Institution Press.

Lee, R. B. 1968. What hunters do for a living, or how to make out on scarce resources. In R. B. Lee and I. Devore, eds., *Man the Hunter,* 30–48. Chicago: Aldine.

Lee, R. B., and R. Daly. 1999. Introduction: Foragers and others. In R. B. Lee and R. Daly, eds., *The Cambridge Encyclopedia of Hunters and Gatherers,* 1–19. Cambridge, U.K.: Cambridge University Press.

Meggers, B. J. 1957. Environment, history, and culture in the Amazon basin: An appraisal of the theory of environmental determinism. In *Studies in Human Ecology,* 71–89. Social Science Monographs no. 111. Washington, D.C.: Pan American Union.

——. 1973. Some problems of cultural adaptation in Amazonia, with emphasis on the pre-European period. In B. Meggers, E. Ayensu, and W. Duckworth, eds., *Tropical Forest Ecosystems in Africa and South America: A Comparative Review,* 311–320. Washington, D.C.: Smithsonian Institution Press.

——. 2003. Revisiting Amazonia circa 1492. *Science* 302:2067–2070.

Métraux, A. 1942. *The Native Tribes of Eastern Bolivia and Western Matto Grosso.* Bureau of American Ethnology Bulletin no. 134. Washington, D.C.: Smithsonian Institution Press.

———. 1963. Tribes of eastern Bolivia and the Madeira headwaters. In *The Tropical Forest Tribes,* 381–454, vol. 3 of J. H. Steward, ed., *Handbook of South American Indians.* Bureau of American Ethnology Bulletin no. 143. Washington, D.C.: U.S. Government Printing Office.

Milton, K. 1984. Protein and carbohydrate resources of the Maku Indians of northwest Amazonia. *American Anthropologist* 86:7–27.

———. 1991. Comparative aspects of diet in Amazonian forest dwellers. *Philosophical Transactions of the Royal Society,* Series B, 334:253–263.

Nimuendaju, C. 1948. The Guajá. In *The Tropical Forest Tribes,* vol. 3 of J. H. Steward, ed., *Handbook of South American Indians,* 135–136. Bureau of American Ethnology Bulletin no. 143. Washington, D.C.: U.S. Government Printing Office.

Noronha, J. M. 1856. Roteiro da viagem da cidade do Pará até ás ultimas colonias dos dominios Portuguezes em os rios Amazonas e Negro. In *Collecção de noticias para a história e geografia das Naçoes ultramarinas que vivem nos dominios Portuguezes, ou lhes são Visinhas,* 1–102. Lisbon: Academia Real das Sciencias.

Parise, F. 1988. *Relatório histórico dos grupos Guajá contatados e sua situação atual.* Brasília: Ministério da Justiça, Fundação Nacional do Índio.

Pollock, D. 2002. Partible paternity and multiple maternity among the Kulina. In S. Beckerman and P. Valentine, eds., *Cultures of Multiple Fathers: The Theory and Practice of Partible Paternity in South America,* 42–61. Gainesville: University of Florida Press.

Rival, L. M. 1998. Domestication as a historical and symbolic process: Wild gardens and cultivated forests in the Ecuadorian Amazon. In W. Balée, ed., *Advances in Historical Ecology,* 232–250. New York: Columbia University Press.

———. 2002. *Trekking Through History: The Huaorani of Amazonian Ecuador.* New York: Columbia University Press.

Rodrigues, A. D. 1984–85. Relações internas na família lingüistica Tupí-Guaraní. *Revista de Antropologia* 27–28:33–53.

Rosch, E. 1978. Principles of categorization. In E. Rosch and B. Lloyds, eds., *Cognition and Categorization,* 27–48. Hillsdale, N.J.: Lawrence Erlbaum Associates.

Schwartz, M. 1997. *A History of Dogs in the Early Americas.* New Haven, Conn.: Yale University Press.

Shepard, G. H., Jr. 2002. Primates in Matsigenka subsistence and world view. In A. Fuentes and L. D. Wolfe, eds., *Primates Face to Face: The Conservation Implications of Human-Nonhuman Primate Interconnections,* 101–136. Cambridge, U.K.: Cambridge University Press.

Silverwood-Cope, P. 1972. A contribution to the ethnography of the Columbian Maku. Ph.D. diss., Selwyn College, Cambridge University.

Stearman, A. M. 1984. The Yuquí connection: Another look at Sirionó deculturation. *American Anthropologist* 86:630–650.

———. 1989. *The Yuquí: Forest Nomads in a Changing World.* Orlando, Fla.: Holt, Rinehart, and Winston.

———. 1995. Neotropical foraging adaptations and the effects of acculturation on sustainable resource use. In L. E. Sponsel, ed., *Indigenous Peoples and the Future of Amazonia: An Ecological Anthropology of an Endangered World,* 207–224. Tucson: University of Arizona Press.

Sweet, D. G. 1975. A rich realm of nature destroyed: The middle Amazon Valley, 1640–1750. Ph.D. diss., University of Wisconsin.

Trigger, D.S. 1999. Hunter-gatherer peoples and nation-states. In R.B. Lee and R. Daly, eds., *The Cambridge Encyclopedia of Hunters and Gatherers,* 473–479. Cambridge, U.K.: Cambridge University Press.

Viveiros de Castro, E. 1998. Cosmological deixis and Amazonian perspectivism. *Journal of the Royal Anthropological Institute* 4:469–488.

——. 1999. The transformation of objects into subjects in Amerindian ontogenies. Paper presented at the annual meeting of the American Anthropological Association, Chicago.

——. 2001. Gut feelings about Amazonia: Potential affinity and the construction of sociality. In N.L. Whitehead and L.M. Rival, eds., *Beyond the Visible and Material: The Amerindianization of Society in the Work of Peter Rivière,* 19–43. Oxford: Oxford University Press.

Wagley, C. [1940] 1974. The effects of depopulation upon social organization as illustrated by the Tapirapé Indians. In P.J. Lyon, ed., *Native South Americans: Ethnology of the Least Known Continent,* 373–376. Prospect Heights, Ill.: Waveland Press.

12

LANDSCAPES OF THE PAST, FOOTPRINTS OF THE FUTURE

Historical Ecology and the Study of Contemporary Land-Use
Change in the Amazon

EDUARDO S. BRONDÍZIO

THE HISTORICAL ECOLOGICAL approach that emerged during the 1990s is contributing to a growing awareness of the long-term and processual interactions between human populations and environment. It has challenged the recurrent simplification of culture-nature interactions as dichotomous and deterministic and the perception that natural and anthropogenic landscapes are mutually exclusive (Balée 1998; Crumley 1994). This chapter reflects on the potential contribution of an applied historical ecology for the analysis of contemporary land-use change in the Amazon. Land use in contemporary Amazonia does not occur in a historical vacuum despite the overwhelming changes taking place in the region recently. Contemporary land-use change in the region reflects variations in regional historical conditions defining land tenure; migration and access to resources; ethnicity; social organization and class; and demands from external markets and policies. My rationale for integrating historical ecology and land-use studies in examining the region is based on three related points.

First, since colonial times there has been a growing complexity of social groups, economic strategies, and forms of land use and resource ownership in the region. Although substitution of land-use systems and social groups has occurred, forms of land use have coevolved cumulatively, resulting in growing intraregional variability. Indigenous systems now coexist with large-scale industrial enterprises; urban-rural networks mingle with a variety of farming systems and settlement arrangements of multiple sizes. Contemporary land use in the region reflects the historical interaction between macrolevel processes and place-specific conditions underlying land tenure, infrastructure, demographic and social organization, technology and knowledge of resource use, and market-economic arrangements. A historical ecological perspective has much to contribute to the understanding of processes creating intraregional variation in social and economic conditions and thus in land-use systems in the region today.

Second, contemporary land-use analysis and historical ecology focus on overlapping theoretical concerns regarding human-environment interactions—such as the implications of changes in settlement pattern and resource-management strategies to the formation of anthropogenic landscapes—thus opening the opportunity to build bridges toward theoretical cross-fertilization. Land use has been actively studied from different perspectives in the Amazon for decades. Examples include models of settlement location (Chibnik 1994; Denevan 1996; Hiraoka 1985; Sternberg 1956); swidden cultivation (Albuquerque 1969; Beckerman 1983; Brabo 1979; Denevan 1998; Denevan and Padoch 1987; Hames 1983); environment and adaptation (Clarke 1976; Hames and Vickers 1983; Moran 1981, 1995); soil fertility, population density, and environmental circumscription (Carneiro 1961 and Meggers 1971, to cite just two). However, theoretical tools implicitly or explicitly used in analytical models explaining patterns of deforestation in the region today—such as by Boserup, Von Thünen, and central-place models—are also helpful in understanding various forms of relationships between Amazonian populations and environmental resources (Geist and Lambin 2001; Kaimowitz and Angelsen 1998; Wood and Porro 2002). Although there are variations in language and terminology in the two approaches, a common interest exists between historical ecology and land-use analysis regarding conceptual models utilized to explain human-environment interaction in the region.

Third, historical ecology and land-use analysis have similar concerns regarding the role of units of analysis and spatial-temporal scales in understanding human-environment interactions. Understanding social and environmental change in the region today requires attention to national and international factors interacting with regionally and locally defined conditions and histories. Historical ecologists are familiar with the recurrent tension between considering the region as an organic entity and envisioning it as a mosaic of microrealities resulting from population dynamics, environmental variability, and historical events. A historical ecological approach to land use may help to avoid generalizations and to improve sampling across historically diverse communities and regions, including areas of recent colonization.

The high rates of deforestation in the region since the 1970s, growing awareness of the global implications of Amazonian environmental change, and demand for integrated social environmental policies have produced numerous analysis and prognostic models for interpreting factors affecting land-use change in the region (Carvalho et al. 2002; Fearnside 1984; Goldenberg 1989; INPE 1988–2001; Kaimovitz and Angelsen 1998; Laurence et al. 2001; Nepstad and Uhl 2000; Nepstad et al. 2002; Skole and Tucker 1993; Verissimo, Cochrane, and Souza 2002; Wood and Porro 2002). Development projects, rural and urban population growth, changing infrastructure, and national and international market developments draw attention to macrolevel processes that cannot be ignored in any sensible analysis (Ab'Saber 1997; Browder and Godfrey 1997; Dincão and Silveira 1994; Lená and Oliveira 1992; Moran 1993b; Schmink and Wood 1992; Wood and Porro 2002). Furthermore, the growing integration of the region with global markets will increase regional complexity as these new corridors meet up with a long history of regional occupation.[1] However, the focus on variables of regional and global relevance should not lead to disregarding complex local differences on the basis of their unmanageability or "irrelevance" (Brondízio 2005).

Although macrolevel socioeconomic processes continue to be important, land-use change cannot be generalized because of a growing spatial diversity of inter- and intraregional conditions—for instance, differences in land tenure and in sociocultural, technological, demographic, and environmental conditions. The human decisions—that is to say, the intentionality—central to shaping the regional environment have occurred at micro- and mesoscales, although they have had basinwide cumulative consequences. In this context, variability is a condition increasingly inherent to the region, and accounting for it is necessary to make any research finding useful to policy. Striking a balance between contemporary and historical, local and macroprocesses—in the interplay among national and international forces, regional conditions, and interregional variability—is and will be increasingly necessary for understanding the present and future of the Amazon region.

The term *"intraregional" variability* is used in this paper as a heuristic tool and an analytical unit of research defined empirically to accommodate human populations in relation to their historical, cultural, biophysical, economic, and institutional environment. Nested units of analysis can be defined according to one's research question and scale of analysis and according to regional socioenvironmental conditions: for instance, households in relation to a community, rural communities in relation to a county (*município* in Portuguese), farm lots in relation to a settlement, settlements in relation to a network of settlements, communities in relation to a conservation unit. Defining the level of detail necessary to capture intraregional variability and related historical differences depends on a given study's scale (spatial and temporal), question, and goals. The point here is that attention to a "region's" historical occupation—variation in settlement time, differences in social groups, forms of access to resources and resource-use rights, land-tenure arrangements, and past economic cycles—may contribute to improving research design and sampling and to avoiding generalizations across diverse social realities within the region.

This chapter builds on empirical analysis of multiple rural and indigenous communities in regions representative of contemporary Amazonia, including rural and periurban riverine communities, indigenous territories, colonization zones, and conservation areas. Data are discussed in the context of historical processes influencing land-use and settlement pattern and the formation of humanized landscapes (Clement, chapter 6, and Erickson and Balée, chapter 7, this volume) in the region. Based on case studies, the chapter relies on ethnographic and archival surveys as well as on multitemporal remote sensing data and geographical information system (GIS) analysis to discuss the variability of factors affecting changing land use and their footprint on the landscape. The chapter is structured to address and discuss each of the three main points presented earlier. It concludes with reflections on the analysis of land-use trajectories in the region today, with a focus mostly on the Brazilian Amazon.

HISTORICAL FORMATION OF AMAZONIAN LAND-USE SYSTEMS: CUMULATIVE STRATEGIES AND GROWING COMPLEXITY

In contrast to other areas of Brazil where large-scale out-migration of rural populations, substitution of land-use systems, and homogenization of landscapes have occurred, a significant part of the Amazon region has increasingly moved toward greater social and land-use complexity. There is no such thing as an "average" cultural landscape in Amazonia either in indigenous areas or in recent colonization settlements. Besides the diversity of environments, variations in social, economic, and cultural history have embedded their footprints in the region. Figure 12.1 illustrates general historical trends in agrarian land-use systems in the region. By and large, Amazonian land use can be seen as evolving according to successive economic and political phases while maintaining some continuity of pre-European indigenous land-use systems and technology. In general terms, land-use systems in the region have evolved along with phases of regional occupation by different migrant groups, government policies, forms of land-tenure grants, and demands from external markets. Along with various new forms deriving from indigenous land-use systems, historically dominant forms of land use have included export-oriented extractivism, cycles of cash crop expansion, and, more recently, large-scale logging, monocrop agriculture, and cattle ranching, together with the implementation of conservation units. Regional settlement patterns were until recently dominated by dispersed rural communities and regional urban centers along the main river networks. Aside from groups such as the Tikuna, occupying the upper Solimões River floodplains, indigenous groups have been characterized as having dispersed but interconnected upland settlements of various sizes, usually associated with the region's main tributaries and transition areas or between types of forests and savannas. During the past 30 years, a complex network of roads connecting urban areas and interspersed by rural settlements evolved. Today, indigenous areas coexist, commonly in conflict, with a growing number of colonization settlements, logging and mining concessions, large urban centers, and a variety of conservation areas of direct- and indirect-use categories.

Although much remains to be learned about pre-Columbian forms of land use, some level of continuity in swidden and agroforestry cultivation systems and environmental management technologies is usually assumed and ethnographically documented among contemporary populations, indigenous and *caboclo*.[2] The scope and kind of land-use change occurring during the first two centuries of European colonization (Denevan 1998, 2001) is still little understood, but this period of transition was crucial to subsequent land-use systems in the region. Among the processes influencing land use during this period were the scaling down of agriculture under conditions of sociocultural chaos, labor shortage, and migration to new environments. Depopulation, migration, and settlement change as well as diffusion of and experimentation with new technological practices brought to the region led to regional variations in forms of land use. The introduction of new crops into the indigenous agricultural repertoire, such as rice, banana, and sugarcane, further diversified land-use systems based on various forms of swidden agriculture. Responses to colonial demands, new crop varieties, and variation in regional environmental conditions made possible the emergence of different forms of swidden agriculture and its widespread adoption not only among indigenous groups, but arguably among all subsequent rural populations since the seventeenth century (Denevan 1998, 2001).

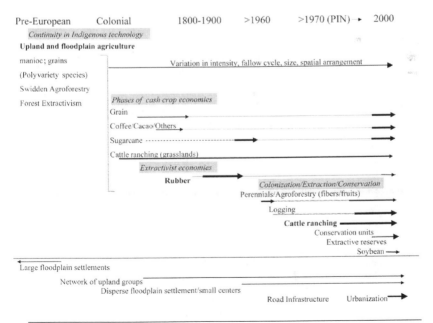

FIGURE 12.1 Historical overview of agrarian and forest use in Amazonia.

Technologies of indigenous origin continue to play a key role within the Amazonian agrarian economy, from the supply of manioc (*Manihot esculenta*) (see, for instance, Albuquerque 1969; Murrieta 2001) and grain staples for rural and urban populations to the dominant açaí (*Euterpe oleracea*) palm agroforestry economy of the Amazon estuary (for a historical review, see Brondízio 2004a). Strategies of swidden agroforestry and resource-management systems are now practiced across indigenous, caboclo, and colonist farming sectors. Continuous diffusion, experimentation, and adoption of land-use technologies underlie variation in land-use practices in different parts of the region today.

The demand for forest products and cash crops intensified during the seventeenth and eighteenth centuries, particularly in the eastern part of Amazonia, which was characterized by Portuguese settlements and missionary villages congregating the remaining indigenous populations. Cacao (*Theobroma cacao*) represents one important example of intensifying production under external demand early in regional history. Although arguments about the level of cultivation versus extractivism are still unanswered, cacao represented an important export for the region during the eighteenth century (R. Anderson 1976; Balée 2003; Santos 1980). Coffee entered Brazil in the seventeenth century through what are now the states of Amapá and Pará. Although modest in production, coffee probably contributed to new forms of perennial agriculture in the region. Also significant to regional land use was the introduction of cattle ranching among Jesuit missions during the seventeenth century and later intensified during the late nineteenth and early twentieth centuries with the introduction of buffalo to Marajó Island. By the mid-1700s, the cattle herd of Marajó Island was estimated at 400,000 (Baena 1969). Sugarcane, already present during the eighteenth century, also experienced a significant cycle of expansion during the nineteenth and twentieth centuries along the floodplains of the estuary with the use of tidal-powered technology to support processing mills. Sugarcane cultivation declined significantly during the latter half of the twentieth century (S. Anderson 1992). Maize was also important among indigenous groups and continues to be among rural populations. Rice has been present at least since the eighteenth century, and, later occupying larger extents of floodplains, it opened the way to mechanization of the floodplains (Baena 1969; Barata 1915; Lima 1956).

It was, however, the rubber economy that definitively shaped land use by articulating merchant capital, land tenure, and social control of labor throughout the Amazon. Rubber attracted new migrants and diverted most of the regional agricultural labor force by shifting attention from the organization of plantation and subsistence agriculture as initially organized during the Directorate period starting after 1750 (see also Cormier, chapter 11, this volume). The arrival of new migrant groups, occupation and claim of forest land, and the opening of colonization settlements represented important changes during this period. After the drastic decline of the rubber economy circa 1910,[3] agricultural systems, in particular cultivation of rice and sugarcane, emerged for specific periods of time. Regional penetration of colonization settlements along the Madeira-Mamoré and the Bragantina rail lines ended because of economic decline and failures in early colonization efforts. The sociocultural, economic, and land-tenure legacy of the so-called rubber boom, however, is still imprinted across large portions of the region and continues to have great influence on contemporary social organization and land uses, particularly along the Amazon River floodplain and penetrating as far upstream as the hinterlands of Acre, Rondônia, Pará, Amapá, and Amazonas states (R. Anderson 1976; Barata 1972; Becker 1997; Cleary 2001; Dean 1987; Derby 1897; Santos 1980; Weinstein 1983).

After the rubber boom, Amazonian land use retracted into small-scale regional systems with a variety of crop combination, episodic forest and wildlife extractivist cycles (e.g., fur and forest essences), and localized expansion of cash crops. Two significant cycles are noteworthy for their role in the diffusion of land-use technology. The first was the growth of the black pepper economy among Japanese migrants in Tomé-Açu County in the state of Pará during the 1940s and 1950s. The intensive cultivation of black pepper production was unprecedented for the region and assumed international importance until its decline due to plant disease and price changes (Tsunoda 1988; Yamada 1999). The decline of black pepper in the Tomé-Açu area prompted the second cycle: the experimentation with and development of agroforestry systems of fruit production based on Amazonian species such as cupuaçu (*Theobroma grandiflorum*), taperebá (*Spondias mombin*), and graviola (*Annona* sp.), as well as on nonnative fruits such as acerola (*Malpighia glabra*). Few developments in agricultural systems have been as significant for the region today as the cultivation of regional fruits for supplying both internal and external markets (Subler and Uhl 1990). The impact of agroforestry cultivation and particularly of the opening of new markets for regional fruit species and products on today's small-scale agriculture throughout the region is paramount (Brondízio 2004a, 2004b). A second land-use economy of regional significance was the jute cycle, in particular that seen from the 1950s to the 1970s along the lower Amazonian floodplains. Influenced by Japanese migrants, jute cultivation expanded to almost all floodplain populations from Gurupá to Manaus (Gentil 1988). In another vein, cattle ranching along the seasonal floodplains of the lower Amazon River grew in importance considerably before post-1970 road construction. In the Santarém region, cattle ranching increased after the opening of the uplands for colonization. Because of limited water supply in large parts of the upland area and seasonal availability of grazing areas in the floodplains, cattle ranching developed strong linkages between the floodplain and upland areas of the region.

As widely documented, the opening of the Belém-Brasília highway in the 1960s, followed by the TransAmazon, Cuiaba-Santarém, and later Porto Velho–Cuiabá and Manaus–Rio Branco highways, underwrote new phases of agricultural expansion and land use in the entire region. Small-scale cash cropping promoted by government agencies, subsidies for perennial agriculture, and subsidized (as well as nonsubsidized) small- and large-scale cattle ranching have been widespread in the region during the past 30 years (Aragón and Mougeot 1986; Hecht 1993; Lená and Oliveira 1992; Mahar 1988; Schmink and Wood 1992). At the same time, colonist farmers have adopted swidden and newer technologies for local production (Browder 1989; Caviglia 1999; Moran 1981; Muchagata 1997; Smith et al. 1996). During the past two decades, logging has expanded considerably to large-scale and extensive exploitation. However, several precedents exist in the region. Localized exploitation has been reported for the Santarém region since the 1930s (C. Sena, curator, Centro Cultural Boanerge Sena Archives, personal communication, July 2000), but the floodplains of the Amazon estuary have been subjected to intensive logging since the 1950s (Anderson, Mousasticoshvily, and Macedo 1993; Barros and Verissimo 1996). Pinedo-Vasquez and colleagues (2001) provide a fascinating account of the boom of the 1950s and the postboom logging in parts of Amapá State. In this area, large-scale selective exploitation of prime wood was followed by management of secondary-value species by farmers and communities targeting the local market. Paralleling agricultural change and the granting of logging and mining concessions, various incipient urban settlements have appeared in the region and been interconnected by a network of planned and unplanned roads and waterways. Large municipalities, previously administered by single urban centers, have been divided into several municipalities, thus creating new forms of institutional and political arrangements underlying land use in the region (Aragón and Mougeot 1986; Browder and Godfrey 1997).

Similar to the "original" highway system (e.g., Programa de Integração Nacional [PIN] I and II, POLONOROESTE), the current opening or reopening of new export routes is defining new forms of land use, agricultural expansion, and regional occupation, such as the case of soybean expansion since the year 2000. At the same time, indigenous, caboclo, and colonist groups have intensified their social and political organization. They have gained access to new forms of resource ownership, such as demarcation of territories, extractivist reserves, and privatization of previously unoccupied areas through agrarian reform. Several analogous processes can be observed, such as the rise of the rubber tappers as a social and political movement since the mid-1970s, the organization of indigenous groups in pan-regional coalitions (e.g., in the Negro River basin, in the Xingu-Iriri basin, and in parts of southern Pará State) (ISA 2000). Also relevant are the regional coalitions of rural unions claiming more incentives and land-tenure rights for small-scale producers (e.g., *grito da terra*)[4] (Tura and Costa 2000). Together, these social groups have created a mosaic of stakeholders varying in political strength; these stakeholders inhabit forest reserves, indigenous areas, and colonization settlements, and thus represent multiple forms of land-tenure and institutional arrangements. These stakeholders are trying to have a voice, a history, together with land claims and visions for land use. On another level, state- and national-level economic-ecological zoning is also lending recognition to different stakeholders, but is not always able to accommodate all groups, so conflict develops.

The northward movement of soybean cultivation from central Brazil through the BR-163 Cuiabá-Santarém highway (and the Cuiabá–Porto Velho highway to the West) to the new Cargill harbor in Santarém marks the definitive presence of large-scale, mechanized, monocultural plantations as a significant land-use system. Although the socioeconomic and environmental impacts are substantial and the rate of rural out-migration high, soybean plantations will in the future likely coexist with rural communities refusing to give up their land and their small-scale production of fruits and other crops, as well as with fishing communities on the floodplain-upland interface. The success of these communities' resistance, however, will depend on local and regional political and economic forces, such as their own internal social organization, access to similar economic incentives as provided to large-scale farmers, and access to market and technologies. Conservation areas, indigenous reserves, logging areas, and unopened forest reserves also coexist with these diverse forms of land use, thus further increasing the spatial complexity of land-use systems even across short distances (Brondízio 2005; Nepstad et al. 2002; Verissimo, Cochrane, and Souza 2002).

In contrast to the Atlantic Forest of Brazil, the scale of deforestation resulting from "cycles" of land use since the seventeenth century in Amazonia was limited until recently (figure 12.2). Cattle ranching has taken advantage of grassland areas in Marajó Island, and rubber extraction was based on native stands. Because widespread swidden agriculture existed under small population density, the proportion of forest to nonforest areas has remained high. The only significant deforestation before 1970 occurred during successive years of colonization of the Bragantina region (east of Belém) starting at the turn of the twentieth century as migrants were settled with the goal of increasing agricultural production to supply the urban population of a growing rubber economy (Penteado 1967). Deforestation in the Bragantina region before 1970, however, pales in comparison to rates documented since the 1970s and exacerbated during the 1990s and 2000s by lower inflation and the economic incentives of governmental development programs, such as Avança Brasil. In this context, the temporal and spatial articulation of a long history of regional occupation creates differential conditions upon which land-use change takes place within and across regions today.

This summary highlights elements relevant to the understanding of processes and units of analysis that account for regional variations in land use. Historically, variability within the region reflects the arrival and emergence of social groups, colonization policies, and external market demand, as well as the opening of new access routes (waterways and roads) and the organization of land

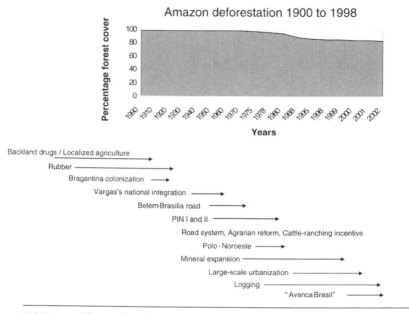

FIGURE 12.2 Historical outline of political and economic events and deforestation in the Amazon.

tenure and control of resources. The implications for research design in historical ecology and analysis of land use are clear because these factors are relevant in defining units of analysis necessary for capturing different processes underlying current transformations of regional landscapes (for prehistoric transformations, see Erickson and Balée, chapter 7; Erickson, chapter 8; Neves and Peterson, chapter 9; and Heckenberger, chapter 10, this volume).

THEORETICAL AND CONCEPTUAL MODELS EXPLAINING SETTLEMENT, LAND USE, AND LANDSCAPE FORMATION

The study of land use has provided themes that integrate the social and environmental sciences for a number of reasons, but particularly because it seeks to understand spatial and temporal dimensions of human behavior related to environmental and socioeconomic problems of local and global interest.[5] In this context, *land use* refers to the purposes and intent of human activities that directly affect and are affected by the biophysical environment (LUCC 1994). Although long recognized as a relevant subject within anthropology, geography, agronomy, ecology, and other fields, land use emerged as a distinctive field of research during the 1990s. International programs such as Land-Use and Cover

Change (LUCC) and Human Dimensions of Global Environmental Change developed by the International Geosphere-Biosphere Program (IGBP) have contributed to consolidating land use as a research field focusing on bridging micro- and macroscales of analysis. In the Amazon, where a long history of land-use studies in the social sciences exists, contemporary research on local and regional land-use change has become one component of the Brazilian-led international program Large Scale Biosphere-Atmosphere Experiment in Amazonia (LBA). In conjunction with theoretical developments in historical ecology, land-use studies harbor a heuristic framework for those interested in articulating factors affecting and mediating both micro- and macrodimensions of social and environmental change.

The goal of most contemporary analysis of land-use change is to understand factors underlying variation in rate, extent, and direction of environmental change. In particular, such analysis attempts to understand the relationship between the so-called underlying causes ("the initial conditions and fundamental forces that underpin human action towards the environment" [Geist and Lambim 2001:8]) and the proximate sources ("direct human activities affecting the biophysical environment" [Geist and Lambin 2001:5–16; also see LUCC 1994, 1999; Turner, Meyer, and Skole 1994]). In most cases, underlying causes include broadly defined demographic, economic, technological, political, institutional, and sociocultural variables, whereas proximate sources refer to the set of transformation activities broadly defined as agricultural expansion, logging, and infrastructure development (Geist and Lambin 2001). One may argue that in Amazonia, as in most other places, the understanding of factors mediating the interaction between underlying conditions and proximate causes requires attention to historical events and economic cycles, which have influenced patterns of land tenure and resource ownership, demographic changes, and the social and political institutions emerging at different periods and places.

Land-use studies in the Amazon have been approached from different theoretical and methodological perspectives depending on the type of question and scale of analysis. Independent of perspective, there has been a continuous concern with integrating the mosaic of microrealities and local conditions with regional patterns of human-environment interactions. Several conceptual models correlating population dynamics and forms of land use have paid attention to the roles of settlement location, soil fertility, population density, management technology, land tenure, environmental conditions, and market and institutional incentives (see Balée 1989; Brondízio, Safar, and Siqueira 2002; Brondízio et al. 2002; Carneiro 1961; Coomes and Burt 1997; Denevan 1996; Denevan and Padoch 1987; Fisher 2000; Futemma and Brondízio 2003; Hiraoka 1985, 1994a; Lathrap 1970; Moran 1981; Muñiz-Miret et al. 1996; Padoch et al. 1985; Posey and Balée 1989; Roosevelt 1989; WinklerPrins 2002a, 2002b). The study of peasant (e.g., rural caboclos and colonists) economy has also contributed significantly to understanding Amazonian land use. It includes attention to the participation of small farmers and agricultural workers in commodity production and economic cycles (historical and contemporary), migration and settlement, and land tenure and labor arrangements (Brondízio and Siqueira 1997; Bunker 1985; Chibnik 1994; Nugent 1993; Sawyer 1986; Schmink and Wood 1992; Tura and Costa 2000). No less important has been the role of ethnobotanical studies to local land use analysis (Balick 1988; Nepstad and Schwartzman 1992; Prance and Kallunki 1984).

The bulk of land-use studies over the past decade has focused on measuring and modeling variables to explain deforestation, ranging from the microscale (such as a farm lot) to the macroscale (such as the Brazilian Amazon). Emphasis has been placed on variables of demography; political economy; political institutions and infrastructure (McCracken et al. 1999; Pichón and Bilsborrow 1992; Wood and Skole 1998); colonization programs; fiscal incentives and inflation (Mahar 1988; Ozório Almeida 1972); and disarticulated urbanization (Browder and Godfrey 1997), among others. Attention to infrastructure and the spatial articulation of colonization areas, in particular access to roads, has also been important, even central, to analysis and formulation of predictive models of land-use change (Alves 2002; Batistela, Robeson, and Moran 2003; Laurence et al. 2001; Walker, Moran, and Anselin 2000). In general, however, prognostic and causal models of land-use change have paid insufficient attention to intraregional variability in conditions that seem to be underlying land-use change (Brondízio 2005). Theoretical models supporting these analyses have included, for example, central-place and Von Thünen theories, particularly with the growing importance of urbanization in the region (Browder and Godfrey 1997). Implicitly or explicitly, Boserupian models of land use intensification have been used to look at rates of regrowth, fallow cycle, and crop frequency in relation to population size, labor, technology, and land circumscription (Brondízio and Siqueira 1997; Scatena et al. 1996). In colonization areas, Chayanovian and other models of household cycles, labor arrangements, and land use have also become increasingly used in studies focused on farm-level dynamics (Marquette 1998; McCracken et al. 2002).

There are parallels between theoretical perspectives used within the literature aiming to explain contemporary causes of land-use change[6] and that used to explain long-term human-environment interactions in the Amazon. Historical ecology—with its roots in archaeology, history, anthropology, ecology, and geography—contributes various models to explain linkages between patterns of indigenous settlement, migration and demographic change, and environmental management in different parts of the region (Balée 1994, 1998; Erickson 2000; Heckenberger et al. 2003; E. Neves 1998). To some extent, historical ecology originated out of the necessity to refute models based on environmental determinism, to engage in long timescales, and to provide cultural context for the study of human-environment interactions. These same problems are also faced in the study of land use in the region today and tend to be dominated by simple and deterministic causality (for example, between road construction and consequent deforestation) and simplified solutions (such as standardized credit programs). Bringing a historical ecological approach to the study of contemporary land use will certainly lead to more emphasis on the role of long-term, processual, spatial, and temporal dimensions of human-environment interactions in the region and help account for the past and present diversity of human experiences and forms of environmental management.

UNITS OF ANALYSIS IN STUDYING HUMAN-ENVIRONMENT INTERACTIONS IN THE AMAZON

As the previous discussion suggests, few aspects are as relevant to the study of human-environment interactions and land-use change as defining units of analysis in space and time. Across disciplines as diverse as anthropology, geography, and ecology, the notion of a unit of analysis has evolved through a long history of definitions and forms of conceptualizing boundaries within which to envision the relationship between human populations and the environment. These concepts include cultural area, community, population, household, niche, ecosystem, landscape, and biome (Geertz 1963; Golley 1992; Hardesty 1975; Kroeber 1939; Moran 1990; Odum 1971; Steward 1946–50, 1955, 1956; Turner, Meyer, and Skole 1994; Vayda and McCay 1975; Vayda and Rappaport 1968).

There is no single way of predefining units of analysis in land-use studies because the exercise in question depends on, among other things, the type of question, data available, time frame for analysis, and the unit of observation within which the data are collected. During the past decade, the emergence and common use of remote sensing data have yielded various levels of spatial and temporal coverage that have contributed to better definition of spatial boundaries and units of analysis. GIS supplies tools for manipulation and querying of spatial data. Global positioning systems furnish location-specific information. Depending on the type of question and level of analysis, one may find a range of possibilities in organizing and nesting spatial units of different categories—for instance, political, institutional, biophysical, sociocultural, demographic, and contextual units (Behrens, Baksh, and Mothes 1994; Evans and Moran 2002; Fox et al. 2003; Liverman et al. 1998; McConnell 2001; McCracken, Boucek, and Moran 2002; Moran and Brondízio 2001; Wood and Porro 2002).[7]

Accounting for the underlying factors that influence land-use change requires integrating field approaches to remote sensing and GIS, including ethnography, survey, and archival work. Rare are "land users" who do not rely on multiple resource and economic strategies, diverse markets, varied forms of resource ownership (private, communal, government), and multiple labor arrangements. For instance, colonization settlements where family and entrepreneurs arrive from different places and at different times often evolve into complex, nonclustered social and economic networks shaped by historical experiences particular to a given context. In regions where occupation dates to the Colonial period, these arrangements are even more complex and embedded in different forms of power relation and class. Throughout the Amazon estuary, for instance, one finds a range of sharecropping arrangements sometimes spanning multiple generations. But a whole sharecropping community may also be the neighbor of a highly organized, cooperative-based community. Although performing different roles and carrying different political and social agendas, middlemen and merchant capital as well as churches, banks, nongovernmental organizations, and development projects often coexist in influencing stakeholders' land use within the same region. Furthermore, intraregional variation among land users may include previous conflicts and power relations underlying local land-tenure arrangements, ascendancy and ethnic differences, varied experience and knowledge of forest resources, cultural preferences in resource use and consumption, and the evolution of social organization and political leadership as well as institutional arrangements underlying local norms, rules, and sanctions regarding land use. Therefore, in moving from the local to a regional scale, where land-use analysis and prognostic models are often developed, the intersection of new colonization areas, older rural communities, urban centers, and indigenous reserves renders focus on the historical context of social and economic differences ever more relevant.

The case studies presented in the next section illustrate this point. These cases illustrate the importance, during research design and sampling, of paying attention to historical processes such as settlement formation, evolving land-tenure arrangement, differences in access to resources among users, and past impacts of environmental management on soil and vegetation, as well as the need for contextualizing the analysis of land-use change observed in a given region today. Even more important, attention to these processes may help to avoid modeling exercises that assume similar conditions and land-use behavior over large socially and environmentally diverse areas. At any level one selects to analyze land-use change, emphasis on the historical differences underlying management of resources may help to avoid unnecessary and flawed generalizations that decontextualize causal relationships between land users and land-use systems.

INTERREGIONAL AND INTRAREGIONAL VARIATION IN LAND-USE TRAJECTORIES: IMPLICATIONS FOR THE FORMATION OF CULTURAL LANDSCAPES

I present three examples here to illustrate intraregional variations in land-use systems and in rates of land-cover change resulting from differences in historical and contemporary processes. I selected these examples to represent variation in scale (from larger to smaller areas); types of occupation and social groups (indigenous peoples, caboclos, small-holder colonists, and large farmers); types of land-tenure arrangements (government, communal, sharecropping, and private systems); and forms of access (roads and riverways) at different distances from regional urban centers.

HISTORICAL STRATIFICATION OF COMPLEX AND LARGE REGIONS: FROM THE XINGU RIVER TO THE TAPAJÓS RIVER

An example of regional complexity is illustrated in figure 12.3. The figure presents a satellite image of the region covering the Xingu River in the east to the Tapajós River in the west. On the southern part, one can follow the TransAmazon highway from Altamira to Itaituba and from its intersection (city of Rurópolis) with the Cuiabá-Santarém highway to the city of Santarém at the confluence of the Tapajós and Amazon rivers (ACT 2003). Many different regional realities and historical depth of regional occupation are represented in this image, illustrating the issues discussed here. In simple terms, the region can be divided into at least two parts based on two highway systems: the Santarém-Belterra-Aveiros region to the northwest and the Altamira region to the east. A more comprehensive analysis would also account for occupation along the Xingu, Tapajós, and Amazon rivers, which differ substantially from the two highway areas.

The region of Santarém-Belterra along the BR-163 highway (Cuiabá-Santarém) illustrates well the complexity of land-use conditions found in an area currently undergoing significant changes with the recent arrival of large-scale soybean cultivation. A site recognized for its large pre-European chiefdoms, Santarém has been subjected to successive colonization and occupation for the past 300 years, but more intensively for the past 150 years, including cacao (ca. 1800–1860), rubber exploitation and commerce (ca. 1850–1910), early logging corridors (1930s), jute cycle (1940s–1980s), migration to rural areas (1960s), agrarian reform (since the 1970s), land grants and rural out-migration to cities (1970s and 1980s), creation of conservation and extractivist reserves (1980s–1990s), and more recently soybean expansion (2000–) and new indigenous areas (2003).[8] Along the Tapajos River, Belterra and Fordlândia were created after 1928 as part of rubber plantations and settlements established by the Ford Motor Company. As much as 8,000 hectares

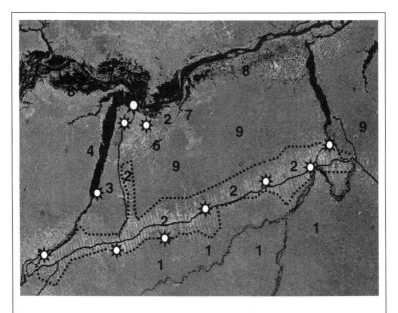

1. Indigenous reserves
2. Colonization settlements (>1970s)
3. Flona Tapajos ("traditional" and migrant communities, logging concessions)
4. Extractive reserve ("traditional" communities)
5. Various rural settlements (>1930s), soybean expansion zone (>2000)
6. Recent farm expansion (>1990s)
7. Government reserve/research area
8. Floodplain communities, lake-management areas, cattle ranching
9. Logging and farm expansion zones

 Urban centers and towns

FIGURE 12.3 Regional complexity in the Xingu-Tapajós region. This map is only illustrative; it does not include all conservation units and roads, and it includes only settlement areas shown in figures 12.4 and 12.5.

of forest were cleared with this purpose. Following uneven success, the area was later transferred to the Brazilian government (ca. 1945) as "land of the union." In the late 1990s, as Belterra became a municipality, part of the old settlement was subdivided and distributed as land grants (around 16 hectares) to old and new residents. Buildings, water utilities, and old patches of rubber trees persist and remind one of the area's past. Still, most of the county remains "land of the union" despite its de facto occupation and use by county administration and residents.

The remaining part of Belterra falls mostly within the boundaries of Flona-Tapajos, where more than rural communities exist. Defining the county as a study area—for instance, to analyze deforestation processes—lumps together a variety of groups and processes, such as colonization areas settled at different time periods. However, phases of regional occupation correspond roughly to the opening of key access routes, the immigration and migration of different social groups, and the organization of different land-tenure arrangements. These phases of occupation include, for instance, the settlement of riverine populations along the Tapajós and Amazon rivers (seventeenth to the twentieth centuries), earlier migrants occupying the Curuá-Una and Mojuí dos Campos roads (early twentieth century), the creation of a national forest encompassing dozen of communities along the Tapajós River (1974–), the creation of the Ford Company rubber plantation (1930–), settlement of colonists and large ranchers along the Cuiabá-Santarém highway (1972–), the formation of Belterra County (1997–), and the current arrival of newcomer soybean farmers.

In the example presented here, the region under study has been stratified according to subregions representing phases of historical occupations (including road construction), arrival of social groups, and dominant forms of land use—all based on archival, ethnographic, and remote sensing research,. Figure 12.4 illustrates an example of stratifying the region according to areas representing different periods and forms of occupation (ACT 2000–2003).[9] Stratifying a region according to historical occupation permits flexibility and robustness in sampling for land use and stakeholders, as well as in accounting for demographic, sociocultural, economic, and environmental factors affecting land-cover change in the here and now. Although histories are not units of analysis, the institutions, social groups, and forms of resource ownership and use created through time can be just that. Accounting for variations in land use may help to avoid comparing deforestation rates across regions occupied during different times and in different settlement areas and farm lots of different age and undergoing different stages of occupation.

In contrast to Santarém, Altamira to the east represents a county that has undergone significant colonization and settlement only since the 1970s, despite its earlier history as a riverine settlement early in the twentieth century and as an area important to indigenous communities. In both cases, the temporal depth is a defining factor if one wants to understand variation in deforestation trajectories. The TransAmazon highway west of the town of Altamira was one of most important foci of the Brazilian government colonization program during the 1970s. Altamira grew from a small riverine town based on rubber collection into a booming town of 85,000 due to agropastoral production stimulated by the highway and subsequent colonization along it. Several counties and planned agrarian villages (*agrovilas*) were created along the highway and its feeder roads. Some planned agrarian villages disappeared, while others formed; large numbers of lots have been

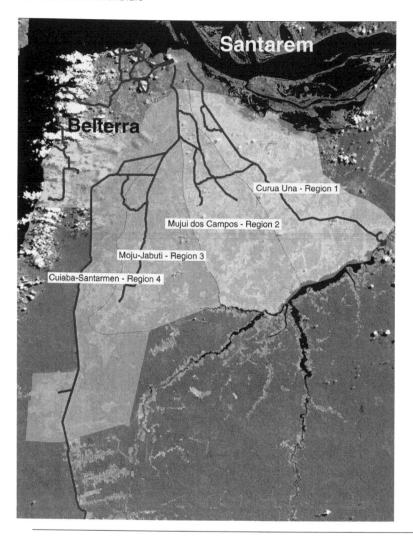

FIGURE 12.4 Example of intraregional stratification representing different periods and forms of occupation. Research example of the Anthropological Center for Training and Research on Global Environmental Change, Indiana University (ACT field data 2000–2003). Floodplain areas not represented. (Map prepared by Scott Hetrick of ACT)

successively sold. Although in the Santarém-Belterra area, which is marked by long-term occupation, subregions indeed can be stratified according to different occupation "routes" (both by roads and rivers) as well as by different social groups (caboclos, immigrants), the recent history of occupation and structure of settlement in the TransAmazon region renders colonization settlements and cohorts of farm lots a suitable choice (Brondízio et al. 2002; McCracken et al. 1999). Such an arrangement is shown in figure 12.5, a group of approximately 3,800 farm lots arranged according to adjacent settlement projects implemented by the National Agrarian Reform Institute (INCRA) during the past 30 years, which cuts across the counties of Altamira and newly created Brasil Novo and Medicilândia, in the state of Pará (Fearnside 1986; Moran 1981; Smith 1982).

Remote sensing data collected at different times since the 1970s (aerial photography, Landsat MSS, and Landsat TM) have allowed researchers to reconstruct the history of the colonization of the study area (figure 12.5). In this way, farm lots can be classified by the duration of occupation and thus organized into cohorts of arrival time (i.e., groups of farms being established during given periods of time) in the region. Demographic concepts may be particularly helpful in accomplishing this task. Such concepts include *period effects,* such as fluctuations in migration, different credit policies, and inflation; *cohort effects,* associated with the arrival and occupation of farm lots by groups of families; and *age effects,* related to the transformations of households and their farms over time (Brondízio et al. 2002; McCracken, Boucek, and Moran 2002; McCracken et al. 1999; McCracken et al. 2002; Moran, Brondízio, and McCracken 2002; Siqueira et al. 2003). Figure 12.5 shows the history of regional occupation and deforestation since 1970 (*top*), which was used to stratify farm lots (*bottom*) according to cohorts of arrival.

Along this stretch of the TransAmazon highway, deforestation rates vary according to age of settlement, location in the region, and environmental characteristics of the lot (Brondízio et al. 2002; McCracken, Boucek, and Moran 2002; McCracken et al. 1999; McCracken et al. 2002; Moran, Brondízio, and McCracken 2002). As in most colonization areas, "old settlers" coexist with new ones, the latter being recent migrants or second-generation colonists taking over new lots or acquiring them from previous colonists. Although land-use trajectories are shaped by economic and social conditions and by the local environment, understanding them within and across farm lots also requires the researcher to note the time of colonists' arrival, their age of occupation, and the type of settlement being examined. This knowledge is necessary for untangling different types of temporal and spatial factors underlying land-use change.

This temporally and spatially sensitive strategy allows understanding and comparison of deforestation rates and land-use trajectories at the level of farm lot, cohort of farm lots, and entire settlements. The rate and extent of deforestation vary significantly when calculated at each spatial level and when units of analysis are stratified according to their history of occupation (time in the area). My colleagues and I found that though a significant, positive correlation between rate of deforestation and age of settlement obtains, this correlation is conditioned by regional period effects, such as changes in economic, institutional, and infrastructural conditions that motivated or inhibited particular land-use behaviors. We also found that deforestation rates and land-use trajectories

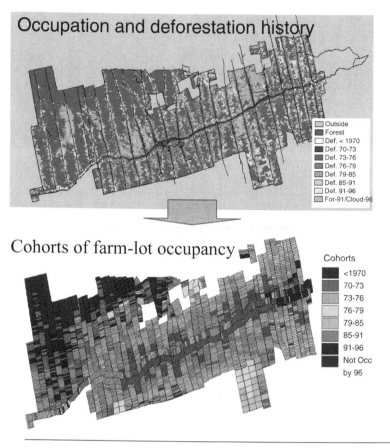

FIGURE 12.5 Reconstitution of occupation/deforestation history and farm lot by cohort of arrival (1970–1996), TransAmazon hghway from Altamira to Medicilândia. Mapped deforestation sequences from 1970 to 1996. (Aerial photographs, Landsat MSS and TM)

are to some extent age dependent in colonization lots, but not necessarily so in indigenous and rural communities predating colonization during and after the 1970s. We have observed deforestation rates characterized by phases of expansion (clearing) and consolidation of colonists' farm lots. However, the magnitude of deforestation during each of these phases varies by cohort (time of arrival and age of occupation) as a function of period effects (regional events) affecting different groups of farmers arriving at different times. Thus, groups of families occupying the region during the past 30 years (the TransAmazon highway and parts of the Cuiabá-Santarém highway) or the past 70 years (Santarém) have encountered different levels of incentives and constraints (period effects) under which to make land-use decisions (see Brondízio et al. 2002; Futemma and Brondízio 2003 for detailed discussion; McCracken, Boucek, and Moran 2002; McCracken et al. 1999, 2002; Moran, Brondízio, and McCracken 2002; Siqueira et al. 2003). Neighboring riverine communities (e.g., along the Tapajós River), floodplain communities (e.g., along the Amazon River), and upland indigenous communities that reflect long histories of land occupation, distinct kinds of social organization (compared to colonist families), and participation in the regional economy present different and lower deforestation rates and different spatial patterns. A far more complex social and environmental scenario exists in this large region, to be sure, but this illustration instantiates the value added and the importance of accounting for ever-increasing regional complexity and historical depth of occupation, both being concerns of historical ecology.

VARIATION IN LAND-USE TRAJECTORIES IN THE AMAZON ESTUARY

The region of Ponta de Pedras offers a microcosm in which to capture land-use changes taking place during the past 30 years in the larger Amazon estuary. Urban population growth and market demand in the nearby state capital of Belém, local development projects and subsidies, and a diversity of plant-utilization forms and local land-use strategies create a constantly evolving landscape. This study area is located in the estuarine region of the Amazon River, on Marajó Island, and in Ponta de Pedras County, Pará (Brondízio 1999; Brondízio et al. 1994, 1996). Occupation dating from the seventeenth century includes *sesmarias* land grants (land concessions by the Portuguese Crown) to individuals and religious missions as well as Directorate land policies (circa 1750) (see Cormier, chapter 11, this volume). The region's current pattern of riverine settlement of dispersed individual households dates back at least to the rubber boom cycle (mid-1800s to 1910). During the past 30 years, development projects, government incentives, and strong market demand for locally produced food products, the açaí palm fruit (*Euterpe oleracea* Mart.) in particular, have resulted in incentives and opportunities for estuarine farmers and communities to intensify their land use in floodplain areas. However, local producers and communities' differential responses to regional market demand and development projects have resulted in diverse land-use systems, even within short distances.

Both household and community are important levels of social organization in this region. Factors affecting these trajectories include historical conditions defining the location of settlement and land tenure, participation in development projects during the past 30 years, and variation in forms of social organization of local communities such as cooperative and sharecropping arrangements. Six rural communities are analyzed in this example (figure 12.6); three are located in upland areas, and three are located adjacent to the floodplains. Their settlement patterns and locations result from land inheritance and sharecropping systems dating back to colonial times (the floodplain), as well as from more recent Catholic Church–based land acquisition and distribution to communities since the 1960s (the uplands).

Population growth in urban areas has created markets for regionally preferred food sources such as the açaí fruit, which is a key regional staple consumed by rural migrants living in urban centers. Impressive intensification of açaí production followed increased market demand, thereby changing the regional economic profile. Market demand, however, has had differential influence on land users' decisions with regard to intensification depending on several factors affecting production, in particular land tenure and availability of floodplain areas, dependency on middlemen, and access to market centers (Brondízio 2004a, 2004b; Brondízio, Safar, and Siquiera 2002; Brondízio and Siqueira 1997).

Analyzing the expansion of açaí agroforestry at the current time helps, in a broader context, to conceptualize declining deforestation rates, widespread

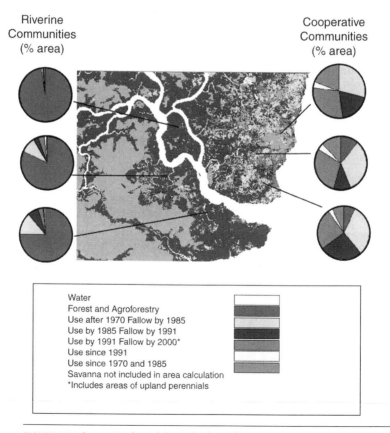

Riverine
Communities
(% area)

Cooperative
Communities
(% area)

Water
Forest and Agroforestry
Use after 1970 Fallow by 1985
Use by 1985 Fallow by 1991
Use by 1991 Fallow by 2000*
Use since 1991
Use since 1970 and 1985
Savanna not included in area calculation
*Includes areas of upland perennials

FIGURE 12.6 Intraregional variability in land-use allocation and trajectories 1970–2000: Ponta de Pedras region, Marajó Island, Pará State, Brazil.

formation of secondary vegetation, and relative decline in swidden agriculture even among riverine communities. Remote sensing data from 1969 on show that coinciding with the growth of açaí agroforestry, there has been virtually no deforestation in the area since the mid-1980s, in sharp contrast to the deforestation of other areas of the Amazon basin in Brazil. As illustrated in figure 12.6, a variety of land-use patterns emerges from the differential spread of market incentives as producers cope with land tenure, availability of resources, and access to external incentives, such as credit (Brondízio 2004a).

Analysis of land-cover change since the 1970s reveals different strategies of land-use intensification and extensification among communities (figure 12.6). The different proportions of forest cover, cropland allocation, and secondary vegetation (representing differing lengths of fallow cycles) reflect the variation in dissimilar contemporary and historical conditions within which these communities and households operate. These conditions include availability of resources (e.g., floodplain forest areas), land ownership, subsidies from development projects, and agricultural technology. Riverine communities illustrated in figure 12.6 offer a view of land allocation and fallow-cycle management based on a combination of agroforestry land use and small-scale swidden agriculture. In these communities, forest cover tends to correspond to more than 75 percent of the area. Older secondary vegetation, fallow for 15 to 50 years, tends to occur in larger amounts when compared to younger vegetation. Families tend to maintain up to 2 hectares in production yearly, utilizing a cycle of rotation that depends on the availability of particular types of soil, secondary vegetation, and household needs. This scenario differs radically in upland communities that underwent cooperative development projects. Forests cleared to allocate areas for mechanized cultivation of annual crops, pasture, and coconuts during the 1970s and 1980s can today be encountered in several stages of regrowth. Contrary to land cover in riverine communities, secondary vegetation regrowth of different ages (up to 20 years) tends to represent at least 75 percent of the area of these formerly cleared areas. Several communal projects implemented during the 1970s and 1980s were not successful in the long run. Reasons vary widely, including the failure of cattle and pasture management. For instance, many pastures could not compete with secondary plant species and were therefore deemed to be inadequate as cattle fodder, thus leading to stakeholders' abandonment of the pastures. As a result, one can observe large areas of pasture abandoned by the late 1980s after numerous attempts to manage and weed out secondary growth. During the 1990s, these communities were cultivating smaller areas using mechanization and implementing agroforestry projects in floodplain and upland areas, while letting most of the formerly forested areas recover fully to forest.

In summary, regional history has created variations in land tenure, social organization, and access to resources, markets, and infrastructure influencing land-use change today across different rural communities located at similar distances to urban areas and sharing in a similar regional context.

TUKANOAN SETTLEMENT HISTORY, SOIL DISTRIBUTION, AND SPATIAL DIMENSIONS OF LAND USE

Understanding the relationship between settlement distribution and the spatial pattern of deforestation is central to many land-use studies. Despite the relative absence of factors common today in many other Amazonian areas, such as urban markets, road access, and colonist populations, the Tukanoan community exemplifies variation in the spatial pattern of land use between two neighboring villages of the northwest Amazon. The areas of Tukanoan-speaking populations are located on the Vaupés basin between Colombia and Brazil in the northwest Amazon. Settlement pattern has been influenced by historical variations in regional migration and missionary occupation, leading to nucleation of the population in village centers (Castro et al 2002; Wilson 1997; Wilson and Dufour 2002). The area is composed of large patches of nutrient-poor Spodosols covered by Amazon "caatinga" (scrubland) intermixed with stretches of Oxisols covered by upland forest. A manioc-based swidden agricultural system characterized by long-fallow cycles is the dominant land-use system in these communities. In this context, land-use choices are closely related to access to appropriate soils (Oxisols), but also influenced by historical events defining the location of each village. Interactions of historical and environmental factors are of paramount importance for understanding the Tukanoans' land use in the present as well as in the recent past. The two neighboring Tukanoan villages are of a relatively similar size (figure 12.7).

The research problem presented here is illustrative of what explains different spatial patterns of land use and land cover between neighboring villages that are otherwise characterized by similar sociocultural conditions and agricultural systems. Although part of the explanation rests on the fact that Community B is surrounded by Spodosols and Community A by areas of Oxisols, their respective locations derive from differences in each settlement's history. In contrast to Community A, Community B moved to this particular site through missionary incentives in previous decades (Castro et al. 2002; Wilson 1997; Wilson and Dufour 2002). In brief, the history of settlement and availability of better agricultural soils has made it possible for Community A to minimize the distance traveled to gardens (farm lots) by opening garden areas near the village center, whereas Community B has to seek appropriate soils by accessing areas via waterways. Consequently, even in the absence of roads, development projects, and private land-tenure arrangements, the same land-use system (i.e., long-fallow, manioc-based swidden agriculture) effects different spatial patterns of land cover (Castro et al. 2002).

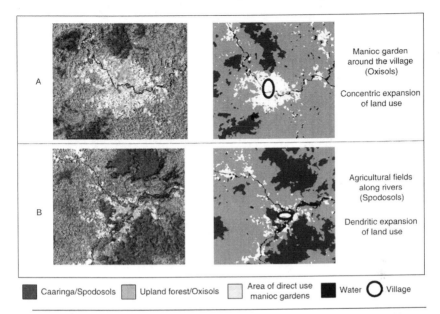

Caaringa/Spodosols   Upland forest/Oxisols   Area of direct use manioc gardens   Water   Village

FIGURE 12.7 Biophysical and access factors influencing land-use patterns between neighboring Tukanoan villages (Community A, *top*, and Community B, *bottom*). Adapted from Castro and Brondízio 2000.

Although land-cover patterning may have been impacted by the distribution of soils around these two villages, it is actually the historical fact of the arrival of missionaries, who for their part encouraged village nucleation and relocation, that in the final analysis underlies contemporary differences in spatial patterns found in these two communities. Methodological lessons taken from this case enrich land-use analysis and explanatory models of land-use change, such as those originating in the historical ecology research program. The amount of land-cover change between the two villages may be similar (e.g., percentage of cleared forest and secondary vegetation), but their spatial patterns differ strongly. This difference suggests that intraregional variation in land use and land-cover change is not restricted to recently colonized areas, but is present throughout the Amazon basin, even in areas such as indigenous reserves where land use is considered to be relatively homogeneous. These results represent the importance of accounting for historical factors and intraregional variations in environmental conditions when interpreting regional patterns of deforestation and when designing explanatory, predictive models of land use and land-cover change.

DISCUSSION

The growing complexity in the forms of human-environment interaction in the Amazon region is expressed in the coexistence of diverse social groups and land-use systems across the region. A challenge for land-use analysis and policy is to take into account local factors influencing human land-use behavior, the materialization of these factors at larger spatial scales, and macrolevel political economic forces underlying these processes. Integrating historical, ethnographic, and spatial analysis offers an opportunity to address these issues while promoting theoretical and methodological cross-fertilization between historical ecology and land-use studies in the study of human-environment interactions in the region. Potential research topics include more attention to historical trajectories of forest-cover change and variability in land-use systems within the region.

HISTORICAL TRAJECTORIES OF FOREST-COVER CHANGE

Contemporary land-use change in the region does not occur in a vacuum; land use is history written onto the landscape (Erickson and Balée, chapter 7, this volume). Multiple "historical depths" coexist in different parts of the Amazon basin—from pre-Columbian and colonial times to recent colonization settlements. The region carries the footprints of successive phases of sociocultural and economic change, territorial occupation and agrarian history, and commodity markets of forest resources. These processes of change have created different forms of access to resources and land tenure, different forms of social organization, and different land-use technologies.

Different models explaining long-term use of forest resources in the region have generally emphasized temporal phases based on successive stages of decline and rebound of forest resources coinciding with expansion and retraction of human activities. This emphasis is reflected, for instance, in discussions of the process leading to the formation of anthropogenic forests, of the trajectories of extractivist economic cycles, and of the impact of economic development in the region today (figure 12.8A). The literature on the sociopolitical formation of pre-Columbian Amazonian populations has debated the extent to which human management practices and forms of organization have been used to overcome the region's environmental limitations (e.g., low soil fertility, sparse concentration of resources) (Carneiro 1961, 1995; Denevan 2001; Lathrap 1970; Meggers 1971; W. Neves 1989; Roosevelt 1989). Balée's seminal paper "The Culture of Amazonian Forests" (1989) represents the most significant paradigm shift in this discussion. It argues that a significant portion of the Amazon forest results from different forms of human land uses, including species concentration, fire management, and large-scale cultivation resulting from dense and widespread populations in parts of the region. Concentration and management of environmental resources allowed for an increase in sedentary populations and for the development of political complexity and multitier settlement patterns. Anthropogenic forests in this case result from successive uses and transformations of forest resources, followed by population decline or migration (as after the European conquest) and by subsequent rebounds of forest cover. Heckenberger and colleagues' (2003) archaeological findings in the upper Xingu River (see also Brondízio 2003; Heckenberger, chapter 10, this volume), showing the articulation of multiple settlements associated with large-scale spread of anthropogenic forests, and previous ethnographic evidence of forest management (e.g., see Anderson and Posey 1989; Brondízio and Siqueira 1997; Denevan and Padoch 1987) seem to corroborate this model (figure 12.8A).

Studies of the region's extractivist economies (e.g., Homma 1993) have suggested that forest resource use goes through phases of large-scale extensive exploitation followed by periods of decline, substitution, and abandonment. Forest resources, depending on market and merchant capital, are exploited to a degree close to exhaustion or to its limited productive capacity until a market decline or a substitute product winds up shifting attention to other areas or resources. Retraction of land users and reduction in exploitation are then followed by eventual regeneration of the resource. However, these trajectories vary according to the type of resource. Although some resources continue to be exploited extensively and continuously (for example, the Brazil nut [*Bertholettia excelsa*] and the babaçu palm [*Attalea speciosa* = *Orbygnia phalerata*]), others became cultivated (such as the açaí palm) (see Brondízio and Siqueira 1997; Clement, chapter 6, this volume). In short, different forms, trajectories, and levels of intensity of forest extractivism and management tend to coexist in the region today (figure 12.8A–C).

A third model discussing the trajectory of forest resources in the region relates to the spread of certain economic development policies during the past four decades. Several explanations for the spread of human occupation and consequent deforestation as a result of this process have been proposed for parts of the region. The region is experiencing a rapid decline in forest due to deforestation resulting from a perception of endless availability of resources, low land value, incentives for forest clearing, and agrarian development policies (Nepstad and Uhl 2000). Although we still do not know the long-term outcome of these processes, national and regional economic policies have used the argument that a period of regional occupation based on high rates of deforestation will lead to economic development, increase in land value, and intensification of land use, which in combination with conservation measures may eventually decrease pressure on remaining forests and potentially allow regeneration of abandoned areas (figure 12.8A). Regional models frequently used as examples to support economic development arguments include the "greening" of parts of the world ranging from Europe and the eastern United States to the Atlantic Coastal

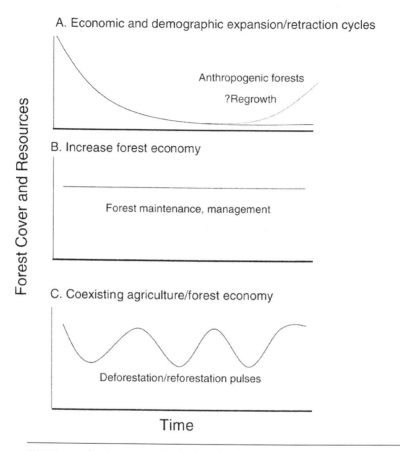

A. Economic and demographic expansion/retraction cycles

Anthropogenic forests

?Regrowth

Forest Cover and Resources

B. Increase forest economy

Forest maintenance, management

C. Coexisting agriculture/forest economy

Deforestation/reforestation pulses

Time

FIGURE 12.8 Coexisting intraregional trajectories of forest cover and resources.

Forest of Brazil (Dean 1996). In all cases, economic development follows the retraction of agricultural frontiers and the increase in industrialization and agricultural intensification. Different arguments explaining these trends occur in academic articles, newspapers, and political discourse (Angelsen and Kaimovitz 2001; Laurence et al. 2001; Nepstad and Uhl 2000; Nepstad et al. 2002; Silveira 2001; Verissimo, Cochrane, and Souza 2002; Wood and Porro 2002). A good example of the political power of the economic development argument is illustrated by the ongoing political debate over the expansion of corporate soybean cultivation into the Amazon (see Sant'ana 2003 and Rohter 2003).

Although it is not my goal here to discuss the explanatory value of these models, they help to elucidate changes in forest cover associated with different historical periods of the region. I have argued that contemporary land use in the region is marked by a high level of intraregional variability resulting from the intersection of historical and contemporary regional occupation, arrival and transformation of social groups, and cumulative forms of resource ownership and land tenure. Consequently, multiple historical conditions and forms of forest use are nested within any given landscape in the region. The institutional, social, economic, and environmental complexity of the region today requires students of land use to examine multitudinous forms of forest-resource use and to analyze the coexistence of different trajectories of forest resource use. The value of the forest and diverse types of land use vary significantly among different social groups occupying the region today, hence the existence of incentives to maintain the forests. Parts of a given region may undergo decline in forest cover, but others will continue to sustain forest-based economies, and still others will experience episodic changes, such as "pulses" of deforestation, followed by regrowth (figure 12.8A–C). The coexistence of different social groups, occupation corridors, agricultural markets, conservation policies, and forms of agrarian reform within parts of the region create different conditions underlying the ways forest resources can be used. Taking into account historical conditions underlying this variability is necessary to comprehend land-use change in the region today and to account for alternative, prognostic models that can inform regional policies.

DIVERSIFYING LAND-USE ALLOCATION: MINIMIZING RISK, MAXIMIZING RETURNS

Historical and ethnographic accounts of Amazonian land use have consistently highlighted the coexistence of multiple land-use and economic strategies among indigenous and nonindigenous Amazonian populations (Browder 1989; Denevan 1984; Moran 1989; Posey and Balée 1989; Redford and Padoch 1992). This coexistence is also observed among colonist populations (Brondízio et al. 2002; Muchagata 1997). Amazonian farmers are increasingly faced with challenging conditions. Climate and environmental change, stronger competition for resources, and also stronger market fluctuations will offer even greater challenges to Amazonian land users to minimize risks while tapping into new economic opportunities. Knowing the factors influencing historical shifts in land-use economies may contribute to a better understanding of land use in the context of local livelihood strategies.

Variability in environmental resources, market opportunities, and forms of resource ownership are factors that have led land users, in particular small-scale ones, to rely on multiple economic strategies in order to achieve consumption needs and to capture available capital while minimizing the associated risks. The various degrees of engagement in cattle ranching among small and large colonist farmers also illustrate this process (Hecht 1993). Coexisting and shifting land-use strategies continue to be and will increasingly be a characteristic of local land-use systems. For instance, decline in the price of the annual crops and increase in the external market for Amazonian fruits and wood products have led to a shift from annual agriculture to forest-based economies in different parts of the Amazon estuary (Brondízio 1999; Hiraoka 1994b; Pinedo-Vasquez et al 2001).

Understanding shifts in land allocation occurring in synchrony with local decision making and external factors are at the very core of land-use analysis. The potential cross-fertilization between land-use analysis and historical ecology may help to provide more sophisticated understanding of the range of livelihood and economic strategies across social groups living side by side in the region today.

METHODOLOGICAL IMPLICATIONS

Several practical applications of a historical ecological approach to the study of land use in the Amazon ensue from the foregoing discussion. Historical analysis of settlement formation and forms of regional occupation contributes directly to sampling design, not only in areas experiencing change since colonial times, but in areas recently colonized. A focus on intraregional historical process may even help to bridge tools such as remote sensing, on the one hand, and ethnographic research, on the other. Remote sensing data capture large regions that represent different environmental and historical conditions and often encompass landscapes that display dissimilar spatial patterns resulting from these conditions. Broadly speaking, ethnographic work, in contrast, contributes only to understanding the life histories of local land use systems. Image data may thus be used to inform fieldwork, ethnographic interviews, and survey design, and vice versa. No less important, knowledge of intraregional variability may inform ecological analysis aimed at elucidating the formation of anthropogenic forests. These examples and others (such as Brondízio et al. 1994, 1996; Castro et al. 2000) illuminate the value brought to the table by incorporating historical concerns into regional and local analyses of land-use change.

Accounting for intraregional variability in land use and land cover resulting from historical occupation also helps to facilitate image classification. Integrating remote sensing and ethnographic work on land-use history and management practices helps to fine-tune classification parameters to different parts of a scene where variation in land-cover types may have originated in differences among land-use systems, including ancient, prehistoric ones (see Erickson and Balée, chapter 7; Erickson, chapter 8; Neves and Peterson, chapter 9; Heckenberger, chapter 10, this volume). Different lengths of occupation and forms of land use result in variations in land-cover types, forest structure, and species composition. When digital satellite data are used, these parameters become important for image classification. Although spectral data provide an initial indication of the main dissimilarities in land-cover structure, spatial patterns, and environmental conditions—such as contiguity and fragmentation of forest cover, shape, and size of patches of the dominant land cover—fieldwork on land-use history, including vegetation inventories, helps to inform and characterize these differences (Brondízio 2005). In summary, cross-fertilization of methods and tools of analysis offers an advance in the daunting task of integrating local and regional levels of analysis for comprehending land-use change.

CONCLUDING REMARKS

The growing complexity in the land uses of the Amazonian region poses challenges and opportunities for understanding human-environment interaction, analyzing land-use change, and contributing to debates and policies on regional development. As the Amazon basin takes a central stage in global and regional environmental change scenarios, analyzing human-environment interactions affecting these processes requires balancing the role of macrolevel and geopolitical forces vis-à-vis local environmental and historical conditions underlying local land-use change. In this context, the current trend in modeling future scenarios of Amazonian land use carries important political implications for different populations in the region. Modeling predictions have an eminent political application in negotiating both the economics of global change at the international level—such as carbon emissions—and national and regional priorities for development policies. The research community thus faces the political and ethical implications of defining causal relationships, developing prognostic models, and informing policy in order to alter or support particular forms of land use and particular land users or stakeholders. Attention to intraregional land-use diversity and the historical dimensions of these uses is critical to minimize misinterpretations and negative long-term consequences of national and international policies for regional development.

This chapter suggests an approach to study land use grounded on the long-term concerns of historical ecology, on the spatial dimensions of remote sensing and other tools for spatial analysis, and on the strength of integrating survey, ethnographic, and ecological tools for capturing local conditions. It is essentially an argument for an applied historical ecology of Amazonia in the immediate context of today and tomorrow. A historical and intraregional perspective to land use among Amazonian populations supplies insight into the complexity of factors affecting social and environmental change in the region. Rather than creating unmanaged complexity for macroregional analysis and policy, a better understanding of regional socioenvironmental diversity will highlight the different needs of regional populations and perhaps address the real problems of deforestation, loss of biodiversity, and poverty in the region.

ACKNOWLEDGMENTS

This chapter benefits from case studies and examples developed as part of several research projects and field campaigns and involving several colleagues from ACT and the Center for the Study of Institutions, Populations, and Environmental Change (CIPEC) at Indiana University; and collaborating Brazilian institutions such as Empresa Brasileira de Pesquisas Agropecuárias (EMBRAPA), Museu Paraense Emilío Goeldi, and the Instituto Nacional de Pesquisas Espaciais (INPE). Much of the historical remote sensing work presented here benefited from the map archives of the Superintendência de Desenvolvimento da Amazônia (SUDAM), which I hope are taken care of for years to come. I am thankful for the support from SUDAM's librarians and personnel. I am particularly thankful for William Balée's and Clark Erickson's comments and for Andrea Siqueira's feedback on earlier versions of this manuscript. I am grateful for Patti Torp's editing and Scott Hetrick's assistance with figures. An earlier version of this chapter was presented at the 2002 Symposium on Neotropical Historical Ecology at Tulane University. The views expressed here are my sole responsibility.

NOTES

1. Existing and planned corridors include the "Pacific route" connecting the region through Brazil, Bolivia, and Peru to Asian markets; the "Guianas route" via Amapá State; the "Caribbean route" via the Manaus-Caracas highway; and the "soybean route" via the Santarém/Cargill port, which links central Brazil and western Amazonia to the Atlantic Ocean.

2. For a detailed discussion of the term *caboclo,* see Brondízio 2004b.

3. During the Second World War, another important, though smaller, period of rubber exploitation began through a program of financial cooperation between Brazil and the United States, aiming at overcoming the stalemate of the international rubber market due to Japanese occupation of rubber-production areas in Asia. Motivated by government recruiting, a significant migration movement of Brazilian northeasterners, the Soldados da Borracha (Soldiers of the Rubber), took off particularly to western Amazonia (now the states of Rondônia and Acrê). This process had significant demographic, social, and economic influence on the formation of this part of the Amazon. Today, more than ever, the descendants of these migrants, best represented by the legacy of Chico Mendes, are shaping politically contemporary forms of occupation and land use throughout the region.

4. An approximate translation is "voice of the land"; the phrase refers to a social movement of smallholders and rural workers.

5. Many precedents in the social and biophysical sciences motivated the growth of land-use studies during the 1990s (Brondízio and Siqueira 1997). Ecologically oriented anthropologists and geographers have moved toward scaling up their local unit of analysis due to the need to understand local agriculture and economy on a more encompassing regional scale (e.g., see the human ecology volume Behrens 1994; Conant 1990; Guyer and Lambin 1993; Mertens et al. 2000; Moran and Brondízio 2001; Nyerges and Green 2000; Wilkie 1987). Anthropology in general and environmental anthropology

in particular share this task through their interest in agrarian studies, political ecology, and studies of consumption and markets. In contrast, ecological and biophysical scientists working at global and regional scales have perceived the need to scale down in order to understand the impact of local land-use strategies on large-scale processes, such as on biogeochemical cycles and climate (Dale et al. 1993; Liverman et al. 1998; Skole and Tucker 1993).

6. For an up-to-date, good overview of theories used in land-use studies, see Vanwey Ostrom, and Merestisk 2005.

7. One example involves the advances in the analysis of Amazonian deforestation. The literature today shows examples of deforestation analysis at the basin, state, and national levels (e.g, INPE 1988–2001; Moran 1993a; Skole and Tucker 1993; Wood and Skole 1998), studies of vegetation type (INPE 1988–2002), and studies illustrating the articulation of even the smallest units such as the farm lots as well as settlements and municipalities (e.g., Batistela, Robeson, and Moran 2003; Brondízio, McCracken et al. 2002; McCracken, Siqueira et al. 2002; Wood and Porro 2002). Until recently, it was common to see reports of deforestation based on a Landsat scene itself or arbitrarily defined areas.

8. The region was also influenced by a significant "gold cycle" of the 1970s and 1980s, particularly close to Itaituba along the Tapajós River.

9. This map represents examples of research taking place at the Anthropological Center for Training and Research on Global Environmental Change (ACT) in this region since 2000. Further spatial stratification of the region has been developed for the purpose of sampling farms and households appearing during different periods of occupation.

REFERENCES

Ab'Saber, A. 1997. *Amazônia: Do discurso a praxis.* São Paulo: EDUSP.

Albuquerque, M. 1969. *A Mandioca na Amazônia.* Belém, Brazil: SUDAM.

Alves, D. S. 2002. An analysis of the geographical patterns of deforestation in the Brazilian Amazon in the period 1991–1996. In C. H. Wood and R. Porro, eds., *Deforestation and Land Use in the Amazon,* 95–106. Gainesville: University of Florida Press.

Anderson, A. B., I. Mousasticoshvily, and D. S. Macedo. 1993. *Impactos ecológicos e sócio-econômicos da exploração seletiva de Virola no estuário Amazônico.* Brasília: World Wildlife Fund.

Anderson, A. B., and D. A. Posey. 1989. Management of a tropical scrub savanna by the Gorotire Kayapó of Brazil. In D. A. Posey and W. Balée, eds., *Resource Management in Amazônia: Indigenous and Folk Strategies,* 159–173. Advances in Economic Botany no. 7. Bronx: New York Botanical Garden.

Anderson, R. L. 1976. Following Curupira: Colonization and migration in Pará 1758–1930 as a study in settlement of the humid tropics. Ph.D. diss., University of California, Davis.

Anderson, S. 1992. Engenhos de várzea: uma análise de declínio de um sistema de produção tradicional na Amazônia. In P. Léna and A. E. Oliveira, eds., *Amazônia: A fronteira agrícola 20 anos depois,* 101–124. Belém, Brazil: CEJUP.

Angelsen, A., and D. Kaimowitz, eds. 2001. *Agricultural Technology and Tropical Deforestation.* Oxfordshire, U.K.: CAB International and Center for International Forestry Research.

Anthropological Center for Training and Research on Global Environmental Change (ACT). 2000–2003. *Field Data and Map Product Generated by the Anthropological Center* for Training and Research on Global Environmental Change. Bloomington: Indiana University.

Aragon, L. E., and L. J. A. Mougeot, eds., 1986. Migrações internas na Amazônia: Contribuições teóricas e metodológicas. Belém, Brazil: UFP/NAEE and Falangola Editora.

Baena, A. L. M. 1969. Compêndio das eras da Província do Pará. Belém, Brazil: Universidade Federal do Pará.

Balée, W. 1989. The culture of Amazonian forests. In: D. A. Posey and W. Balée, eds., Resource Management in Amazonia: Indigenous and Folk Strategies, 1–21. Advances in Economic Botany no. 7. Bronx: New York Botanical Garden.

———. 1994. Footprints of the Forest: Ka'apor Ethnobotany—the Historical Ecology of Plant Utilization by an Amazonian People. New York: Columbia University Press.

———, ed. 1998. Advances in Historical Ecology. New York: Columbia University Press.

———. 2003. Historical-ecological influences on the word for cacao in Ka'apor. Anthropological Linguistics 45 (3): 259–280.

Balick, M. J., ed. 1987. The Palm—Tree of Life: Biology, Utilization, and Conservation. Advances in Economic Botany no. 6. Bronx: New York Botanical Garden.

Barata, M. 1915. A antiga produção e exportação do Pará. Belém, Brazil: Livraria Gillet.

———. 1972. Formação histórica do Pará. Belém, Brazil: Universidade Federal do Pará.

Barros, M. C., and A. Verissimo, eds. 1996. A expansão Madeireira na Amazônia. Belém, Brazil: Imazon.

Batistella, M., S. Robeson, and E. F. Moran. 2003. Settlement design, forest fragmentation, and landscape change in Rodônia, Amazonia. Photogrammetiric Engineering and Remote Sensing 69 (7): 805–812.

Becker, B. K. 1990. Amazônia. São Paulo: Ática.

Beckerman, S., ed. 1983. Does the swidden ape the jungle? Human Ecology 11 (1): 1–12.

Behrens, C. A., ed. 1994. Recent Advances in the Regional Analysis of Indigenous Land Use and Tropical Deforestation. Special issue of Human Ecology 22 (3).

Behrens, C. A., M. G. Baksh, and M. Mothes. 1994. A regional analysis of Bari land use intensification and its impact on landscape heterogeneity. Human Ecology 22 (3): 279–316.

Brabo, M. J. C. 1979. Os roceiros de Muaná. Publicações Avulsas. Belém, Brazil: MPEG.

Brondízio, E. S. 1999. Agroforestry intensification in the Amazon estuary. In T. Granfelt, ed., Managing the Globalized Environment: Local Strategies to Secure Livelihoods, 88–113. London: IT.

———. 2003. Letter: Revisiting Amazonia circa 1492. Science 302:2067–2068.

———. 2004a. Agricultural intensification, economic identity, and shared invisibility in Amazonian peasantry: Caboclos and colonists in comparative perspective. Culture and Agriculture 26 (1–2): 1–24.

———. 2004b. From staple to fashion food: Shifting cycles, shifting opportunities, and the case of açaí fruit (Euterpe oleracea Mart.) of the Amazon estuary. In D. J. Zarin, J. R. R. Alavalapati, F. E. Putz, and M. Schmink, eds., Working Forests in the American Tropics: Conservation Through Sustainable Management? 348–361. New York: Columbia University Press.

Brondízio, E. S. 2005. Intra-regional analysis and land-use trajectories in Amazônia. In E. F. Moran and E. Ostrom, eds., Seeing the Forest and the Trees: Human-Environment Interactions in Forest Ecosystems, 223–252. Cambridge, Mass.: MIT Press.

Brondízio, E. S., S. McCracken, E. F. Moran, A. D. Siqueira, D. R. Nelson, and C. Rodriguez-Pedraza. 2002. The colonist footprint: Towards a conceptual framework of deforestation trajectories among small farmers in frontier Amazônia. In C. H. Wood and R. Porro, eds., Deforestation and Land Use in the Amazon, 133–161. Gainesville: University of Florida Press.

Brondízio, E. S., E. F. Moran, P. Mausel, and Y. Wu. 1996. Changes in and cover in the Amazon estuary: Integration of thematic mapper with botanical and historical data. *Photogrammetric Engineering and Remote Sensing* 62 (8): 921–929.

Brondízio, E. S., E. F. Moran, A. D. Siqueira, P. Mausel, Y. Wu, and Y. Li. 1994. Mapping anthropogenic forest: Using remote sensing in a multi-level approach to estimate production and distribution of managed palm forest (*Euterpe oleracea*) in the Amazon estuary. *International Archives of Photogrammetry and Remote Sensing* 30 (7a): 184–191.

Brondízio, E. S., C. Safar, and A. D. Siqueira. 2002. The urban market of açaí fruit (*Euterpe oleracea* Mart.) and rural land use change: Ethnographic insights into the role of price and land tenure constraining agricultural choices in the Amazon estuary. *Urban Ecosystems* 6 (1–2): 67–98.

Brondízio, E. S., and A. D. Siqueira. 1997. From extractivists to forest farmers: Changing concepts of caboclo agroforestry in the Amazon estuary. *Research in Economic Anthropology* 18:234–279.

Browder, J., ed. 1989. *Fragile Lands of Latin America: Strategies for Sustainable Development.* Boulder, Colo.: Westview Press.

Browder, J., and B. Godfrey. 1997. *Rainforest Cities: Urbanization, Development, and Globalization of the Brazilian Amazon.* New York: Columbia University Press.

Bunker, S. G. 1985. *Underdeveloping the Amazon: Extraction, Unequal Exchange, and the Failure of the Modern State.* Chicago: Chicago University Press.

Carneiro, R. L. 1961. Slash-and-burn cultivation among the Kuikuru and its implications for cultural development in the Amazon basin. In J. Wilbert, ed., *The Evolution of Horticultural Systems in Native South America: Causes and Consequences,* 47–67. Anthropological Supplement no. 2. Caracas, Venezuela: Sociedad de Ciencias Naturales La Salle.

——. 1995. The history of ecological interpretations of Amazônia: Does Roosevelt have it right? In L. Sponsel, ed., *Indigenous Peoples and the Future of Amazônia: An Ecological Anthropology of an Endangered World,* 45–70. Tucson: University of Arizona Press

Carvalho, G., D. Nepstad, D. McGrath, M. del C. Vera Diaz, M. Santilli, and A. C. Barros. 2002. Frontier expansion in the Amazon: Balancing development and sustainability. *Environment* 44:34–45.

Castro, F. de, and E. S. Brondízio. 2000. The role of ecological patchiness in forest fragment patterns: The upper Negro River case. Poster presented at the First International Open Science Meeting of the Large Scale Biosphere-Atmosphere Experiment in Amazonia. Belém, Brazil, June 21–26.

Castro, F., M. C. Silva-Forsberg, W. Wilson, E. S. Brondízio, and E. F. Moran. 2002. The use of remotely-sensed data in rapid rural assessment. *Field Methods* 14 (3): 243–310.

Caviglia, J. L. 1999. *Sustainable Agriculture in Brazil: Economic Development and Deforestation.* Cheltenham, U.K.: Edward Elgar.

Chibnik, M. 1994. *Risky Rivers: The Economics and Politics of Floodplain Farming in Amazonia.* Tucson: University of Arizona Press.

Clarke, W. C. 1976. Maintenance of agriculture and human habitats within the tropical forest ecosystem. *Human Ecology* 4 (3): 247–259.

Cleary, D. 2001. Towards an environmental history of the Amazon: From prehistory to the nineteenth century. *Latin American Research Review* 36 (2): 65–96.

Conant, F. P. 1990. 1900 and beyond: Satellite remote sensing and ecological anthropology. In. E. F. Moran ,ed., *The Ecosystem Approach in Anthropology: From Concept to Practice,* 357–388. Ann Arbor: University of Michigan Press.

Coomes, O. T., and G. J. Burt. 1997. Indigenous market-oriented agroforestry: Dissecting local diversity in western Amazônia. *Agroforestry Systems* 37:27–44.

Crumley, C. L., ed. 1994. *Historical Ecology.* Santa Fe: School of American Research.

Dale, V. H., R. V. O'Neill, M. Pedlowski, and F. Southworth. 1993. Causes and effects of land-use change in central Rondônia, Brazil. *Photogrammetric Engineering and Remote Sensing* 59 (6): 997–1005.

Dean, W. 1987. *Brazil and the Struggle for Rubber.* Cambridge, U.K.: Cambridge University Press.

——. 1996. *A ferro e fogo: A história da devastação da Mata Atlântica Brasileira.* São Paulo: Cia. de Letras.

Denevan, W. M. 1984. Ecological heterogeneity and horizontal zonation of agriculture in the Amazon floodplain. In M. Schmink and C. Wood, eds., *Frontier Expansion in Amazonia,* 311–336. Gainesville: University of Florida Press.

——. 1996. A bluff model of riverine settlement in prehistoric Amazônia. *Annals of the Association of American Geographers* 86 (4): 654–681.

——. 1998. Comments on prehistoric agriculture in Amazônia. *Cultural and Agriculture* 20:54–59.

——. 2001. *Cultivated Landscapes of Native Amazônia and the Andes.* New York: Oxford University Press.

Denevan, W. M., and C. Padoch, eds. 1987. *Swidden-Fallow Agroforestry in the Peruvian Amazon.* Advances in Economic Botany no. 5. Bronx: New York Botanical Garden.

Derby, O. 1897. A Ilha de Marajó. *Boletim do Museu Paraense Emílio Goeldi* 4:163–173.

Dincão, M. A., and I. M. Silveira, eds. 1994. *A Amazônia e a crise da modernização.* Belém, Brazil: Museu Paraense Emílio Goeldi.

Erickson, C. L. 2000. An artificial landscape-scale fishery in the Bolivian Amazon. *Nature* 408: 190–193.

Evans, T. P., and E. F. Moran. 2002. Spatial integration of social and biophysical factors related to landcover change. In W. Lutz, A. Frskawetz, and W. C. Sanderson, eds., *Population and Environment: Methods of Analysis,* supplement to *Population and Development Review* 28:165–186.

Fearnside, P. M. 1984. A floresta vai acabar? *Ciência Hoje* 2 (10): 42–52.

——. 1986. *Human Carrying Capacity of the Brazilian Rainforest.* New York: Columbia University Press.

Fisher, W. H. 2000. *Rainforest Exchanges: Industry and Community on an Amazonian Frontier.* Washington, D.C.: Smithsonian Institution Press.

Fox, J.; V. Mishra, R. Rindfuss, and S. Walsh. 2003. *People and the Environment: Approaches to Linking Household and Community Surveys to Remote Sensing and GIS.* Boston: Kluwer Academic.

Futemma, C., and E. S. Brondízio. 2003. Land reform and land use changes in the lower Amazon: Implications to agricultural intensification. *Human Ecology* 31 (3): 369–402.

Geertz, C. 1963. *Agricultural Involution: The Process of Ecological Change in Indonesia.* Berkeley: University of California Press.

Geist, H., and E. F. Lambin. 2001. *What Drives Tropical Deforestation: A Meta-Analysis of Proximate and Underlying Causes of Deforestation Based on Subnational Case Study Evidence.* Land-Use and Cover Change (LUCC) Report Series no. 4. Brussels: CIACO Printshop. Available at: www.geo.ucl.ac.be/LUCC.

Gentil, J. 1988. A juta na agricultura de várzea na área de Santarém-Médio Amazonas. Série Antropologia. *Boletim do Museu Paraense Emílio Goeldi* 4 (2): 1–50.

Goldenberg, J., ed. 1989. *Amazônia: Facts, Problems, and Solutions.* São Paulo: Universidade de São Paulo.

Golley, F. B. 1992. *A History of the Ecosystem Concept in Ecology: More Than the Sum of the Parts.* New Haven, Conn.: Yale University Press.

Guyer, J., and E. Lambin. 1993. Land use in an urban hinterland: Ethnography and remote sensing in the study of African intensification. *American Anthropologist* 95 (4): 839–859.

Hames, R. 1983. Monoculture, polyculture, and polyvariety in tropical forest swidden cultivation. *Human Ecology* 11 (1): 13–34.

Hames, R., and W. Vickers, eds. 1983. *Adaptive Responses of Native Amazonians.* New York: Academic Press.

Hardesty, D. L. 1975. The niche concept: Suggestions for its use in human ecology. *Human Ecology* 3 (2): 71–85.

Hecht, S. 1993. The logic of livestock and deforestation in Amazônia. *Bioscience* 43 (10): 687–695.

Heckenberger, M. J., A. Kuikuruo, U. T. Kuikuro, J. C. Russell, M. Schmidt, C. Fausto, and B. Franchetto. 2003. Amazônia 1492: Pristine forest or cultural parkland? *Science* 301:1710–1713.

Hiraoka, M. 1985. Zonation of mestizo riverine farming systems in northeast Peru. *National Geographic Research* 2 (3): 354–371.

———. 1994a. Mudanças nos padrões econômicos de uma população ribeirinha do estuário do Amazonas. In L. Furtado, A. F. Mello, and W. Leitão, eds., *Poros das aguas: Realidade perspectivas na Amazônia,* 133–137. Belém, Brazil: MPEG/UFPa.

———. 1994b. The use and management of "Miriti" (*Mauritia flexuosa*) palms among the ribeirinhos along the Amazon estuary. Paper presented at the conference "Whitewater Várzeas: Diversity, Development, and Conservation of Amazonian Floodplain," Macapá, Amapa, Brazil, December 12–14.

Homma, A. 1993. *Extrativismo vegetal na Amazônia: Limites e oportunidades.* Brasília: Empresa Brasileira de Produção Agrícola e Agropecuária.

———. 2003. *História da agricultura na Amazônia: Da era pré-Columbiana ao terceiro milênio.* Belém, Brazil: EMBRAPA.

Instituto Nacional de Pesquisas Espaciais (INPE). 1988–2001. *INPE Deforestation Reports 1988–2001.* São José dos Campos, Brazil: Diretoria de Observação da Terra, INPE. Available at: www.inpe.br/.

Instituto Socioambiental (ISA). 2000. *Povos indígenas no Brasil 1996/2000.* São Paulo: ISA.

Kaimowitz, D., and A. Angelsen. 1998. *Economic Models of Tropical Deforestation: A Review.* Bogor, Indonesia: Center for International Forestry Research.

Kroeber, A. 1939. *Cultural and Natural Areas of Native North America.* Berkeley: University of California Press.

Land-Use and Cover Change (LUCC). 1994. *Land-Use and Land-Cover Change, Science/ Research Plan.* International Geosphere-Biosphere Program (IGBP) Report no. 35 and Human Dimensions Program (HDP) Report no. 7. Stockholm: IGBP, HDP. Available at: www.igbp.kva.se/publicat.html.

Land-Use and Cover Change (LUCC). 1999. *Land-Use and Land-Cover Change Implementation Strategy.* Prepared by the LUCC Scientific Steering Committee and International Project Office: E. F. Lambin, X. Baulies, N. Bockstael, G. Fischer, T. Krug, R. Leemans, E. F. Moran, R. R. Rindfuss, Y. Sato, D. Skole, B. L. Turner II, and C. Vogel. Edited by C. Nunes and J. I. Augé. International Geosphere-Biosphere Program (IGBP) Report no. 48 and International Human Dimensions Program (IHDP) Report no. 10. Stockholm: IGBP and HDP. Available at: www.igbp.kva.se/publicat.html.

Lathrap, D. W. 1970. *The Upper Amazon.* London: Thames and Hudson.

Laurence, W. F., M. A. Cochrane, S. Bergen, P. M. Fearnside, P. Delamonica, C. Barber, S. D'Angelo, and T. Fernandes. 2001. The future of the Brazilian Amazon. *Science* 291: 438–442.

Lená, P., and A. Oliveira, eds. 1992. *Amazônia: A fronteira agrícola 20 anos depois*. Belém, Brazil: Edições CEJUP.

Lima, R. R., 1956. Agricultura nas várzeas do estuário do Amazonas. *Boletim Técnico do Instituto Agronômico do Norte* 33:1–164.

Liverman, D., E. F. Moran, R. Rindfuss, and P. Stern, eds. 1998. *People and Pixels: Linking Remote Sensing and Social Science*. Washington, D.C.: National Academy Press.

Mahar, D. J. 1988. *Government Policies and Deforestation in the Brazilian Amazon Region*. Washington, D.C.: World Bank.

Marquette, C. 1998. Land use patterns among small farmer settlers in the northeastern Eucadorian Amazon. *Human Ecology* 26 (4): 573–598.

McConnell, W. 2001. Meeting in the middle: The challenge of meso-level integration, a report. *Land Use Policy* 19 (2): 99–101.

McCracken, S., B. Boucek, and E. F. Moran. 2002. Deforestation trajectories in a frontier region of the Brazilian Amazon. In S. Walsh and K. Crews-Meyer, eds., *Linking People, Place, and Policy: A GIScience Approach*, 215–234. Boston: Kluwer Academic.

McCracken, S., E. S. Brondízio, D. Nelson, E. F. Moran, A. D. Siqueira, and C. Rodriguez-Peraza. 1999. Remote sensing and GIS at farm property level: Demography and deforestation in the Brazilian Amazon. *Photogrammetric Engineering and Remote Sensing* 65 (11): 1311–1320.

McCracken, S., A. D. Siqueira, E. F. Moran, and E. S. Brondízio. 2002. Land use patterns on an agricultural frontier in Brazil: Insights and examples from a demographic perspective. In C. H. Wood and R. Porro, eds., *Deforestation and Land Use in the Amazon*, 162–192. Gainesville: University Press of Florida.

Meggers, B. J. 1971. *Amazonia: Man and Culture in a Counterfeit Paradise*. Chicago: Aldine, Atherton.

Mertens, B., W. D. Sunderlin, O. Ndoye, and E. F. Lambin. 2000. Impact of macroeconomic change on deforestation in South Cameroon: Integration of household survey and remotely sensed data. *World Development* 28 (6): 983–999.

Moran, E. F. 1981. *Developing the Amazon*. Bloomington: Indiana University Press.

———. 1989. Models of native and folk adaptation in Amazonia. In D. A. Posey and W. Balée, eds., *Resource Management in Amazonia: Indigenous and Folk Strategies*, 22–29. Advances in Economic Botany no. 7. Bronx: New York Botanical Garden.

———, ed. 1990. *The Ecosystem Approach in Anthropology: From Concept to Practice*. Ann Arbor: University of Michigan Press.

———. 1993a. Deforestation and land use in the Brazilian Amazon. *Human Ecology* 21 (1): 1–21.

———. 1993b. *Through Amazonian Eyes: The Human Ecology of Amazonian Populations*. Iowa City: University of Iowa Press.

Moran, E. F. 1995. Disaggregating Amazonia: A strategy for understanding biological and cultural diversity. In L. Sponsel, ed., *Indigenous Peoples and the Future of Amazonia*, 71–95. Tucson: University of Arizona Press.

Moran, E. F., and E. S. Brondízio. 2001. Human ecology from space: Ecological anthropology engages the study of global environmental change. In M. Lambek and E. Messer, eds., *Ecology and the Sacred: Engaging the Anthropology of Roy A. Rappaport*, 64–87. Ann Arbor: University of Michigan Press.

Moran, E. F., E. S. Brondízio, and S. McCracken. 2002. Trajectories of land use: Soils, succession, and crop choice. In C. H. Wood and R. Porro, eds., *Deforestation and Land Use in the Amazon*, 193–217. Gainesville: University of Florida Press.

Muchagata, M. 1997. Forests and people: The role of forest production in frontier farming systems in eastern Amazônia. Development Studies Occasional Paper no. 36. Norwich, U.K.: University of East Anglia, School of Development Studies.

Muñiz-Miret, N., R. Vamos, M. Hiraoka, F. Montagnini, and R. O. Mendelsohn. 1996. The economic value of managing the açaí palm (*Euterpe oleracea* Mart.) in the floodplains of the Amazon estuary, Pará, Brazil. *Forest Ecology and Management* 87 (1–3): 163–173.

Murrieta, R. S. S. 2001. Dialética do sabor: Alimentação, ecologia e vida cotidiana em comunidades ribeirinhas da Ilha de Ituqui, Baixo Amazonas, Pará. *Revista de Antropología (USP)* (São Paulo) 44 (2): 39–88.

Nepstad, D. C., A. McGrath, A. Alencar, C. Barros, G. Carvalho, M. Santilli, and M. Del C. Vera Diaz. 2002. Frontier governance in Amazonia. *Science* 295:629–632.

Nepstad, D. C., and S. Schwartzman, eds. 1992. *Non-timber Products from Tropical Forests: Evaluation of a Conservation and Development Strategy.* Advances in Economic Botany no. 9. Bronx: New York Botanical Garden.

Nepstad, D. C., and C. Uhl. 2000. Amazonia at the millennium. *Interciência* 25 (3): 159–164.

Neves, E. G. 1998. Twenty years of Amazonian archeology in Brazil (1977–1997). *Antiquity* 71:625–632.

Neves, W. A., ed. 1989. *Biologia e ecologia humana na Amazonia: Avaliacao e perspectives.* Belém, Brazil: MPEG.

Nugent, S. 1993. *Amazonian Caboclo Society: An Essay on Invisibility and Peasant Economy.* Oxford: Oxford University Press.

Nyerges, A. E., and G. M. Green. 2000. The ethnography of landscape: GIS and remote sensing in the study of forest change in West African Guinea savanna. *American Anthropologist* 102 (2): 272–290.

Odum, E. 1971. *Fundamentals of Ecology.* 3rd ed. Philadelphia: Saunders.

Ozório Almeida, A. L. 1992. *The Colonization of the Amazon.* Austin: University of Texas Press.

Padoch, C., J. Chota Inuma; W. de Jong, and J. Unruh. 1985. Amazonian agroforestry: A market oriented system in Peru. *Agroforestry Systems* 3:47–58.

Penteado, A. R. 1967. *Problemas de colonização e de uso de terra da Região Bragantina do Estado do Pará.* Belém, Brazil: Editora da Universidade Federal do Pará.

Pichón, F., and R. Bilsborrow. 1992. Land use systems, deforestation, and associated demographic factors in the humid tropics. In R. Bilsborrow and D. Hogan, eds., *Population and Deforestation in the Humid Tropics,* 175–207. Paris: International Union for the Scientific Study of Population.

Pinedo-Vasquez, M., D. J. Zarin, K. Coffey, C. Padoch, and F. Rabelo. 2001. Post-boom logging in Amazônia. *Human Ecology* 29 (2): 219–239.

Posey, D. A., and W. Balée, eds. 1989. *Resource Management in Amazônia: Indigenous and Folk Strategies.* Advances in Economic Botany no. 7. Bronx: New York Botanical Garden.

Prance, G. T., and J. A. Kallunki, eds. 1984. *Ethnobotany in the Neotropics.* Advances in Economic Botany no. 1. Bronx: New York Botanical Garden.

Redford, K. H., and C. Padoch, eds. 1992. *Conservation of Neotropical Forests: Working from Traditional Resource Use.* New York: Columbia University Press.

Rohter, L. 2003. Relentless foe of the Amazon jungle: Soybeans. *New York Times,* September 17.

Roosevelt, A. C. 1989. Resource management in Amazônia before the Conquest: Beyond ethnographic projection. In D. A. Posey and W. Balée, eds., *Resource Management in Amazonia: Indigenous and Folk Strategies,* 30–62. Advances in Economic Botany no. 7. Bronx: New York Botanical Garden.

Sant'ana, L. 2003. Amazônia: A soja avança na floresta. *O Estado de São Paulo,* October 26, B6–B7.

Santos, R. 1980. *Historia econômica da Amazônia (1800–1920).* Sao Paulo: T. A. Queiroz.

Sawyer, D. R. 1986. A fronteira inacabada: Industrializacao da agricultura brasileira e debilitacao da fronteira Amazônica. In L. E. Aragon and L. J. A. Mougeot, eds., *Migrações internas na Amazônia: Contribuições teóricas e metodológicas,* 54–90. Belém, Brazil: UFP/ NAEE e Falangola Editora.

Scatena, F., A. Walker, A. Homma, C. Couto, C. Ferreira, R. Cavalho, A. Rocha, A. Santos, and P. de Oliveira. 1996. Cropping and fallowing sequences of small farms in the *terra firme* landscape of the Brazilian Amazon: A case study from Santarém, Pará. *Ecological Economics* 18:29–40.

Schmink, M., and C. H. Wood. 1992. *Contested Frontiers in Amazonia.* New York: Columbia University Press.

Silveíra, J. P. 2001. Development of the Brazilian Amazon. *Science* 292:1651–1654.

Siqueira, A. D, S. McCracken, E. S. Brondízio, and E. F. Moran. 2003. Women in a Brazilian agricultural frontier. In G. Clark, ed., *Gender at Work in Economic Life,* 243–267. Society for Economic Anthropology (SEA) Monograph Series. Lanham, Md.: AltaMira Press.

Skole, D., and C. J. Tucker. 1993. Tropical deforestation and habitat fragmentation in the Amazon: Satellite data from 1978 to 1988. *Science* 260:1905–1910.

Smith, N. J. H. 1982. *Rainforest Corridors.* Berkeley: University of California Press.

Smith, N. J. H., I. C. Falesi, P. T. Alvim, and E. A. S. Serrão. 1996. Agroforestry trajectories among smallholders in the Brazilian Amazon: Innovations and resiliency in pioneer and older settled areas. *Ecological Economics* 18:15–27.

Sternberg, H. O. 1956. *A água eo homem na várzea do Carreiro.* Rio de Janeiro: Universidade do Brasil.

Steward, J. H., ed. 1946–50. *Handbook of South American Indians.* 7 vols. Bureau of American Ethnology Bulletin no. 143. Washington, D.C.: U.S. Government Printing Office.

——. 1955. *Theory of Culture Change.* Urbana: University of Illinois Press.

——. 1956. *The People of Puerto Rico: A Study in Social Anthropology.* Urbana: University of Illinois Press.

Subler, S., and C. Uhl. 1990. Japanese agroforestry in Amazônia: A case study in Tomé-Acu, Brazil. In A. B. Anderson, ed., *Alternatives to Deforestation: Steps Toward Sustainable Use of the Amazon Rain Forest,* 152–166. New York: Columbia University Press.

Tsunoda, F. 1988. *Canção da Amazônia.* Rio de Janeiro: Editora Francisco Alves.

Tura, L. R., and R. A. Costa, eds. 2000. *Campesinato e estado na Amazônia.* Belém, Brazil: Brasilia Jurídica.

Turner, B. L., W. B. Meyer, and D. L. Skole. 1994. Global land-use/land-cover change: Towards an integrative study. *Ambio* 23 (1): 91–95.

Vanwey, L, E. Ostrom, and V. Merestisk. 2005. Theory in human-environment interaction studies. In E. F. Moran and E. Ostrom, eds., *Seeing the Forest and the Trees: Human-Environment Interactions in Forest Ecosystems,* 23–56. Cambridge, Mass.: MIT Press.

Vayda, A. P., and B. MacCay. 1975. New directions in ecology, cultural and ecological anthropology. *Annual Review of Anthropology* 4:93–306.

Vayda, A. P., and R. Rappaport. 1968. Ecology, cultural and noncultural. In J. A. Clinton, ed., *Introduction to Cultural Anthropology,* 476–497. Boston: Houghton Mifflin.

Verissimo, A., M. A. Cochrane, and C. Souza Jr. 2002. National forests in the Amazon. *Science* 297:1478.

Walker, R., E. F. Moran, and L. Anselin. 2000. Deforestation and cattle ranching in the Brazilian Amazon: External capital and household processes. *World Development* 28 (4): 683–699.

Weinstein, B. 1983. *The Amazon Rubber Boom, 1850–1920.* Stanford, Calif.: Stanford University Press.

Wilkie, D. 1987. Cultural and ecological survival in the Ituri Forest: The role of accurately monitoring natural resources and agricultural land use. *Cultural Survival Quarterly* 11 (2): 72–74.

Wilson, W. M. 1997. Why bitter cassava (*Manihot esculenta* Crantz)? Productivity and perception of cassava in a Tukanoan Indian settlement in the northwest Amazon. Ph.D. diss., University of Colorado, Boulder.

Wilson, W. M., and D. L. Dufour. 2002. Why bitter cassava? Productivity of bitter and sweet cassava in a Tukanoan Indian settlement in the northwest Amazon. *Economic Botany* 56 (1): 49–57.

WinklerPrins, A. M. G. A. 2002a. House-lot gardens in Santarém, Pará, Brazil: Linking rural with urban. *Urban Ecosystems* 6: 43–65.

———. 2002b. Recent seasonal floodplain-upland migration along the lower Amazon River, Brazil. *Geographical Review* 92 (3): 415–431.

Wood, C. H., and R. Porro, eds. 2002. *Deforestation and Land Use in the Amazon.* Gainesville: University of Florida Press.

Wood, C. H., and D. Skole. 1998. Linking satellite, census, and survey data to study deforestation in the Brazilian Amazon. In D. M. Liverman, E. F. Moran, R. R. Rindfuss, and P. C. Stern, eds., *People and Pixels: Linking Remote Sensing and Social Science,* 70–93. Washington, D.C: National Academy Press.

Yamada, M. 1999. Japanese immigrant agroforestry in the Brazilian Amazon: A case study of sustainable rural development in the tropics. Ph.D. diss., University of Florida.

INDEX

açaí palm (*Euterpe oleracea*), 171–72, 284–85, 370, 385–86
acerola (*Malpighia glabra*), 371
Aché (Indians, Paraguay), 283, 358n
achira (*Canna edulis*), 105, 111–13, 115–16, 117–18
Açutuba (site, Amazon), 288, 296, 300, 320
Adams, William Y., 4
agouti (*Dasyprocta* sp.), 144
agricultural regression, 156, 238, 351, 357; agricultural or cultural devolution, 357
agrovilas, 381
aguaí (*Pouteria* sp.), 215
Aguiar, Jaime P. L., 176
alpha diversity 3, 30, 52, 208 (defined), 216–17
Altamira (city, Brazil), 381
Amanayé (Indians, Brazil), 349
Amazon basin, 77, 93, 95, 96, 103, 264, 395
Amazon Polychrome Ceramic tradition, 321
Amazon River, 157, 281, 371, 379
Amazonia, 312, 336
Amazonian anthropogenic forest, 172
Amazonian Dark Earth (ADE), 57–79, 156, 200, 245, 263, 280, 290–92, 294, 299–301
Amazonian forest, 95, 153, 154
Anambé (Indians, Brazil), 349
Ancon Bay, 111
Anderson, Edgar, 97, 268
Andes, 90, 94

Andors, Allison, 146
animism, 353
anteater (*Tamandua* sp.), 144
anthropic (defined), 70
anthropogenic (defined), 70; anthropogenic clearance 130, 144; anthropogenic forest, 172, 209, 217, 391; anthropogenic landscape, 3, 93, 94, 96, 145, 165, 189, 244–45, 366; anthropogenic soil horizons, 187
Apere River (Bolivia), 189, 251
apête, 55 (*see also* forest island)
Apinayé (Indians, Brazil), 160
apple snail (*Pomacea* sp. or *Ampularia* sp.), 199
applied knowledge, 67
Araracuara (site, Colombian Amazon), 158
Arawak (Indians, Neotropics), 317, 318, 327, 328, 329, 330; diaspora of, 317–19
Arawak (language family), 319
archaeofauna (Jama, Ecuador), 127–149
archaeology of landscapes, 127, 262
armadillo (*Dasypus* spp.), 144
Arnason, John T., 59
arroyos. *See* meander scars
Asia (site), 115, 116
assacu, 172, 177; *see also* ochoó
Atlantic Coastal Forest (Mata Atlântica), 301, 373
Avá-Canoeiro (Indians, Brazil), 358
Avança Brasil (program), 373

avocado (*Persea americana*), 106, 115–17
Ayacucho Valley (Peru), 111–13, 116

babaçu palm (*Attalea speciosa* = *Orbignya phalerata*), 171, 212, 343, 356, 391
bacaba palm (*Oenocarpus* sp.), 284; Mart. see Bailey, Robert G., 238
bacuri (*Platonia insignis*), 172
bajo clays, 70
Balée, William, 1–17, 72, 77, 93, 94, 99, 100, 103, 121, 145, 172, 176, 187–233, 237, 238, 334, 341, 343, 349, 350, 353, 356, 357, 358, 390
Bandurria (site, Peru), 115
Barasana (Indians, Colombia), 99, 101–102
Barrancoid Manacapuru Phase, 292
barrow pits, 190, 192, 195–96
Baure (Indians), 327
Baures (region Bolivia), 200, 258, 255, 259, 260
Baures Hydraulic Complex, 255
"beater syndrome," 130
Beck, Stephan G., 208
Beckerman, Stephen, 166, 173, 351, 352, 354
begonia, 105, 110, 111
behavior (human), 69
Belém (city, Brazil), 168
Belize, 21, 58, 73, 74, 75, 76
Bella Vista (site, Bolivia), 258
Belterra (site), 156, 379, 380–81
Bender, Barbara, 103
Beni (Bolivia), 199, 200
Bernal, Rodrigo, 171, 284
beta diversity, 30; (defined), 52, 208
bi. *See* genipapo
"big models," 65
biodiversity, 245, 254, 345
biribá (*Rollinia mucosa*), 173
black mangrove (*Avicennia nitida*), 75
black pepper, 371
Bladen Nature Reserve (BNR), 32
Blake, Michael, 97
bluff cultivation, 158
boa constrictor (*Boa* sp.), 144
Bolivia, 7, 94, 312
Bolivian Amazon, 258, 260, 263, 265–67
bone (human), 199, 295–96
Boom, Brian, 207
Bororo (Indians, Brazil), 160, 329, 330

Bottega, Ricardo, 190, 269
bottle gourd (*Lageneria siceraria*), 105, 109, 111, 113, 115, 117
Bragantina region (Brazil), 373
Braidwood, Robert J., 103
Brazil nut (*Bertholletia excelsa*), 172, 391, 394
Brewer, Steven W., 32
brocket deer (*Mazama* spp.) 133, 144
Brondizio, Eduardo, 8, 365–405
Browder, John O., 336
Brown, Cecil, 92
built environment (defined), 245
Bunchosia armeniaca, 106, 115
buriti (palm (*Mauritia flexuosa*), 171–72, 251, 261, 269n, 284
burning (practice), 154–56; in Bolivian Amazon, 250–51
Bustos, Victor, 190, 269

caatinga, 388
Cabanagem Civil War (Brazil), 350
caboclo, 300–301, 369, 379, 396n
cacao (*Theobroma cacao*), 172, 260, 370, 379
Cahokia (site, Illinois), 333
cahoon. *See* corozo palm
caiaué palm (*Elaeis oleifera*), 172
Calandra, Horacio, 190, 269
Calathea allouia, 183
calcite, 63
California, 92
Callejon de Huaylas (Peru), 108
calzadas. See causeways
Campbell, David G., 7, 21–56, 207
campina, 288
campinarana, 288
Campo Alegre (site, Brazil), 300
canals, 190, 196–98, 247, 250, 254–57
cane mouse (*Zygodontomys* sp.), 144
Cannaceae, 183
cañadas. See meander scars
Capim River, 349
capitão, 350
Caquetá River (Colombia), 158, 263
carbon, 62, 157, 293–94; emissions, 395
Caribbean, 58
Carjaval, Gaspar de, 157
Carneiro, Robert, 319, 323
Caru River (Brazil), 349
Casarabe (site, Bolivia), 190, 199, 203, 204

cattle, 370, 373
causeways, 190, 192, 196–98, 219, 247, 250,
    254–57
Cayalo Mound (site, Bolivia), 206
Cayo District (forest site, Belize), 22, 32, 33
Cecropia concolor, 213
Central Amazon Project (CAP), 286–91,
    300–2
central place theory, 322, 276
ceramics, 62, 62, 64, 67, 75, 158, 189, 198–
    99, 218, 230n, 239, 257, 282, 288–89, 290,
    292, 301, 317, 321
cerritos, 201
cerros, 201
chaco. See swidden
Chan Chan (site, Peru), 333
charcoal, 61, 155, 156, 188, 200, 202, 295
Chase, A. K., 241, 242, 243, 268
chert, 62
chicha, 198
Chilca River Valley (Peru), 106, 108,
    110, 111, 112
chili peppers (*Capsicum frutescens, C.
    baccatum, C. pubscens*), 102, 105,
    106, 111, 113, 115, 116, 118
Chillon Valley (Peru), 111, 116, 117
China, 78
Chirabaya (Chile), 118
chocolate. *See* cacao
chonta palm (*Astrocaryum murumuru* var.
    murumuru), 209, 212, 215
Classic Period (Mesoamerica), 31;
    (Amazonia), 320–21
Clastres, Pierre, 319, 328
Clement, Charles R., 8, 121, 165–85,
    268n, 351
climbing rat (*Rhipidomys* sp), 144
Coastal Salish (Indians, Canada), 348n
coca (*Erythroxylum coca*), 106, 111–13
cocoyam (*Xanthosoma* sp.), 166, 183n, 252
coffee, 370
cohort effects, 383
collared peccary (*Tayassu tajacu*), 133, 144
Colombia, 101
Colonial period (Brazil), 378; (Bolivian
    Amazon), 257
Colorado, 92
common bean (*Phaseolus vulgaris*), 105, 106,
    109, 110, 118

Companhia Geral do Comércio do Grão
    Pará e Maranhão, 349
complexity (of society), 8, 90, 91, 103, 249–
    50, 290–91, 301–2, 311, 321, 322, 324–25,
    327–28, 335
Comunidade Terra Preta (site, Brazil), 158
conch shells, 72
Condamine, Charles-Marie La, 316
Conklin, Beth A., 355
conservation biology, 4
Conservation International, 260
Convolvulaceae, 183
coquino (*Pouteria* sp. 1), 215
Cordia alliodora Oken, 213
Cormier, Loretta A., 8, 92, 98, 341–63
corozo palm (*Attalea cohune*), 31
cotton (*Gossypium barbadense*), 105, 106, 111,
    113, 115, 117
cotton rat (*Sigmodon* sp.), 144
craft annuals, 105
crookneck squash (*Cucurbita moschata*), 106
crops: diffusion, 88, 96; domestication, 93
Crumley, Carole L., 2
c-transforms (defined), 71
Cuba, 74
cubiu (*Solanum sessiliflorum*), 173
Cuiabá–Santarém Highway, 379, 384
Culebras Valley (Peru), 114
cultivation (in Amazonian forests), 153–63
cultural ecology, 12n
cultural forests, 11
culture, 2, 66, 72, 73, 78–79; (defined), 79
culture areas, 302
cupuaçu (*Theobroma grandiflorum*), 371
curiche (bajio), 192

dark earth, 156, 158; *see also* Amazonian
    Dark Earth
DBH (diameter at breast height, of vegeta-
    tion), 22, 23, 24, 208, 211
De Aguirre, Pedro de Ursúa-Lope, 157
deep history, 316–22
Denevan, William, 2, 10, 153–64, 189, 237,
    247, 255, 268n, 269, 186, 300, 351, 352,
    356, 369
Descola, Philippe, 353
diaspora (linguistic, in Amazonia), 317–18,
    319, 327, 329, 330
dicotyledons, 169

diffusion (of plants), 96
Dioscoreaceae, 166, 183
Distel, Alicia F., 269
Dobyns, Henry, 316
domestic plant, 87, 90, 93
domesticated food, 87
domesticated landscape (Bolivian Amazon), 235–78
domestication (defined), 235; 239–40 , 268n; of plants, 87–91, 96, 97, 166, 171, 173; of fruit trees, 165–86; of landscapes, 265, (defined), 241; two-way, 91; of nature, 334; of neotropical environments, 239
domestication of landscape, 255; see also anthropogenic; domestication; landscape
domiculture, 241
Dona Stella (site, Amazon), 288
Dos Islas (site, Bolivia), 206
Dougherty, Bernardo, 190, 269
Douglas, Mary, 103
Dufour, Darna, 301

Early Archaic Period, 95
Early Archaic-Preceramic III Phase (Western South America), 106
ecofacts, 62
ecological anthropology, 6
ecological epistéme, 9
ecologically ignoble savage (Homo devastans), 10, 57, 78
ecologically noble savage (Homo ecologicus), 10, 70, 235
ecosystem (concept), 6
ecotone (of eastern edges of Andes), 90; aquatic, 254
Ecuador, 7, 106, 109, 110, 111, 113, 114, 115, 128, 133, 135
Edwards, Clinton R., 21
Efe (Pygmies, Central Africa), 345
El Aspero (site, Peru), 114, 115
El Niño, 94, 104, 109
El Paraiso (site, Peru), 116, 117
El Pilar (forest site, Belize), 22, 25, 31
enculturation, 91
endogamy, 303
Engels, Friedrichederick, 242
engine (pump) models, 129
environmental circumscription, 366
environmental history, 6

Erickson, Clark L., 1–17, 89, 121, 146, 176, 187–278, 341
Erikson, Philippe, 351
Esmeraldas Province (Ecuador), 144
Esperanza II Mound (Bolivia), 206
etic rationalism, 4
Euphorbiaceae, 183
European World System Period (Amazonia), 321–22
eurytopic taxa, 128, 133
Evans, Clifford, 324
evolutionary ecology (behavioral ecology), 5, 238

Fairhead, James, 249
Faldín, Juan, 190, 269
Faron, Louis, 328, 335
Farrington, Ian S., 97
Fedick, Scott L., 21
felids (Felis sp., Panthera sp.), 144
fermentation, 170, 171
fields. See swidden
fire, 200, 260; managed, 11, 31, 22; fire histories, 263; see also burning; slash-and-burn; slash and char; swidden
fish weirs, 258, 260–62, 266
Flannery, Kent V., 333
floodplains, 157, 172
Flores, José Salvador, 21
food annuals, 105
food production, 165, 174
forager, 92, 238, 241, 283, 343, 344; (defined), 343; traditional forager, 344
foraging exclusion hypothesis, 238, 282
Ford, Anabel, 21–57
Ford Motor Company, 379
Fordlândia (Brazil), 379
forest: fragmentation, 130–31
forest island, 11, 248, 218, 258
forest isolates, 131
Forline, Louis C., 350
Formative Period (Amazonia), 317–19, 320, 325
frogs, 131
fruit tree domestication, 165–85
fruit-eating bat (Artibeus sp.), 144

galactic clusters, 331
Garcia, E. Emilia, 208

garden. *See* swidden

garden hunting, 130, 144

gardenification of nature (metaphor), 236, 265

Gê diaspora, 318

Geertz, Clifford, 335

genipapo (*Genipa americana*) (genipapo or bi), 214

Gentry, Alwyn, 207, 208

Geographic Information System (GIS), 368, 377

Giddens, Anthony, 325

Global Positioning System (GPS), 377–96

Glover's Reef Atoll, 75

Godfrey, Brian J., 336

gold cycle (Brazil), 397n

Gómez-Pompa, Arturo, 21

Goodenough Island (Papua New Guinea), 103–4

Gould, Stephen J., 237

Gran Chaco (Bolivia, Argentina, Paraguay), 112–13

Graham, Elizabeth, 7, 57–86, 245, 333

grass mouse (*Akodon* sp.), 144

grater bowls, 198

graviola (*Annona* sp.), 173, 179, 371

Greenland, Dennis J., 57

Gross, Daniel R., 166

Guajá (Indians, Brazil), 92, 284, 341–64

gualusa. *See* cocoyam

Guarayo (Indians, Bolivia), 213

Guarita phase (or subtradition), 289, 293, 294, 296, 321

Guatemala, 21, 73

guava (*Psidium guajava*), 96, 104, 106, 110, 111, 113, 115

guayusa (*Ilex* sp.), 252

Guazuma ulmifolia, 215

Guinea (Africa), 11

Guitarrero Cave (site, Peru), 106, 108

Gurupá (town, Brazil), 371

Gurupi River (Brazil), 349

Haida (Indians, Canada and United States), 358n

Hammond, Norman, 21

Hanke, Wanda, 189, 269

Harlan, Jack, 97, 170

Hastorf, Christine, 7, 87–126

Hatarara (site, Solimões River, Brazil), 291, 293–97, 299, 300

Hayden, Brian, 103

Headland, Thomas N., 345

Hecht, Susanna B., 70

Heckenberger, Michael J., 8, 121n, 311–40, 391

Heiser, Charles B., 97

Heliconia hirsute, 183

Hemming, John, 157, 160

Henderson, Andrew, 212

hermit crabs (*Coenobita clypeatus*), 71

Hetá (Indians, Brazil), 358n

heterotherms, 131

hierarchy (as system of values), 325

Hilbert, Peter, 288

historical ecology, ix, 1–17, 10, 12, 21, 87, 88, 90, 91, 100, 127, 153, 154, 187, 217–19, 236, 238, 240, 243, 249, 262, 267, 313, 336, 342, 352, 356, 365, 366, 377, 378, 395; (defined), 1, 153, 311; research design in, 374; as research program, 1–17, 358

history (as distinguished from historical ecology), 314–16

Hodder, Ian, 268n

hog plum (*Spondias mombin*), 212, 213, 215; *see also* taperebá

Holmberg, Allan, 198

Holocene: Early Holocene, 8, 172, 282; 127, 167, 183n, 247, 262, 280, 281, 282–83

Homo devastans (defined), 10; 57

Homo ecologicus (defined), 10; 78

Hoti (Indians, Venezuela), 246

howler monkey (*Alouatta* sp.), 133, 144

Huaorani (Indians, Ecuador), 98, 99, 103, 283, 346, 348

Huaynuná, (site, Peru), 117

Hugh-Jones, Stephen, 101, 102

human ecology, 6

Humboldt Current, 93

Humboldt, Alexander von, 153

hunter-gatherer. *See* forager

Hynes, Ross A., 241, 242, 243, 268

Ibaré River (Bolivia), 189

Ibiato (site, Sirionó Indian community, Bolivia), 188, 201, 206, 217

Ibibate Mound Complex (Bolivia), 189, 190–98, 201, 204, 205, 207, 210, 217, 230n, 247
Ica (river valley, Peru), 115
igapó, 286; forests, 287
Igarapé Guariba (site, Brazil), 285
igarapés, 287
iguana (*Iguana* sp.), 144
Illinois River valley, 297
Iltis, Hugh H., 97
inajá palm (*Attalea maripa*), 284
India, 78
Inga spp., 106, 111, 215
Ingold, Tim, 97, 99, 101, 242, 315
Initial Period (Andes), 104, 116
intensive shifting cultivation, 159
intraregional variability (Amazonia), 166n, 367
Iranduba phase (Brazil), 289
irrigation, 60
Iruyañez River (Bolivia), 252
Ituqui Island (Brazil), 301
Ix Chel (forest site, Belize), 22, 25, 32

jack bean (*Canavalia* sp.), 106, 115
Jackson, Jean, 345
Jama River (Ecuador), 133, 136, 137, 138, 139, 145
Jamari River (Brazil), 290
Janzen, Dan, 236, 265
Jasiaquiri (site, Bolivia), 258

K-selected species, 129
Ka'apor (Indians, Brazil), 103, 284, 349–50
kannan k'aax ("well-cared for forest"), 21, 31
kapitá, 350
kapok tree (*Ceiba pentandra*), 213
Karajá (Indians, Brazil), 329
Kayapó (Indians, Brazil), 11, 155, 160, 284
Kendal (site, Belize), 62
Kern, Dirse C., 62, 79
Killeen, Timothy J., 208
kitchen garden, 118
Kuhn, Thomas, 3
Kulina (Indians, Brazil), 355
Kumeyaay (Indians, California), 92, 100
Kwakiutl (Indians, Canada), 358

La Galgada (site, Peru), 116, 118, 121n
Lago Grande (site, Brazil), 285, 291–94, 300

Lamanai (site, Belize), 71, 76, 78
Lambert, John D.H., 59
land crab (*Cardisoma* sp), 70
landscape, 87, 244, 268; (defined), 1, 244–45, 341; domestication of, 165, 255; landscape ecology, 2, 4, 165; management, 279, 285–86; landscape history, 301; cultural landscape, 188, 368; (defined), 1; *see also* anthropogenic landscape
land-use, 365–405
Langstroth, Robert, 190, 210, 258, 269
La Paloma (site, Peru), 105, 110, 112
Large Scale Biosphere-Atmosphere Experiment in Amazonia (LBA), 375
Las Haldas (site, Peru), 116
Late Classic Period, 21, 32, 71; in Eastern Guatemala and Western Belize, 22
Late Intermediate Period (Peru), 118
Late Pleistocene, 283
Late Preceramic VI (Peru), 113, 115
Lathrap, Donald W., 90, 116, 238, 239, 281–82, 323, 327, 335
Latin America, 153
Leach, Melissa, 249
leaf-nosed bat (*Phyllostomus* sp.), 144
Lee, Kenneth, 190, 259, 269
Lee, Richard B., 343
Leguminosae (Papilionoideae), 183
Lehmann, Johannes, 79, 200
Lentz, David L., 176
Lévi-Strauss, Claude, 238
Lewontin, Richard, 237
liana, 33
lima bean (*Phaseolus lunatus*), 105, 106, 109, 110, 111, 118
limestone, 62
"limiting factors," 90, 246, 341
Llanos de Mojos (Bolivia), 188–89, 247–48, 268
Loma Alta (site, Ecuador), 112–13
Loma Alta de Casarabe (site, Bolivia), 190, 200, 205, 217
lomas, as seasonal cloud forests, foothills of Andes: 92, 93–95, 121n; as archaeological mounds, Bolivian Amazon: (defined), 188, 257–58
long-fallow shifting cultivation, 153
Loreto (site, Bolivia), 190
Los Buchillones (site, Cuba), 64, 75

lowland Amazon basin population, 160
lowlands (defined), 1
lucuma (*Pouteria lucuma*), 106, 112–13
lupine (*Lupinus mutabilis*), 94

macaúba palm (*Acrocomia aculeata*), 171, 172
Madeira River (Brazil), 282
maize (*Zea mays*), 94, 105, 106, 109, 113–14,
    116, 370
Maku (Indians, Brazil), 283, 345–47
Mamoré River (Bolivia), 189
Manabí Province (Ecuador), 133
Manacapuru phase (Brazil), 288, 289, 294
Manao (Indians, Brazil), 288
Manaus (city, Brazil), 289, 371
mangrove, 75; *see also* red mangrove, black
    mangrove, white mangrove
manioc (*Manihot esculenta*), 96, 105–6, 115,
    166, 183, 253, 301, 370, 388
Marajó (prehistoric people), 320
Marajó Island (Brazil), 281, 282, 285, 312,
    302, 370
Maranhão (Brazil), 209
Maranta arundinacea, 183
Marantaceae, 183
Maranta ruiziana, 183
Massangana phase (Brazil), 290
Matis (Indians, Bolivia), 351
Matisgenka (Indians, Peru), 347
Maya (Indians, Mesoamerica), 7, 21, 25, 26,
    27, 30, 31, 32, 58, 59, 63, 67, 70, 76, 77–79
Maya forest 21, 22, 30–32
Mazzulo, Sal J., 66
Mbuti (Pygmies, Central Africa), 345
McCann, Joe M., 58, 68
meander scars, 195
Meggers, Betty J., 166, 301, 324
Mendoza Mound (site, Bolivia), 201, 204,
    205, 217
Mesoamerica (lowland), 58 109
Messer, Ellen, 88
Métraux, Alfred, 347
Mexico, 106
Meyer, Thomas P., 58, 68
Michel, Marcos, 269n
micromammal, 144
microvertebrate, 137, 138, 145;
    synecology, 128
Middle East, 78

Middle Preceramic Phase, 111
migration (of plants), 96
Milton, Katharine, 346, 347, 348
Mojo (Indians, Bolivia), 260
monkeys, 352–56, 358
monocotyledons, 169
Monte Alegre (site, Brazil), 262–63
Monte Sinai (site, Bolivia), 205
Moran, Emilio F., 237
Morcote-Rios, Gaspar, 171, 284
Mori, Scott A., 207
moriche palm. *See* buriti palm
morphocategories, 23
motacú palm (*Attalea phalerata*), 197, 209,
    211, 212, 215
mounds in Bolivia (lomas), 189–206,
    257–58
mounds (burial), 295–300
mouse opossum (*Caluromys derbianus*), 144
mulefa, 92
Mundurucu (Indians, Brazil), 329, 330
Mura (Indians, Brazil), 288
Murrieta, Rui, 301
murumuru palm (*Astrocaryum aculeatum*),
    172, 177, 284; *see also* tucumã palm
mururé (*Brosimum acutifolium*), 215
museum models, 129

Nasca (valley, Peru), 115
National Science Foundation, 219
nature (defined), 78
Nature Conservancy, 264
NDE ("Neotropic Dark Earth"), 74
Negro River, 286, 288, 296, 298, 320
Neolithic Revolution, 243; in Amazonia
    (questionable existence of), 327
Neotropics (defined), 1, 79, 128, 129, 239,
    262, 264
Neves, Eduardo G., 10, 279–309
new ecology, 236
Nimuendajú, Curt, 349
nine-banded armadillo (*Dasypus
    novemcinctus*), 133
nonequilibrium ecology, 6
Nordenskiöld, Erland, 247, 255, 268–69n
Noronha, José M., 349
Northern Gê (Indians, Brazil), 330
n-transforms (defined), 71
Nukak (Indians, Colombia), 246, 283, 284

oca (*Oxalis* sp.), 105
ochoó tree (*Hura crepitans*), 213–14
Oitavo Bec (site, Brazil), 158
oligarchy (of forests): defined, 7, 25, 26, 30,
    32, 168, 171, 172, 183
omnivore's paradox, 99–100
opossum (*Didelphis* sp.), 144
orange tree (*Citrus maxima*), 215
organics, 66–67
Orinoco basin, 289
Oxisols, 388

paca (*Agouti* sp., *Cuniculus* sp.), 144
pacae (*Inga feuillei*), 106
pacay (*Inga* spp.), 215
Pachyrrhizus, 106, 111; *Pachyrhizus tuberosus*,
    183
paleoecology, 128
palimpsest (cultural imprint), 7, 220
pampa. *See* savanna
Pampa de los Llamas Moxeke (site, Peru),
    116, 117
Papua New Guinea, 101
Parada, Rodolfo Pinto, 190, 257, 269
paradigm, 3
Parakanã (Indians, Brazil), 284, 302,
    303, 347
Pareci (Indians, Brazil) 160, 327
Paredão phase, 289, 294, 296
Paresi. *See* Pareci
pascana (campsite), 195
patauá palm (*Oenocarpus bataua*), 171
Patiño, Victor M., 167, 175
patrilineal (descent), 102
peach palm. *See* pupunha palm
peanut, 106, 115
Pearsall, Deborah M., 89, 105, 106, 108,
    113, 133, 145, 246, 172
Pechichal (site, Ecuador), 136
Pedra Pintada (site, Amazon, Brazil),
    281, 283
Peña Roja (site, Colombia), 262, 282, 284
Pendergast, David M., 71
perennial fruit trees, 105
period effects, 383
personhood (defined), 316
Peru, 88–90, 97, 98, 106, 118
Petén forest, 7, 21, 31
Peters, Charles M., 59, 171

Peterson, James, 10, 279–309
pet kot (Maya anthropogenic forest), 9
Phaseolus spp. beans, 94, 118, 119
phytosociology, 22, 25
Piñedo-Vasquez, Miguel, 372
Piperno, Dolores R., 89, 105, 106, 172, 246
Pires de Campos, Antônio, 327
Pisco (site, Peru), 115
Plafker, George, 247, 268
plant population domestication (defined),
    165–66
Pleistocene, 172, 183n, 247
plural paternity, 354
poisonwood (*Metopium brownei*), 32, 75, 76
Pollock, Donald J., 355
polyandry, 354
Polychrome Guarita Phase, 292
polygyny, 354
Pomona (site, Belize), 62
ponds, 190, 195–96
Ponta de Pedras (site, Brazil), 385
Posey, Darrell A., 237
Postclassic period (Mesoamerica), 71
postforaging society (defined), 344
potato (*Solanum tuberosum* and
    S. andigenum), 94, 105, 106, 117
potsherds. *See* ceramics
pottery. *See* ceramics
pozas. See ponds
Preceramic (Amazonia), 282–83
Preceramic (Peru), 106–15
Preucel, Robert., 268
Prosopis (forests), 93
Prümers, Heiko, 190, 201, 204, 269
Puleston, Dennis E., 59, 60
Pullman, Philip, 92
pupunha palm (*Bactris gaspaes*) 173, 176,
    284–85
Pyne, Stephen J., 10, 263

qochas (sunken gardens), 94
quinoa (*Chenopodium quinoa*), 94;
    Chenopodium spp., 94, 105, 111, 116

r-selected species, 329
rabbit (*Sylvilagus* sp.), 144
radiocarbon dating, 158, 201
raised fields, 246, 249, 250, 251–54,
    266, 269n

ramon tree (*Brosimum alicastrum*), 59, 60
Ranzi, Alceu, 259
Real Alto (site, Ecuador), 115, 121n
red mangrove (*Rhizophora mangle*), 75
Regional Development Period (Amazonia), 319–20
Reid, Lawrence A., 345
relative dominance (defined), 27
remote sensing data, 377–96
rias, 281
rice, 371
rice rats (*Oryzomys caliginosus*), 133, 144
Rimac River (Peru), 111
Rindos, David, 268n
ring ditch sites (Bolivia, Brazil), 258–60
Ring Site (site, Peru), 95
Rio Arapiuns (Brazil), 158
Rio Tapajós (Brazil), 158
Rival, Laura, 92, 98, 103, 346, 347, 348, 351
Roosevelt, Anna C., 157, 246, 324
root crops, 105
Rosa, Nelson A., 207
Rozin, Paul, 99
rubber, 370, 373, 379, 380, 396n
Rydén, Stig, 189, 269n

Salomão, Rafael P., 207
Sanders, William T., 60
San Jacinto (site, Colombia), 282
San Pablo Forest (Bolivia), 190
Santa Cruz (Ambergris Caye, Belize), 66, 75
Santa River (Peru), 116
Santarém (site and city, Brazil), 156, 312, 320, 321, 333, 379, 381
Sapindus saponaria, 214
Sapium glandulosum, 213
Sauer, Carl O., 97, 154, 281
savanna, 200, 207, 208, 209, 210, 211, 254, 255, 258
Schama, Simon, 315
Schiffer, Michael B., 71
Schmidt, Max, 327, 328
Schor, Nicholas, 300
semi-intensive cultivation, 159
semipermanent cultivation, 159
sesmarias, 385
Shannon's index of diversity (defined), 52
Shipek, Florence, 100
short tail opossum (*Monodelphis* sp.), 132

Silva, Fabíola Andréa, 73
Silva, Manoela Ferreira Fernandes da, 207
Simoons, Frederick J., 92
Sirionó (Indians, Bolivia), 188, 190, 191, 195, 212, 214, 215, 220, 283, 347
site (archaeological concept of), 63
slash-and-burn, 250, 263, 258; *see also* swidden
slash and char, 156
slavery, 349–50
sloth (*Bradypus* sp.), 144
small mouse opossum (*Marmosa* spp.), 133, 144
Smalley, John, 97
Smith, C. Earle, 116
snakes, 131
Snead, James, 268n
soils, 7, 57, 68, 155, 200, 257, 366, 388; *see also* Amazonian Dark Earth
Solimões River (Brazil), 286–87, 293, 300, 302, 368
Sombroek, Wim G., 62, 73, 156
Sorenson's index (of similarity), 52n
Sosa, Victoria, 21
South America, 88–91, 93, 95, 103
spiny rats (*Proechimys semispinosus*), 132, 133, 144
Spodosols, 388
squash (*Cucurbita: C. ficifolia, C. maxima, C. moschata*), 104, 105, 106, 109, 111–13, 115, 117, 253; *see also* crookneck squash
squirrel (*Sciurus* sp.), 144
Stahl, Peter, 7, 127–49
stakeholders, 378
Stan Creek District (Belize), 59, 62, 63, 65, 75, 76
Stearman, Allyn, 347–48, 358
stenotopic taxa, 128
Steward, Julian H., 4, 324, 328, 335
stone axe thesis, 156
sugarcane, 370
sumuqué palm (*Syagrus sancona*), 212, 215
Suyá (Indians, Brazil), 329
sweet potato (*Ipomoea batatas*), 105, 106, 115, 166, 183n, 253
Sweet, David G., 349, 350
swidden, 61, 102, 190, 196, 205, 219, 352, 369, 372, 388
swidden agroforestry, 370
synantropic species, 130

Tabebuia serratifolia, 213
Taino (Indians, Cuba), 64–65
Tambiah, Stanley, 331, 335
Tapajós River (Brazil), 156, 158, 281, 302, 379, 381, 397n
taperebá (*Spondias mombin*), 172, 212, 371; see also hog plum
Taperinha (site, Brazil), 281
Tapirapé (Indians, Brazil), 329, 330
taro (*Colocasia* sp.) 166, 242
Tembé (Indians, Brazil), 358n
Teotihuacan (site, Mexico)
Terborgh, John, 130
Terminal Classic period, 71
Terminalia sp., 213
terracing, 60
terra firme (defined), 154, 207, 212, 281, 317
terra mulata, 60–63, 65, 68, 73, 156–59
Terra Nova (forest site, Belize), 22, 25, 26, 32
terraplenes. See causeways
terra preta do índio, 57, 75, 154, 156–57, 263, 280; see also Amazonian Dark Earth
Terrell, John E., 242
Teso dos Bichos (site, Brazil), 285
theater state (Bali), 335
Tigris-Euphrates region, 78
Tikal (site, Guatemala), 60, 73
Tikuna (Indians, Brazil, Peru), 300, 368
tinamou (*Cypturellus* sp.), 137, 144
tiny mouse opossum (*Marmosa* sp.), 144
Tocantins River (Brazil), 349
Towle, Margaret, 104
Townsend, Wendy, 211, 212
Trans-Amazon Highway, 382, 383, 384
Trekker (defined), 343
Tres Ventanas (site, Peru), 108
Trinidad (Bolivia), 189
tucumá palm (*Astrocaryum aculeatum*), 172, 177, 284; see also murumuru palm
Tukanoans (Indians, Brazil, Colombia), 303, 345, 388
Tukunyapé (Indians, Brazil), 349
Tupian Indians, 318
Tupi-Guarani tradition, 318
Tupí-Guaraní (language family), 319, 348–49, 356
Turiwára (Indians, Brazil), 349
Turner, Brian L., 21

turumbúri tree (*Sorocea guilleminiana*), 215, 216

Upper Xingu (region), 313, 320–22, 326, 329–30, 335
urban ecology, 313, 335–36, 337n
urbanism, 333
Urry, James, 97
urucu (*Bixa orellana*), 253

Valentine, Paul, 354
várzea, 31, 172, 286, 317
Vasina, Jan, 314
Vaupés (river basin, Colombia and Brazil), 388
Vilca, Mario, 190, 269
Viru (site, Peru), 114, 115
Viveiros de Castro, Eduardo, 318, 353
Voss, Rob, 146
voucher specimens, 23

Wagley, Charles, 350
Wagner, Roy W., 100
Walker, John, 269n
Waorani (Indians, Ecuador). See Huaorani
Wari (Indians, Brazil), 355
water hyacinth (*Eichhornia azurea*), 253
Webb, Molly A.H., 32
Wenner-Gren Foundation, 219
wetland agriculture, 64
white mangrove (*Laguncularia racemosa*), 75
white poisonwood (*Cameraria* sp.), 76
white-tailed deer (*Odocoileus virginianus*), 133, 144
Wiersum, K. Freerk, 166
Wilson, Peter J., 268n
Winkler, Wilma, 190
Wiseman, Frederick M., 21
Woods, William I., 58, 64, 68, 72
woolly opossum (*Caluromys derbianus*), 144
World Wildlife Fund, 260

Xarae (Indians, Brazil), 160
Xingu River (Brazil), 160, 209, 281, 302, 313, 379
Xinguano polity (prehistory), 330–33
Xinguanos (Indians, Brazil), 8, 320, 333

yam (*Dioscorea* spp.): *Dioscorea trifida*, 183;
   Dioscorea dodecaneura, 183
yellow mombin. *See* hog plum
Yen, Douglas E., 239, 241–43, 268n
Young, Michael, 103
yuca. *See Manihot esculenta*
Yucatán Peninsula, 21
Yucatec Maya forest, 21

yu palm (*Astrocaryum gynacanthum*), 356
Yuquí (Indians, Bolivia), 283,
   347–48, 358n

Zaña Valley (Peru), 111
zanjas. See canals
Zeidler, James A., 133, 145
zooarchaeology (defined), 127